Graduate Texts in Physics

Graduate Texts in Physics publishes core learning/teaching material for graduate- and advanced-level undergraduate courses on topics of current and emerging fields within physics, both pure and applied. These textbooks serve students at the MS- or PhD-level and their instructors as comprehensive sources of principles, definitions, derivations, experiments and applications (as relevant) for their mastery and teaching, respectively. International in scope and relevance, the textbooks correspond to course syllabi sufficiently to serve as required reading. Their didactic style, comprehensiveness and coverage of fundamental material also make them suitable as introductions or references for scientists entering, or requiring timely knowledge of, a research field.

Philipp O. J. Scherer

Computational Physics I

Numerical Methods

Fourth Edition

Springer

Philipp O. J. Scherer
TUM School of Natural Sciences,
Theoretical Physics T38
Technische Universität München
Garching, Bayern, Germany

ISSN 1868-4513 ISSN 1868-4521 (electronic)
Graduate Texts in Physics
ISBN 978-3-032-07855-1 ISBN 978-3-032-07856-8 (eBook)
https://doi.org/10.1007/978-3-032-07856-8

This Springer imprint is published by the registered company Springer Nature Switzerland AG
The registered company address is: Gewerbestrasse 11, 6330 Cham, Switzerland

Preface to the First Volume of the Fourth Edition

Computers have become an integral part of modern physics. They help to acquire, store and process enormous amounts of experimental data. Algebra programs have become very powerful and give the physician the knowledge of many mathematicians at hand. Traditionally, physics has been divided into experimental physics which observes phenomena occurring in the real world and theoretical physics which uses mathematical methods and simplified models to explain the experimental findings and to make predictions for future experiments. But there is also a new part of physics which has an ever growing importance. Computational physics combines the methods of the experimentalist and the theoretician. Computer simulation of physical systems helps to develop models and to investigate their properties. The recent development of machine learning and, more general, artificial intelligence, opens quite new computational possibilities for data analysis and model generalization. A very novel branch of computational physics is the simulation of quantum systems with quantum computers. While a universal digital quantum computer has not yet been realized and analog computing is still limited to very specific prove-of-principle experiments, quantum annealers with thousands of qubits are already commercially available and the simulation of quantum computers on a classical computer is an important aspect of classical computational physics.

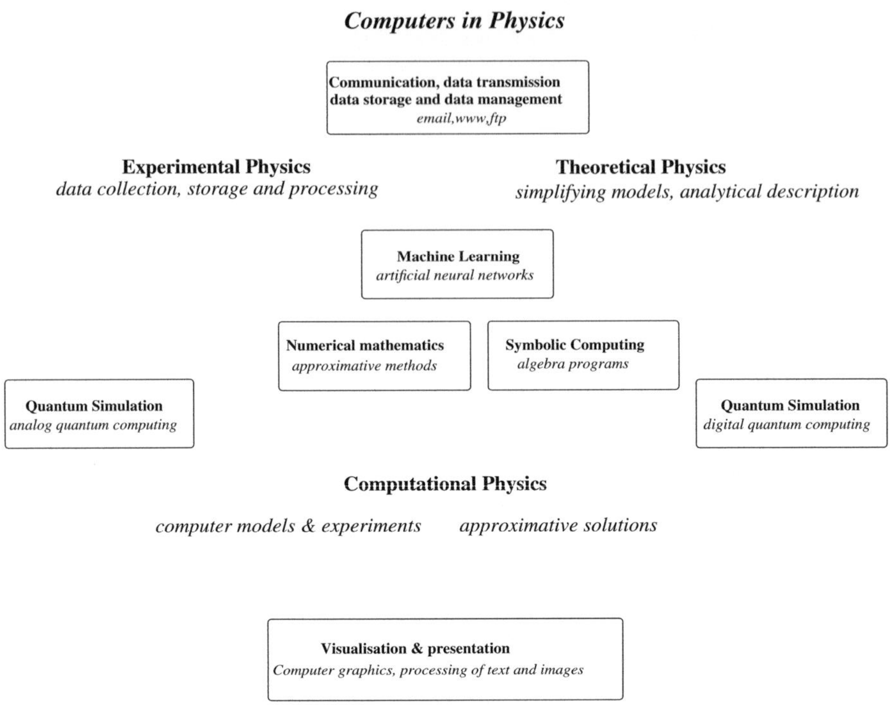

This book attempts to give the undergraduate physics students a profound background in numerical methods and in computer simulation methods but is also very welcome by students of mathematics and computational science who want to learn about applications of numerical methods in physics. This book may also support lecturers of computational physics, numerical analysis, machine learning, and engineering sciences, especially quantum technology.

The first edition of this textbook was based on a one-year course on computational physics which I have given at the Technische Universität München (TUM) for several years on a regular basis. In the later editions, its extent has been increased substantially and is now divided into two volumes representing the two parts of the earlier editions. The first volume gives an introduction into numerical methods and the second volume shows the application of computer simulation methods to a variety of physical systems. This book tries to bridge between simple examples which can be solved analytically and more complicated but instructive applications which provide insight with the help of a large number of computer exercises. For those readers who are unable to try these themselves, the essential results are shown in detailed figures.

In this first volume, I tried to give an overview of essential numerical methods, traditional ones as well as more recent developments. In many cases it is not possible to pin down the "best" algorithm, since this may depend on subtle features of a certain application, the general opinion changes from time to time with new methods appearing and computer architectures evolving, and each author is convinced that his method is the best one. Therefore I concentrated on a discussion of the prevalent methods and a comparison for selected examples. For a comprehensive description I would like to refer the reader to specialized textbooks like "Numerical Recipes" or elementary books in the field of the engineering sciences.

Basic equations are derived in detail and efficient implementations are discussed together with numerical accuracy and stability of the algorithms. Analytical results are given for simple test cases which serve as a benchmark for the numerical methods.

In 14 chapters, this volume explains numerical differentiation and integration, data interpolation and fitting, Fourier transformation and time-frequency analysis including wavelet analysis, systems of linear equations and Eigenproblem, root finding and extremal points, random numbers and Monte Carlo integration and finally discretization and solution of differential equations. Besides classical algorithms, also machine learning methods for algorithmic differentiation, data analysis and differential equations are presented.

Garching, Germany
August 2025

Philipp O. J. Scherer

Contents

Chapter 1
Error Analysis

Several sources of errors are important for numerical data processing:

Experimental uncertainty: Input data from an experiment have a limited precision. Instead of the vector of exact values $\mathbf{x}$ the calculation uses $\mathbf{x} + \Delta\mathbf{x}$, with an uncertainty $\Delta\mathbf{x}$. This can lead to large uncertainties of the calculated results if an unstable algorithm is used or if the unavoidable error inherent to the problem is large.

Rounding errors: The arithmetic unit of a computer uses only a subset of the real numbers, the so called machine numbers $A \subset \Re$. The input data as well as the results of elementary operations have to be represented by machine numbers whereby rounding errors can be generated. This kind of numerical error can be avoided in principle by using arbitrary precision arithmetics[1] or symbolic algebra programs. But this is unpractical in many cases due to the increase in computing time and memory requirements.

Truncation errors: Results from more complex operations like trigonometric functions, numerical differentiation and integration can have even larger errors if series expansions have to be truncated.

Error propagation:
Complex, especially iterative algorithms accumulate the errors of individual steps. This can lead to instability of an algorithm.

[1] For instance the open source GNU MP bignum library.

Supplementary Information The online version contains supplementary material available at https://doi.org/10.1007/978-3-032-07856-8_1.

P. O. J. Scherer, *Computational Physics I*, Graduate Texts in Physics,
https://doi.org/10.1007/978-3-032-07856-8_1

Table 1.1 Binary floating-point formats

Format	Sign	Exponent	Hidden bit	Fraction	Precision ε_M
Float	s	$b_0 \cdots b_7$	1	$a_0 \cdots a_{22}$	$2^{-24} = 5.96\text{E}^{-8}$
Double	s	$b_0 \cdots b_{10}$	1	$a_0 \cdots a_{51}$	$2^{-53} = 1.11\text{E}^{-16}$
Quadruple	s	$b_0 \cdots b_{14}$	1	$a_0 \cdots a_{111}$	$2^{-113} = 9.63\text{E}^{-35}$

Table 1.2 Exponent bias E

Decimal value	Binary value	Hexadecimal value	Data type
127_{10}	1111111_2	\$3F	Single
1023_{10}	1111111111_2	\$3FF	Double
16383_{10}	11111111111111_2	\$3FFF	Quadruple

1.1 Machine Numbers and Rounding Errors

Floating point numbers are internally stored as the product of sign, mantissa and a power of 2. According to the IEEE754 standard [1] single, double and quadruple precision numbers are stored as 32, 64 or 128 bits (Table 1.1):

The sign bit s is 0 for positive and 1 for negative numbers. The exponent b is biased by adding E which is half of its maximum possible value (Table 1.2).[2] The value of a number is given by

$$x = (-)^s \times a \times 2^{b-E}. \tag{1.1}$$

The mantissa a is normalized such that its first bit is 1 and its value is between 1 and 2

$$1.000_2 \cdots 0 \leq a \leq 1.111 \cdots 1_2 < 10.0_2 = 2_{10}. \tag{1.2}$$

Since the first bit of a normalized floating point number always is 1, it is not necessary to store it explicitly (hidden bit or J-bit). However, since not all numbers can be normalized, only the range of exponents from \$001 $\cdots$ \$7FE is used for normalized numbers. An exponent of \$000 signals that the number is not normalized (zero is an important example, there exist even two zero numbers with different sign) whereas the exponent \$7FF is reserved for infinite or undefined results (Table 1.3). The range of normalized double precision numbers is between

$$\text{Min\_Normal} = 2.2250738585072014 \times 10^{-308}$$

[2] In the following the usual hexadecimal notation is used which represents a group of 4 bits by one of the digits 0,1,2,3,4,5,6,7,8,9,A,B,C,D,E,F.

Table 1.3 Special double precision numbers

Hexadecimal value	Symbolic value
\$000 0000000000000	+0
\$080 00000000000000	−0
\$7FF 0000000000000	+inf
\$FFF 0000000000000	−inf
\$7FF 0000000000001 ⋯ \$7FF FFFFFFFFFFFFF	NAN
\$001 0000000000000	Min_Normal
\$7FE FFFFFFFFFFFFF	Max_Normal
\$000 0000000000001	Min_Subnormal
\$000 FFFFFFFFFFFFF	Max_Subnormal

and

$$\text{Max\_Normal} = 1.7976931348623157E \times 10^{308}.$$

Example
Consider the following bit pattern which represents a double precision number:

$$\$4059000000000000.$$

The exponent is $100\,0000\,0101_2 - 011\,1111\,1111_2 = 110_2$ and the mantissa including the J-bit is $1\,1001\,0000\,0000 \cdots_2$. Hence the decimal value is

$$1.5625 \times 2^6 = 100_{10}.$$

Input numbers which are not machine numbers have to be rounded to the nearest machine number. This is formally described by a mapping $\Re \rightarrow A$

$$x \rightarrow rd(x)$$

with the property[3]

$$|x - rd(x)| \leq |x - g| \text{ for all } g \in A. \tag{1.3}$$

For the special case that x is exactly in the middle between two successive machine numbers, a tie-breaking rule is necessary. The simplest rules are to round up always (*round-half-up*) or always down *(round-half-down)*. However, these are not symmetric and produce a bias in the average round-off error. The IEEE-754 standard

[3] Sometimes rounding is replaced by a simpler truncation operation which, however leads to significantly larger rounding errors.

[1] recommends the *round-to-nearest-even* method, i.e. the least significant bit of the rounded number should always be zero. Alternatives are *round-to-nearest-odd,* stochastic rounding and alternating rounding.
The cases of *exponent overflow* and *exponent underflow* need special attention:

Whenever the exponent b has the maximum possible value $b = b_{\max}$ and $a = 1.11 \cdots 11$ has to be rounded to $a' = 10.00 \cdots 0$, the rounded number is not a machine number and the result is $\pm\,\mathrm{inf}$.

Numbers in the range $2^{b_{\min}} > |x| \geq 2^{b_{\min}-t}$ can be represented with loss of accuracy by denormalized machine numbers. Their mantissa cannot be normalized since it is $a < 1$ and the exponent has the smallest possible value $b = b_{\min}$. Even smaller numbers with $|x| < 2^{-t+b_{\min}}$ have to be rounded to ± 0.

The maximum rounding error for normalized numbers with t binary digits

$$a' = s \times 2^{b-E} \times 1.a_1 a_2 \cdots a_{t-1} \tag{1.4}$$

is given by (Fig. 1.1)

$$|a - a'| \leq 2^{b-E} \times 2^{-t} \tag{1.5}$$

and the relative error is bounded by

$$\left| \frac{rd(x) - x}{x} \right| \leq \frac{2^{-t} \times 2^b}{|a| \times 2^b} \leq 2^{-t}. \tag{1.6}$$

The error bound determines the relative machine precision[4]

$$\varepsilon_M = 2^{-t} \tag{1.7}$$

and the rounding operation can be described by

$$rd(x) = x(1 + \varepsilon) \text{ with } |\varepsilon| \leq \varepsilon_M. \tag{1.8}$$

The round-off error takes its maximum value if the mantissa is close to 1. Consider a number

$$x = 1 + \varepsilon.$$

If $\varepsilon < \varepsilon_M$ then $\mathrm{rd}(x) = 1$ whereas for $\varepsilon > \varepsilon_M$ rounding gives $\mathrm{rd}(x) = 1 + 2^{1-t}$ (Fig. 1.2). Hence ε_M is given by the largest number ε for which $\mathrm{rd}(1.0 + \varepsilon) = 1.0$ and is therefore also called *unit roundoff.*

[4] Also known as machine epsilon.

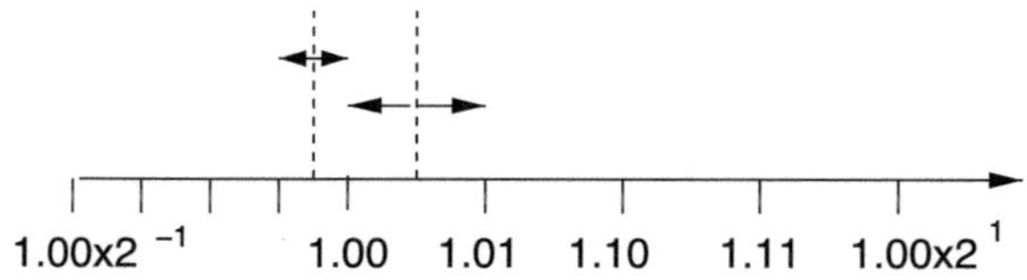

Fig. 1.1 Round to nearest. Normalized machine numbers with $t = 3$ binary digits are shown. Rounding to the nearest machine number produces a round-off error which is bounded by half the spacing of the machine numbers

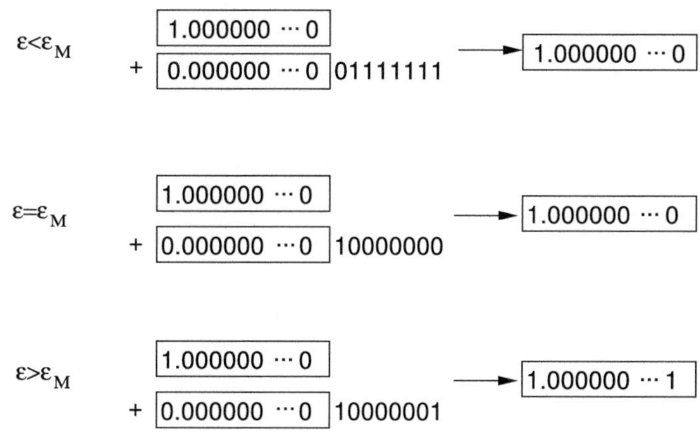

Fig. 1.2 Unit round off

1.2 Numerical Errors of Elementary Floating Point Operations

Even for two machine numbers $x, y \in A$ the results of addition, subtraction, multiplication or division are not necessarily machine numbers. We have to expect some additional round-off errors from all these elementary operations [2]. We assume that the results of elementary operations are approximated by machine numbers as precisely as possible. The IEEE754 standard [1] requires that the exact operations $x + y, x - y, x \times y, x \div y$ are approximated by floating point operations $A \rightarrow A$ with the property:

$$\begin{aligned} fl_+(x, y) &= rd(x + y) \\ fl_-(x, y) &= rd(x - y) \\ fl_*(x, y) &= rd(x \times y) \\ fl_\div(x, y) &= rd(x \div y). \end{aligned} \tag{1.9}$$

1.2.1 Numerical Extinction

For an addition or subtraction one summand has to be denormalized to line up the exponents (for simplicity we consider only the case $x > 0$, $y > 0$)

$$x + y = a_x 2^{b_x - E} + a_y 2^{b_y - E} = (a_x + a_y 2^{b_y - b_x}) 2^{b_x - E}. \tag{1.10}$$

If the two numbers differ much in their magnitude, numerical extinction can happen. Consider the following case:

$$y < 2^{b_x - E} \times 2^{-t} \tag{1.11}$$

$$a_y 2^{b_y - b_x} < 2^{-t}$$

The mantissa of the exact sum is

$$a_x + a_y 2^{b_y - b_x} = 1.\alpha_2 \cdots \alpha_{t-1} 01 \beta_2 \cdots \beta_{t-1}. \tag{1.12}$$

Rounding to the nearest machine number gives

$$rd(x + y) = 2^{b_x} \times (1.\alpha_2 \cdots \alpha_{t-1}) = x \tag{1.13}$$

since

$$|0.01\beta_2 \cdots \beta_{t-1} - 0| \leq |0.011 \cdots 1| = 0.1 - 0.00 \cdots 01$$

$$|0.01\beta_2 \cdots \beta_{t-1} - 1| \geq |0.01 - 1| = 0.11. \tag{1.14}$$

Consider now the case

$$y < x \times 2^{-t-1} = a_x \times 2^{b_x - E - t - 1} < 2^{b_x - E - t}. \tag{1.15}$$

For normalized numbers the mantissa is in the interval

$$1 \leq |a_x| < 2 \tag{1.16}$$

hence we have

$$rd(x + y) = x \text{ if } \frac{y}{x} < 2^{-t-1} = \frac{\varepsilon_M}{2}. \tag{1.17}$$

Especially for $x = 1$ we have

$$rd(1 + y) = 1 \text{ if } y < 2^{-t} = 0.00 \cdots 0_{t-1} 1_t 000 \cdots \tag{1.18}$$

2^{-t} could be rounded to 0 or to 2^{1-t} since the distance is the same $|2^{-t} - 0| = |2^{-t} - 2^{1-t}| = 2^{-t}$.

The smallest machine number with $fl_+(1, \varepsilon) > 1$ is either $\varepsilon = 0.00 \cdots 1_t 0 \cdots = 2^{-t}$ or $\varepsilon = 0.00 \cdots 1_t 0 \cdots 01_{2t-1} = 2^{-t}(1 + 2^{1-t})$. Hence the machine precision ε_M can be determined by looking for the smallest (positive) machine number ε for which $fl_+(1, \varepsilon) > 1$.

1.2.2 Addition

Consider the sum of two floating point numbers

$$y = x_1 + x_2. \tag{1.19}$$

First the input data have to be approximated by machine numbers:

$$\begin{aligned} x_1 &\rightarrow rd(x_1) = x_1(1 + \varepsilon_1) \\ x_2 &\rightarrow rd(x_2) = x_2(1 + \varepsilon_2). \end{aligned} \tag{1.20}$$

The addition of the two summands may produce another error α since the result has to be rounded. The numerical result is

$$\tilde{y} = fl_+(rd(x_1), rd(x_2)) = (x_1(1 + \varepsilon_1) + x_2(1 + \varepsilon_2))(1 + \alpha). \tag{1.21}$$

Neglecting higher orders of the error terms we have in first order

$$\tilde{y} = x_1 + x_2 + x_1\varepsilon_1 + x_2\varepsilon_2 + (x_1 + x_2)\alpha \tag{1.22}$$

and the relative error of the numerical sum is

$$\frac{\tilde{y} - y}{y} = \frac{x_1}{x_1 + x_2}\varepsilon_1 + \frac{x_2}{x_1 + x_2}\varepsilon_2 + \alpha. \tag{1.23}$$

If $x_1 \approx -x_2$ then numerical extinction can produce large relative errors and uncertainties of the input data can be strongly enhanced.

1.2.3 Multiplication

Consider the multiplication of two floating point numbers

$$y = x_1 \times x_2. \tag{1.24}$$

The numerical result is

$$\tilde{y} = fl_*(rd(x_1), rd(x_2)) = x_1(1+\varepsilon_1)x_2(1+\varepsilon_2)(1+\mu) \approx x_1x_2(1+\varepsilon_1+\varepsilon_2+\mu) \tag{1.25}$$

with the relative error

$$\frac{\tilde{y}-y}{y} = 1+\varepsilon_1+\varepsilon_2+\mu. \tag{1.26}$$

The relative errors of the input data and of the multiplication just add up to the total relative error. There is no enhancement. Similarly for a division

$$y = \frac{x_1}{x_2} \tag{1.27}$$

the relative error is

$$\frac{\tilde{y}-y}{y} = 1+\varepsilon_1-\varepsilon_2+\mu. \tag{1.28}$$

1.3 Error Propagation

Consider an algorithm consisting of a sequence of elementary operations. From the set of input data which is denoted by the vector

$$\mathbf{x} = (x_1 \cdots x_n) \tag{1.29}$$

a set of output data is calculated

$$\mathbf{y} = (y_1 \cdots y_m). \tag{1.30}$$

Formally this can be denoted by a vector function

$$\mathbf{y} = \varphi(\mathbf{x}) \tag{1.31}$$

which can be written as a product of r simpler functions representing the elementary operations

$$\varphi = \varphi^{(r)} \times \varphi^{(r-1)} \cdots \varphi^{(1)}. \tag{1.32}$$

Starting with $\mathbf{x}$ intermediate results $\mathbf{x}_i = (x_{i1}, \cdots x_{in_i})$ are calculated until the output data $\mathbf{y}$ result from the last step:

$$\begin{aligned}
\mathbf{x}_1 &= \varphi^{(1)}(\mathbf{x}) \\
\mathbf{x}_2 &= \varphi^{(2)}(\mathbf{x}_1) \\
&\vdots \\
\mathbf{x}_{r-1} &= \varphi^{(r-1)}(\mathbf{x}_{r-2}) \\
\mathbf{y} &= \varphi^{(r)}(\mathbf{x}_{r-1}).
\end{aligned} \tag{1.33}$$

In the following we analyze the influence of numerical errors onto the final results. We treat all errors as small quantities and neglect higher orders. Due to round-off errors and possible experimental uncertainties the input data are not exactly given by $\mathbf{x}$ but by

$$\mathbf{x} + \Delta\mathbf{x}. \tag{1.34}$$

The first step of the algorithm produces the result

$$\tilde{\mathbf{x}}_1 = rd(\varphi^{(1)}(\mathbf{x} + \Delta\mathbf{x})). \tag{1.35}$$

Taylor series expansion gives in first order

$$\tilde{\mathbf{x}}_1 = \left(\varphi^{(1)}(\mathbf{x}) + D\varphi^{(1)}\Delta\mathbf{x}\right)(1 + E_1) + \cdots \tag{1.36}$$

with the partial derivatives

$$D\varphi^{(1)} = \left(\frac{\partial x_{1i}}{\partial x_j}\right) = \begin{pmatrix} \frac{\partial x_{11}}{\partial x_1} & \cdots & \frac{\partial x_{11}}{\partial x_n} \\ \vdots & \ddots & \vdots \\ \frac{\partial x_{1n_1}}{\partial x_1} & \cdots & \frac{\partial x_{1n_1}}{\partial x_n} \end{pmatrix} \tag{1.37}$$

and the round-off errors of the first step

$$E_1 = \begin{pmatrix} \varepsilon_1^{(1)} & & \\ & \ddots & \\ & & \varepsilon_{n_1}^{(1)} \end{pmatrix}. \tag{1.38}$$

The error of the first intermediate result is

$$\Delta\mathbf{x}_1 = \tilde{\mathbf{x}}_1 - \mathbf{x}_1 = D\varphi^{(1)}\Delta\mathbf{x} + \varphi^{(1)}(\mathbf{x})E_1. \tag{1.39}$$

The second intermediate result is

$$\begin{aligned}
\tilde{\mathbf{x}}_2 &= \left(\varphi^{(2)}(\tilde{\mathbf{x}}_1)\right)(1 + E_2) = \varphi^{(2)}(\mathbf{x}_1 + \Delta\mathbf{x}_1)(1 + E_2) \\
&= \mathbf{x}_2(1 + E_2) + D\varphi^{(2)}D\varphi^{(1)}\Delta\mathbf{x} + D\varphi^{(2)}\mathbf{x}_1 E_1
\end{aligned} \tag{1.40}$$

with the error

$$\Delta \mathbf{x}_2 = \mathbf{x}_2 E_2 + D\varphi^{(2)} D\varphi^{(1)} \Delta \mathbf{x} + D\varphi^{(2)} \mathbf{x}_1 E_1. \tag{1.41}$$

Finally the error of the result is

$$\Delta \mathbf{y} = \mathbf{y} E_r + D\varphi^{(r)} \cdots D\varphi^{(1)} \Delta \mathbf{x} + D\varphi^{(r)} \cdots D\varphi^{(2)} \mathbf{x}_1 E_1 + \cdots D\varphi^{(r)} \mathbf{x}_{r-1} E_{r-1}. \tag{1.42}$$

The product of the matrices $D\varphi^{(r)} \cdots D\varphi^{(1)}$ is the matrix which contains the derivatives of the output data with respect to the input data (chain rule)

$$D\varphi = D\varphi^{(r)} \cdots D\varphi^{(1)} = \begin{pmatrix} \frac{\partial y_1}{\partial x_1} & \cdots & \frac{\partial y_1}{\partial x_n} \\ \vdots & \ddots & \vdots \\ \frac{\partial y_m}{\partial x_1} & \cdots & \frac{\partial y_m}{\partial x_n} \end{pmatrix} \tag{1.43}$$

The first two contributions to the total error do not depend on the way in which the algorithm is divided into elementary steps in contrary to the remaining summands. Hence the inevitable error which is inherent to the problem can be estimated as [2]

$$\Delta^{(\text{in})} y_i = \varepsilon_M |y_i| + \sum_{j=1}^{n} \left| \frac{\partial y_i}{\partial x_j} \right| |\Delta x_j| \tag{1.44}$$

or in case the error of the input data is dominated by the round-off errors $|\Delta x_j| \leq \varepsilon_M |x_j|$

$$\Delta^{(\text{in})} y_i = \varepsilon_M |y_i| + \varepsilon_M \sum_{j=1}^{n} \left| \frac{\partial y_i}{\partial x_j} \right| |x_j|. \tag{1.45}$$

Additional errors which are smaller than this inevitable error can be regarded as harmless. If all errors are harmless, the algorithm can be considered well behaved.

1.4 Stability of Iterative Algorithms

Often iterative algorithms are used which generate successive values starting from an initial value $\mathbf{x}_0$ according to an iteration method

$$\mathbf{x}_{j+1} = f(\mathbf{x}_j), \tag{1.46}$$

for instance to solve a large system of equations or to approximate a time evolution $\mathbf{x}_j \approx \mathbf{x}(j\Delta t)$. Consider first a linear iteration equation which can be written in matrix form as

$$\mathbf{x}_{j+1} = A\mathbf{x}_j. \tag{1.47}$$

If the matrix A is the same for all steps we have simply

$$\mathbf{x}_j = A^j \mathbf{x}_0. \tag{1.48}$$

Consider the unavoidable error originating from errors $\Delta\mathbf{x}$ of the start values:

$$\mathbf{x}_j = A^j(\mathbf{x}_0 + \Delta\mathbf{x}) = A^j \mathbf{x}_0 + A^j \Delta\mathbf{x}. \tag{1.49}$$

The initial errors $\Delta\mathbf{x}$ can be enhanced exponentially if A has at least one eigenvalue[5] λ with $|\lambda| > 1$. On the other hand the algorithm is stable if for all eigenvalues $|\lambda| \leq 1$ holds. For a more general nonlinear iteration

$$\mathbf{x}_{j+1} = \varphi(\mathbf{x}_j) \tag{1.50}$$

the error propagates according to

$$\mathbf{x}_1 = \varphi(\mathbf{x}_0) + D\varphi\Delta\mathbf{x}$$

$$\mathbf{x}_2 = \varphi(\mathbf{x}_1) = \varphi(\varphi(\mathbf{x}_0)) + (D\varphi)^2 \Delta x$$

$$\vdots$$

$$\mathbf{x}_j = \varphi(\varphi \cdots \varphi(\mathbf{x}_0)) + (D\varphi)^j \Delta x. \tag{1.51}$$

The algorithm is conditionally stable if all eigenvalues of the derivative matrix $D\varphi$ have absolute values $|\lambda| \leq 1$.

1.5 Example: Rotation

Consider a simple rotation in the complex plane (Fig. 1.3).

The equation of motion

$$\dot{z} = \mathrm{i}\omega z \tag{1.52}$$

obviously has the exact solution

$$z(t) = z_0 \mathrm{e}^{\mathrm{i}\omega t}. \tag{1.53}$$

[5] The eigenvalues of A are solutions of the eigenvalue equation $A\mathbf{x} = \lambda\mathbf{x}$ (10).

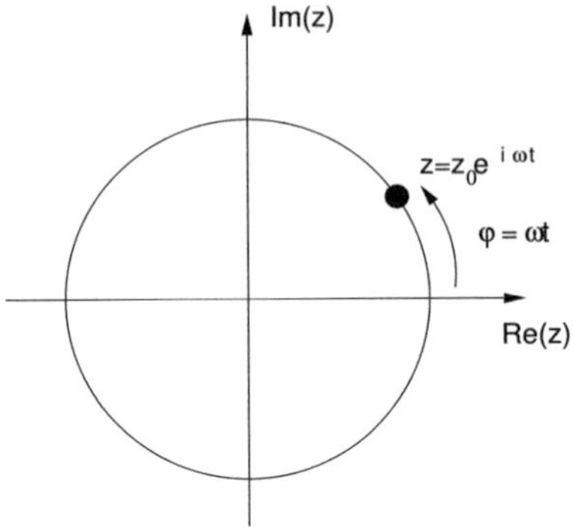

Fig. 1.3 Rotation in the complex plane

As a simple algorithm for numerical integration we use a time grid

$$t_j = j\Delta t \quad j = 0, 1, 2 \ldots \tag{1.54}$$

$$z_j = z(t_j) \tag{1.55}$$

and iterate the function values

$$z_{j+1} = z_j + \dot{z}(t_j) = (1 + \mathrm{i}\omega\Delta t)z_j. \tag{1.56}$$

Since

$$|1 + \mathrm{i}\omega\Delta t| = \sqrt{1 + \omega^2\Delta t^2} > 1 \tag{1.57}$$

uncertainties in the initial condition will grow exponentially and the algorithm is not stable. A stable method is obtained by taking the derivative in the middle of the time interval (Sect. 14.5)

$$\dot{z}(t + \frac{\Delta t}{2}) = \mathrm{i}\omega z(t + \frac{\Delta t}{2})$$

and making the approximation (Sect. 14.5)

$$z(t + \frac{\Delta t}{2}) \approx \frac{z(t) + z(t + \Delta t)}{2}.$$

This gives the implicit equation

$$z_{j+1} = z_j + \mathrm{i}\omega\Delta t\frac{z_{j+1} + z_j}{2} \tag{1.58}$$

which can be solved by

$$z_{j+1} = \frac{1 + \frac{\mathrm{i}\omega\Delta t}{2}}{1 - \frac{\mathrm{i}\omega\Delta t}{2}}z_j. \tag{1.59}$$

Now we have

$$\left|\frac{1+\frac{i\omega\Delta t}{2}}{1-\frac{i\omega\Delta t}{2}}\right| = \frac{\sqrt{1+\frac{\omega^2\Delta t^2}{4}}}{\sqrt{1+\frac{\omega^2\Delta t^2}{4}}} = 1 \tag{1.60}$$

and the calculated orbit is stable.

1.6 Truncation Error

The algorithm in the last example is stable but of course not perfect. Each step produces an error due to the finite time step. The exact solution

$$z(t+\Delta t) = z(t)e^{i\omega\Delta t} = z(t)(1 + i\omega\Delta t - \frac{\omega^2\Delta t^2}{2} + \frac{-i\omega^3\Delta t^3}{6}\cdots) \tag{1.61}$$

is approximated by

$$\begin{aligned} z(t+\Delta t) &\approx z(t)\frac{1+\frac{i\omega\Delta t}{2}}{1-\frac{i\omega\Delta t}{2}} \\ &= z(t)(1+\frac{i\omega\Delta t}{2})(1+\frac{i\omega\Delta t}{2}-\frac{\omega^2\Delta t^2}{4}-\frac{i\omega^3\Delta t^3}{8}+\cdots) \\ &= z(t)(1+i\omega\Delta t-\frac{\omega^2\Delta t^2}{2}+\frac{-i\omega^3\Delta t^3}{4}\cdots) \end{aligned} \tag{1.62}$$

which deviates from the exact solution by a term of the order $O(\Delta t^3)$, hence the *local error* order of this algorithm is $O(\Delta t^3)$ which is indicated by writing

$$z(t+\Delta t) = z(t)\frac{1+\frac{i\omega\Delta t}{2}}{1-\frac{i\omega\Delta t}{2}} + O(\Delta t^3). \tag{1.63}$$

Integration up to a total time $T = N\Delta t$ accumulates a *global error* of the order $N\Delta t^3 = T\Delta t^2$.

Table 1.4 Maximum and minimum integers

Java format	Bit length	Minimum	Maximum
Byte	8	−128	127
Short	16	−32768	32767
Integer	32	−2147483647	2147483648
Long	64	−9223372036854775808	9223372036854775807
Char	16	0	65535

Exercises

Exercise 1.1 Machine Precision

In this computer experiment we determine the machine precision ε_M. Starting with a value of 1.0, x is divided repeatedly by 2 until numerical addition of 1 and $x = 2^{-M}$ gives 1. Compare single and double precision calculations.

Exercise 1.2 Maximum and Minimum Integers

Integers are used as counters or to encode elements of a finite set like characters or colors. There are different integer formats available which store signed or unsigned integers of different length (Table 1.4). There is no infinite integer and addition of 1 to the maximum integer gives the minimum integer.

In this computer experiment we determine the smallest and largest integer numbers. Beginning with $I = 1$ we add repeatedly 1 until the condition $I + 1 > I$ becomes invalid or subtract repeatedly 1 until $I - 1 < I$ becomes invalid. For the 64 bit long integer format this takes to long. Here we multiply alternatively I by 2 until $I - 1 < I$ becomes invalid. For the character format the corresponding ordinal number is shown which is obtained by casting the character to an integer.

Exercise 1.3 Truncation Error

This computer experiment approximates the cosine function by a truncated Taylor series

$$\cos(x) \approx \mathrm{mycos}(x, n_{\max}) = \sum_{n=0}^{n_{\max}} (-)^n \frac{x^{2n}}{(2n)!} = 1 - \frac{x^2}{2} + \frac{x^4}{24} - \frac{x^6}{720} + \cdots \quad (1.64)$$

in the interval $-\pi/2 < x < \pi/2$. The function $\mathrm{mycos}(x, n_{\max})$ is numerically compared to the intrinsic cosine function.

Chapter 2
Interpolation

Experiments usually produce a discrete set of data points $(\mathbf{x}_i, f_i)$ which represent the value of a function $f(\mathbf{x})$ for a finite set of arguments $\{\mathbf{x}_0 \ldots \mathbf{x}_n\}$. If additional data points are needed, for instance to draw a continuous curve, interpolation is necessary. Interpolation also can be helpful to represent a complicated function by a simpler one or to develop more sophisticated numerical methods for the calculation of numerical derivatives and integrals. In the following we concentrate on the most important interpolating functions which are polynomials, splines and rational functions. Trigonometric interpolation is discussed in Chap. 7. An interpolating function reproduces the given function values at the interpolation points exactly (Fig. 2.1). The more general procedure of curve fitting, where this requirement is relaxed, is discussed in Chap. 11.

The interpolating polynomial can be explicitly constructed with the Lagrange method. Newton's method is numerically efficient if the polynomial has to be evaluated at many interpolating points and Neville's method has advantages if the polynomial is not needed explicitly and has to be evaluated only at one interpolation

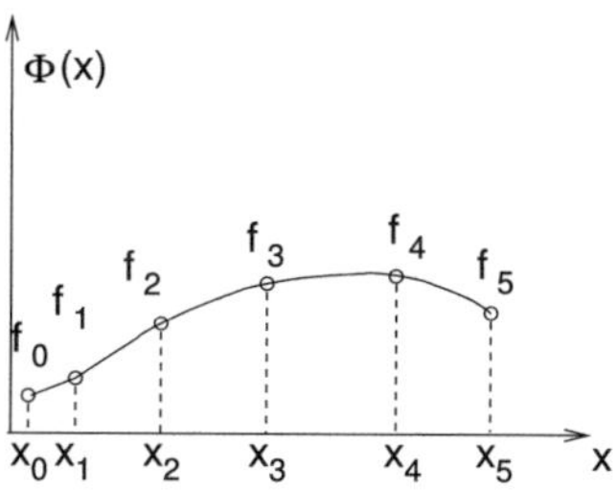

Fig. 2.1 Interpolating function. The interpolating function $\Phi(x)$ reproduces a given data set $\Phi(x_i) = f_i$ and provides an estimate of the function $f(x)$ between the data points

---

Supplementary Information The online version contains supplementary material available at https://doi.org/10.1007/978-3-032-07856-8_2.

P. O. J. Scherer, *Computational Physics I*, Graduate Texts in Physics,
https://doi.org/10.1007/978-3-032-07856-8_2

point. Bernstein polynomials have became popular for Hermite interpolation and the construction of Beziér curves.

Polynomials are not well suited for interpolation over a larger range. Spline functions can be superior which are piecewise defined polynomials. Especially cubic splines are often used to draw smooth curves. Curves with poles can be represented by rational interpolating functions whereas a special class of rational interpolants without poles provides a rather new alternative to spline interpolation.

2.1 Interpolating Functions

Consider the following problem: Given are $n+1$ sample points (x_i, f_i) , $i = 0 \cdots n$ and a function of x which depends on $n+1$ parameters a_i :

$$\Phi(x; a_0 \cdots a_n). \tag{2.1}$$

The parameters are to be determined such that the interpolating function has the proper values at all sample points (Fig. 2.1)

$$\Phi(x_i; a_0 \cdots a_n) = f_i \quad i = 0 \cdots n. \tag{2.2}$$

An interpolation problem is called linear if the interpolating function is a linear combination of functions

$$\Phi(x; a_0 \cdots a_n) = a_0 \Phi_0(x) + a_1 \Phi_1(x) + \cdots a_n \Phi_n(x). \tag{2.3}$$

Important examples are

- polynomials

$$a_0 + a_1 x + \cdots a_n x^n \tag{2.4}$$

- trigonometric functions

$$a_0 + a_1 e^{ix} + a_2 e^{2ix} + \cdots a_n e^{nix} \tag{2.5}$$

- spline functions which are piecewise polynomials, for instance the cubic spline

$$s(x) = \alpha_i + \beta_i (x - x_i) + \gamma_i (x - x_i)^2 + \delta_i (x - x_i)^3 \quad x_i \le x \le x_{i+1}. \tag{2.6}$$

Important examples for nonlinear interpolating functions are

- rational functions

$$\frac{p_0 + p_1 x + \cdots p_M x^M}{q_0 + q_1 x + \cdots q_N x^N} \tag{2.7}$$

- exponential functions

$$a_0 e^{\lambda_0 x} + a_1 e^{\lambda_1 x} + \cdots \tag{2.8}$$

where amplitudes a_i and exponents λ_i have to be optimized.

2.2 Polynomial Interpolation

For n+1 sample points $(x_i, f_i), \quad i = 0 \cdots n, \quad x_i \neq x_j$ there exists exactly one interpolating polynomial of degree n with

$$p(x_i) = f_i, \quad i = 0 \cdots n. \tag{2.9}$$

2.2.1 *Lagrange Polynomials*

Lagrange polynomials [3] are defined as

$$\mathrm{L}_i(x) = \frac{(x - x_0) \cdots (x - x_{i-1})(x - x_{i+1}) \cdots (x - x_n)}{(x_i - x_0) \cdots (x_i - x_{i-1})(x_i - x_{i+1}) \cdots (x_i - x_n)}. \tag{2.10}$$

They are of degree n and have the property

$$\mathrm{L}_i(x_k) = \delta_{i,k}. \tag{2.11}$$

The interpolating polynomial is given in terms of Lagrange polynomials by

$$p(x) = \sum_{i=0}^{n} f_i \mathrm{L}_i(x) = \sum_{i=0}^{n} f_i \prod_{k=0, k \neq i}^{n} \frac{x - x_k}{x_i - x_k}. \tag{2.12}$$

2.2.2 *Barycentric Lagrange Interpolation*

With the polynomial

$$\omega(x) = \prod_{i=0}^{n} (x - x_i) \tag{2.13}$$

the Lagrange polynomial can be written as

$$\mathrm{L}_i(x) = \frac{\omega(x)}{x - x_i} \frac{1}{\prod_{k=0,k\neq i}^{n}(x_i - x_k)} \tag{2.14}$$

which, introducing the Barycentric weights [4]

$$u_i = \frac{1}{\prod_{k=0,k\neq i}^{n}(x_i - x_k)} \tag{2.15}$$

becomes the first form of the barycentric interpolation formula

$$\mathrm{L}_i(x) = \omega(x)\frac{u_i}{x - x_i}. \tag{2.16}$$

The interpolating polynomial can now be evaluated according to

$$p(x) = \sum_{i=0}^{n} f_i \mathrm{L}_i(x) = \omega(x) \sum_{i=0}^{n} f_i \frac{u_i}{x - x_i}. \tag{2.17}$$

Having computed the weights u_i, evaluation of the polynomial only requires $O(n)$ operations whereas calculation of all the Lagrange polynomials requires $O(n^2)$ operations. Calculation of $\omega(x)$ can be avoided considering that

$$p_1(x) = \sum_{i=0}^{n} L_i(x) = \omega(x) \sum_{i=0}^{n} \frac{u_i}{x - x_i} \tag{2.18}$$

is a polynomial of degree n with

$$p_1(x_i) = 1 \quad i = 0 \ldots n. \tag{2.19}$$

But this is only possible if

$$p_1(x) = 1. \tag{2.20}$$

Therefore

$$p(x) = \frac{p(x)}{p_1(x)} = \frac{\sum_{i=0}^{n} f_i \frac{u_i}{x-x_i}}{\sum_{i=0}^{n} \frac{u_i}{x-x_i}} \tag{2.21}$$

which is known as the second form of the barycentric interpolation formula.

2.2.3 Newton's Divided Differences

Newton's method of divided differences [5] is an alternative for efficient numerical calculations [6]. Rewrite

$$f(x) = f(x_0) + \frac{f(x) - f(x_0)}{x - x_0}(x - x_0). \tag{2.22}$$

With the first order divided difference

$$f[x, x_0] = \frac{f(x) - f(x_0)}{x - x_0} \tag{2.23}$$

this becomes

$$f[x, x_0] = f[x_1, x_0] + \frac{f[x, x_0] - f[x_1, x_0]}{x - x_1}(x - x_1) \tag{2.24}$$

and with the second order divided difference

$$\begin{aligned} f[x, x_0, x_1] = \frac{f[x, x_0] - f[x_1, x_0]}{x - x_1} &= \frac{f(x) - f(x_0)}{(x - x_0)(x - x_1)} - \frac{f(x_1) - f(x_0)}{(x_1 - x_0)(x - x_1)} \\ &= \frac{f(x)}{(x - x_0)(x - x_1)} + \frac{f(x_1)}{(x_1 - x_0)(x_1 - x)} + \frac{f(x_0)}{(x_0 - x_1)(x_0 - x)} \end{aligned} \tag{2.25}$$

we have

$$f(x) = f(x_0) + (x - x_0)\, f[x_1, x_0] + (x - x_0)(x - x_1)\, f[x, x_0, x_1]. \tag{2.26}$$

Higher order divided differences are defined recursively by

$$f[x_1 x_2 \cdots x_{r-1} x_r] = \frac{f[x_1 x_2 \cdots x_{r-1}] - f[x_2 \cdots x_{r-1} x_r]}{x_1 - x_r}. \tag{2.27}$$

They are invariant against permutation of the arguments which can be seen from the explicit formula

$$f[x_1 x_2 \cdots x_r] = \sum_{k=1}^{r} \frac{f(x_k)}{\prod_{i \neq k}(x_k - x_i)}. \tag{2.28}$$

Finally we have

$$f(x) = p(x) + q(x) \tag{2.29}$$

with a polynomial of degree n

$$p(x) = f(x_0) + f[x_1, x_0](x - x_0) + f[x_2x_1x_0](x - x_0)(x - x_1) + \cdots$$

$$\cdots + f[x_nx_{n-1}\cdots x_0](x - x_0)(x - x_1)\cdots(x - x_{n-1}) \qquad (2.30)$$

and the function

$$q(x) = f[xx_n \cdots x_0](x - x_0)\cdots(x - x_n). \qquad (2.31)$$

Obviously $q(x_i) = 0 \quad , i = 0 \cdots n$, hence $p(x)$ is the interpolating polynomial. The divided differences can be arranged in the following way:

$$\begin{array}{llllll} f_0 & & & & & \\ f_1 & f[x_0x_1] & & & & \\ \vdots & \vdots & \ddots & & & \\ f_{n-1} & f[x_{n-2}x_{n-1}] & f[x_{n-3}x_{n-2}x_{n-1}] & \cdots & f[x_0\ldots x_{n-1}] & \\ f_n & f[x_{n-1}x_n] & f[x_{n-2}x_{n-1}x_n] & \cdots & f[x_1\cdots x_n] & f[x_0\cdots x_n]. \end{array} \qquad (2.32)$$

Since only the diagonal elements are needed, a one-dimensional data array $t[0]\cdots t[n]$ is sufficient for the calculation of the polynomial coefficients (Algorithms 1 and 2.).

Algorithm 1: Newtons divided differences

```
/* calculates f[x0x1]···f[x0x1···xn-1xn]          */
for i = 0 to N do
    ti ← fi;
    for k = i − 1 downto 0 do
        tk ← (tk+1 − tk)/(xi − xk);
    end
    ai ← t0;
end
```

The polynomial can then be evaluated with algorithm

Algorithm 2: Evaluation of the polynomial

```
/* evaluates the polynomial p(x)                  */
p ← aN;
for i = N − 1 downto 0 do
    p ← p ∗ (x − xi) + ai;
end
```

2.2.4 Neville Method

The Neville method [7] is advantageous if the polynomial is not needed explicitly and has to be evaluated only at one point. Consider the interpolating polynomial for the points $x_0 \cdots x_k$, which will be denoted as $P_{0,1,\cdots k}(x)$. Obviously

$$P_{0,1,\cdots k}(x) = \frac{(x - x_0)P_{1\cdots k}(x) - (x - x_k)P_{0\cdots k-1}(x)}{x_k - x_0} \tag{2.33}$$

since for $x = x_1 \cdots x_{k-1}$ the right hand side is

$$\frac{(x - x_0)f(x) - (x - x_k)f(x)}{x_k - x_0} = f(x). \tag{2.34}$$

For $x = x_0$ we have

$$\frac{-(x_0 - x_k)f(x)}{x_k - x_0} = f(x) \tag{2.35}$$

and finally for $x = x_k$

$$\frac{(x_k - x_0)f(x)}{x_k - x_0} = f(x). \tag{2.36}$$

We use the following scheme to calculate $P_{0,1\cdots n}(x)$ recursively:

$$\begin{array}{lllll} P_0 & & & & \\ P_1 & P_{01} & & & \\ P_2 & P_{12} & P_{012} & & \\ \vdots & \vdots & \vdots & \ddots & \\ P_n & P_{n-1,n} & P_{n-2,n-1,n} & \cdots & P_{01\cdots n}. \end{array} \tag{2.37}$$

The first column contains the function values $P_i(x) = f_i$. The value $P_{01\cdots n}$ can be calculated using a one-dimensional data array $p[0] \cdots p[n]$ (Algorithm 3).

Algorithm 3: Neville Algorithm

```
/* calculate interpolating polynomial P_{01...n}                */
for i = 0 to N do
    p_i ← f_i;
    for k = i − 1 downto 0 do
        p_k ← [p_{k+1} * (x − x_k) − p_k * (x − x_i)]/(x_k − x_i);
    end
end
f ← p_0;
```

2.2.5 Bernstein Polynomials

In 1912, S. Bernstein used a certain class of polynomials for a direct prove of the Weierstrass approximation theorem [8]. He showed that any real valued continuous function $f(x)$ on the interval [0, 1] [1] can be uniformly approximated by the following sequence of polynomials

$$\lim_{n\to\infty} p_n(x) = \sum_{k=0}^{n} f\left(\frac{k}{n}\right)\binom{n}{k} x^k(1-x)^{n-k} = f(x). \tag{2.38}$$

The $n+1$ Bernstein basis polynomials of degree n are defined as [9, 10]

$$B_{k,n}(x) = \binom{n}{k} x^k(1-x)^{n-k} \quad k = 0, 1 \ldots n. \tag{2.39}$$

They have become popular for spline interpolation [11, 12], numerical approximation [13] and integration [14], finite elements [15, 16], higher order differential equations [17] and the construction of Beziér curves.

The Bernstein polynomials are all non-negative over the interval $x \in [0, 1]$. They have a maximum at $x = k/n$ and have boundary values

$$B_{k,n}(0) = \delta_{k,0} \quad B_{k,n}(1) = \delta_{k,n}. \tag{2.40}$$

The n+1 polynomials of degree n form a partition of unity, they are linearly independent and span a basis of the polynomials with order $\leq n$. Their explicit form in terms of the power basis is

$$B_{k,n}(x) = \sum_{i=k}^{n} (-1)^{i-k} \binom{n}{i}\binom{i}{k} x^i \tag{2.41}$$

and the inverse relation is given by

$$x^k = \sum_{i=k-1}^{n-1} \frac{\binom{i}{k}}{\binom{n}{k}} B_{i,n}(x). \tag{2.42}$$

They obey the recursion relation

---

[1] In the current literature, Bernstein polynomials are mostly used on the interval |0, 1], especially in the context of computer graphics. They can easily be extended to any other finite interval $[a, b]$ by substitution $x' = a + x(b-a)$ to obtain the generalized Bernstein polynomials $B_{k,n}(x) = \frac{1}{(b-a)^n}\binom{n}{k}(x-a)^k(b-x)^{n-k}$.

$$B_{k,n}(x) = (1-x)B_{k,n-1}(x) + xB_{k-1,n-1}(x) \tag{2.43}$$

and their derivatives

$$\frac{dB_{k,n}(x)}{dx} = n\left(B_{k-1,n-1}(x) - B_{k,n-1}(x)\right). \tag{2.44}$$

Indefinite integrals (higher order integrals are derived in [17]) can be expanded as

$$\int B_{k,n}(x)dx = \frac{1}{n+1}\sum_{j=k+1}^{n+1} B_{j,n+1}(x) \tag{2.45}$$

from which also definite integrals

$$\int_0^1 B_{k,n}(x)dx = \frac{1}{n+1} \tag{2.46}$$

follow.

The matrix representation of a Bernstein-Beziér polynomial is

$$\begin{aligned} b(x) &= c_0 B_{0,n}(x) + \cdots + c_n B_{n,n}(x) \\ &= \begin{pmatrix} B_{0,n} & \cdots & B_{n.n} \end{pmatrix} \begin{pmatrix} c_0 \\ \vdots \\ c_n \end{pmatrix} \\ &= \begin{pmatrix} x^0 & \cdots & x^n \end{pmatrix} \begin{pmatrix} b_{0,0} & \cdots & 0 \\ \vdots & \ddots & \vdots \\ b_{n,0} & \cdots & b_{n.n} \end{pmatrix} \begin{pmatrix} c_0 \\ \vdots \\ c_n \end{pmatrix}. \end{aligned} \tag{2.47}$$

Lagrange interpolation of f(x) for $x_k = \frac{k}{n}, \quad k = 0, 1 \cdots n$ gives

$$\begin{aligned} p(x) &= \sum_{k=0}^{n} f(\frac{k}{n}) L_{k,n}(x) \\ &= \begin{pmatrix} L_{0,n}(x) & \cdots & L_{n.n}(x) \end{pmatrix} \begin{pmatrix} f_0 \\ \vdots \\ f_n \end{pmatrix} \end{aligned}$$

$$B_{k,n}(x) = \sum_j B_{k,n}(x_j) L_{j,n}(x) \tag{2.48}$$

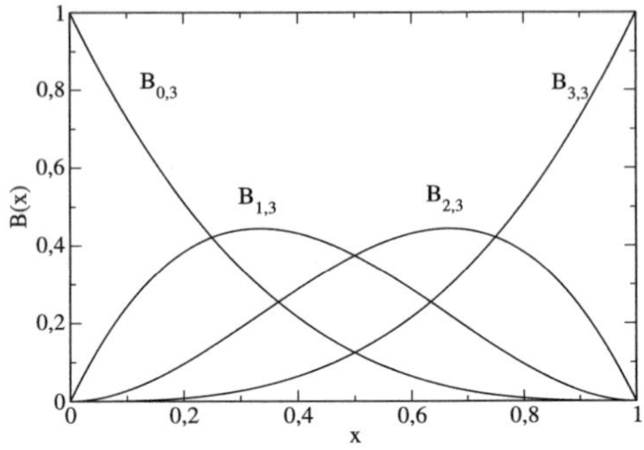

Fig. 2.2 Bernstein polynomials of degree 3

$$\begin{pmatrix} B_{0,n}(x) & \cdots & B_{n,n}(x) \end{pmatrix} = \begin{pmatrix} L_{0,n}(x) & \cdots & L_{n,n}(x) \end{pmatrix} \begin{pmatrix} B_{0,n}(x_0) & \cdots & B_{n,n}(x_0) \\ \vdots & \ddots & \vdots \\ B_{0,n}(x_n) & \cdots & B_{n,n}(x_n) \end{pmatrix} \tag{2.49}$$

$$p(x) = \begin{pmatrix} B_{0,n}(x) & \cdots & B_{n,n}(x) \end{pmatrix} \begin{pmatrix} B_{0,n}(x_0) & \cdots & B_{n,n}(x_0) \\ \vdots & \ddots & \vdots \\ B_{0,n}(x_n) & \cdots & B_{n,n}(x_n) \end{pmatrix}^{-1} \begin{pmatrix} f_0 \\ \vdots \\ f_n \end{pmatrix}. \tag{2.50}$$

Example $n = 3$: (Fig. 2.2)

$$B_{0,3} = (1-x)^3,\ B_{1,3} = 3x(1-x)^2,\ B_{2,3} = 3x^2(1-x),\ B_{3,3} = x^3. \tag{2.51}$$

Bernstein polynomials are well suited for Hermite spline interpolation. To find a cubic polynomial which has given values of function and derivative at the interval boundaries we set

$$p(x) = f_0 B_{0,3} + f_1 B_{3,3} + c_1 B_{1,3} + c_2 B_{2,3} \tag{2.52}$$

and evaluate the derivative

$$p'(x) = -3 f_0 B_{0,2}(x) + 3 f_1 \left(B_{2,2}(x)\right)$$
$$+ 3c_1 \left(B_{0,2}(x) - B_{1,2}(x)\right) + 3c_2 \left(B_{1,2}(x) - B_{2,2,}(x)\right) \tag{2.53}$$

from which

$$p'(0) = -3 f_0 + 3c_1 \tag{2.54}$$

$$p'(1) = 3 f_1 - 3c_2. \tag{2.55}$$

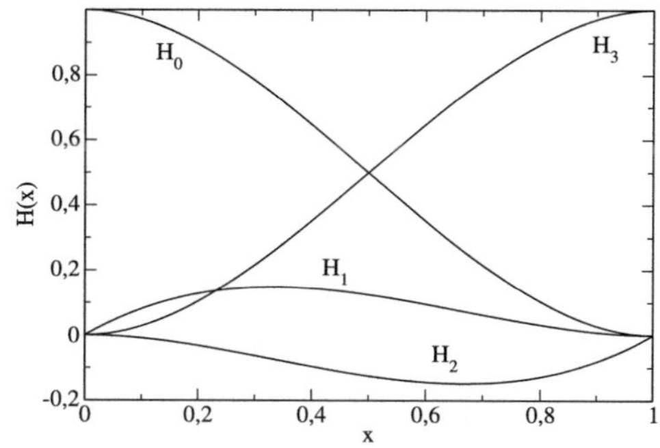

Fig. 2.3 Hermite basis functions of degree 3

The desired polynomial is

$$p(x) = f_0 B_{0,3} + f_1 B_{3,3} + \left(\frac{f_0'}{3} + f_0\right) B_{1,3} + \left(f_1 - \frac{f_1'}{3}\right) B_{2,3} \tag{2.56}$$

which can be written alternatively as

$$\begin{aligned} p(x) &= f_0(B_{0,3} + B_{1,3}) + f_1(B_{3,3} + B_{2,3}) + f_0' \frac{B_{1,3}}{3} + f_1 \frac{-B_{2,3}}{3} \\ &= f_0 H_0(x) + f_0' H_1(x) + f_1' H_2(x) + f_1 H_3(x) \end{aligned} \tag{2.57}$$

with the Hermite basis functions of order 3 (Fig. 2.3)

$$\begin{aligned} H_0 &= (1-x)^2(1+2x) \\ H_1 &= x(1-x)^2 \\ H_2 &= -x^2(1-x) \\ H_3 &= x^2(3-2x). \end{aligned} \tag{2.58}$$

De Casteljau's Algorithm

A polynomial of degree n in Bernstein form

$$p(x) = \sum_{k=0}^{n} p_k B_{k,n}(x) \tag{2.59}$$

can be written with the help of the recursion (2.43) as the sum of two Bernstein polynomials of order $n-1$:

$$p(x) = p_0(1-x)B_{0,n-1} + \sum_{k=1}^{n-1} p_k\left[(1-x)B_{k,n-1}(x)\right.$$

$$\left. + xB_{k-1,n-1}(x)\right] + p_n x B_{n-1,n-1}$$

$$= (1-x)\sum_{k=0}^{n-1} p_k B_{k,n-1}(x) + x\sum_{k=1}^{n} p_k B_{k-1,n-1}(x). \tag{2.60}$$

Hence the following tableau is useful to build up $p(x)$:

$p_0^{(0)} = p_0$	$p_1^{(0)} = p_1$	$\cdots$	$p_n^{(0)} = p_n$
	$p_1^{(1)} = (1-x)p_0^{(0)} + xp_1^{(0)}$	$\cdots$	$p_n^{(1)} = (1-x)p_{n-1}^{(0)} + xp_n^{(0)}$
		$\ddots$	$\vdots$
			$p_n^{(n)} = (1-x)p_{n-1}^{(n-1)} + xp_n^{(n-1)}$

De Casteljau's recursive algorithm [18] is numerically more stable than the direct method and needs only a one-dimensional array p_k (Algorithm 4).

Algorithm 4: De Casteljau's algorithm

```
for k=1 to n do
    p_k^(0) = p_k
end
for j=1 to n do
    for k=n downto j do
        p_k^(j) ← p_{k-1}^(j-1) * (1 - x) + p_k^(j-1) * x
    end
end
p(x) ← p_n^(n)
```

2.2.6 Error of Polynomial Interpolation

The error of polynomial interpolation [19] can be estimated with the help of the following theorem:

If $f(x)$ is $n+1$ times differentiable then for each $\overline{x}$ there exists ξ within the smallest interval containing $\overline{x}$ as well as all the x_i with

$$q(\overline{x}) = \prod_{i=0}^{n}(\overline{x} - x_i)\frac{f^{(n+1)}(\xi)}{(n+1)!}. \tag{2.61}$$

From a discussion of the function

$$\omega(x) = \prod_{i=0}^{n}(x - x_i) \tag{2.62}$$

it can be seen that the error increases rapidly outside the region of the sample points (extrapolation is dangerous!). As an example consider the sample points (Fig. 2.4).

$$f(x) = \sin(x) \quad x_i = 0, \frac{\pi}{2}, \pi, \frac{3\pi}{2}, 2\pi. \tag{2.63}$$

The maximum interpolation error is estimated by ($|f^{(n+1)}| \leq 1$)

$$|f(x) - p(x)| \leq |\omega(x)|\frac{1}{120} \leq \frac{35}{120} \approx 0.3 \tag{2.64}$$

whereas the error increases rapidly outside the interval $0 < x < 2\pi$ (Fig. 2.5).

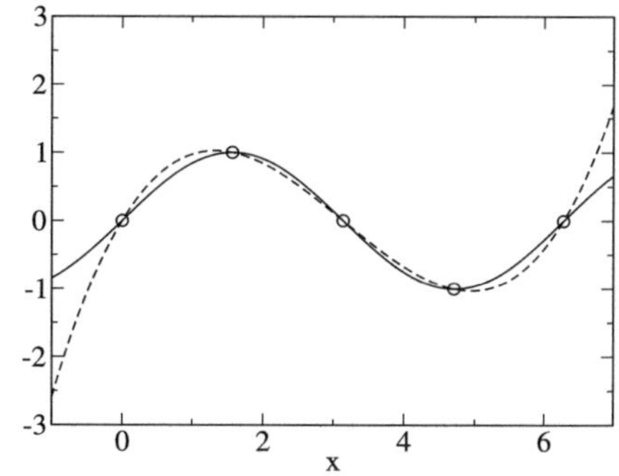

Fig. 2.4 Interpolating polynomial. The interpolated function (solid curve) and the interpolating polynomial (broken curve) for the example (2.63) are compared

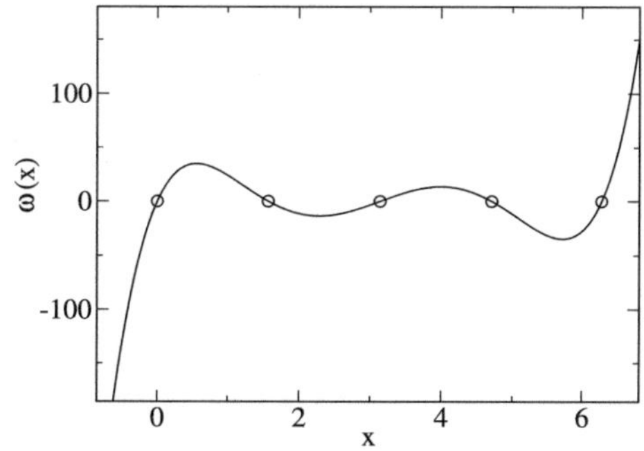

Fig. 2.5 Interpolation error. The polynomial $\omega(x)$ is shown for the example (2.63). Its roots x_i are given by the x values of the sample points (circles). Inside the interval $x_0 \cdots x_4$ the absolute value of ω is bounded by $|\omega(x)| \leq 35$ whereas outside the interval it increases very rapidly

2.3 Spline Interpolation

Polynomials are not well suited for interpolation over a larger range. Often spline functions are superior which are piecewise defined polynomials [20, 21]. The simplest case is a linear spline which just connects the sampling points by straight lines:

$$p_i(x) = y_i + \frac{y_{i+1} - y_i}{x_{i+1} - x_i}(x - x_i) \tag{2.65}$$

$$s(x) = p_i(x) \text{ where } x_i \leq x < x_{i+1}. \tag{2.66}$$

The most important case is the cubic spline which is given in the interval $x_i \leq x < x_{i+1}$ by

$$p_i(x) = \alpha_i + \beta_i(x - x_i) + \gamma_i(x - x_i)^2 + \delta_i(x - x_i)^3. \tag{2.67}$$

We want to have a smooth interpolation and assume that the interpolating function and their first two derivatives are continuous. Hence we have for the inner boundaries:

$$i = 0, \cdots n - 1$$

$$p_i(x_{i+1}) = p_{i+1}(x_{i+1}) \tag{2.68}$$

$$p_i'(x_{i+1}) = p_{i+1}'(x_{i+1}) \tag{2.69}$$

$$p_i''(x_{i+1}) = p_{i+1}''(x_{i+1}). \tag{2.70}$$

We have to specify boundary conditions at x_0 and x_n. The most common choice are natural boundary conditions $s''(x_0) = s''(x_n) = 0$, but also periodic boundary conditions $s''(x_0) = s''(x_n)$, $s'(x_0) = s'(x_n)$, $\quad s(x_0) = s(x_n)$ or given derivative values $s'(x_0)$ and $s'(x_n)$ are often used. The second derivative is a linear function [2]

$$p_i''(x) = 2\gamma_i + 6\delta_i(x - x_i) \tag{2.71}$$

which can be written using $h_{i+1} = x_{i+1} - x_i$ and $M_i = s''(x_i)$ as

$$p_i''(x) = M_{i+1}\frac{(x - x_i)}{h_{i+1}} + M_i\frac{(x_{i+1} - x)}{h_{i+1}} \qquad i = 0 \cdots n - 1 \tag{2.72}$$

since

$$p_i''(x_i) = M_i\frac{x_{i+1} - x_i}{h_{i+1}} = s''(x_i) \tag{2.73}$$

$$p_i''(x_{i+1}) = M_{i+1}\frac{(x_{i+1} - x_i)}{h_{i+1}} = s''(x_{i+1}). \tag{2.74}$$

Integration gives with the two constants A_i and B_i

$$p_i'(x) = M_{i+1}\frac{(x-x_i)^2}{2h_{i+1}} - M_i\frac{(x_{i+1}-x)^2}{2h_{i+1}} + A_i \tag{2.75}$$

$$p_i(x) = M_{i+1}\frac{(x-x_i)^3}{6h_{i+1}} + M_i\frac{(x_{i+1}-x)^3}{6h_{i+1}} + A_i(x-x_i) + B_i. \tag{2.76}$$

From $s(x_i) = y_i$ and $s(x_{i+1}) = y_{i+1}$ we have

$$M_i\frac{h_{i+1}^2}{6} + B_i = y_i \tag{2.77}$$

$$M_{i+1}\frac{h_{i+1}^2}{6} + A_i h_{i+1} + B_i = y_{i+1} \tag{2.78}$$

and hence

$$B_i = y_i - M_i\frac{h_{i+1}^2}{6} \tag{2.79}$$

$$A_i = \frac{y_{i+1}-y_i}{h_{i+1}} - \frac{h_{i+1}}{6}(M_{i+1}-M_i). \tag{2.80}$$

Now the polynomial is

$$\begin{aligned} p_i(x) &= \frac{M_{i+1}}{6h_{i+1}}(x-x_i)^3 - \frac{M_i}{6h_{i+1}}(x-x_i-h_{i+1})^3 + A_i(x-x_i) + B_i \\ &= (x-x_i)^3\left(\frac{M_{i+1}}{6h_{i+1}} - \frac{M_i}{6h_{i+1}}\right) + \frac{M_i}{6h_{i+1}}3h_{i+1}(x-x_i)^2 \\ &+ (x-x_i)\left(A_i - \frac{M_i}{6h_{i+1}}3h_{i+1}^2\right) + B_i + \frac{M_i}{6h_{i+1}}h_{i+1}^3. \end{aligned} \tag{2.81}$$

Comparison with

$$p_i(x) = \alpha_i + \beta_i(x-x_i) + \gamma_i(x-x_i)^2 + \delta_i(x-x_i)^3 \tag{2.82}$$

gives

$$\alpha_i = B_i + \frac{M_i}{6}h_{i+1}^2 = y_i \tag{2.83}$$

$$\beta_i = A_i - \frac{h_{i+1}M_i}{2} = \frac{y_{i+1}-y_i}{h_{i+1}} - h_{i+1}\frac{M_{i+1}+2M_i}{6} \tag{2.84}$$

$$\gamma_i = \frac{M_i}{2} \tag{2.85}$$

$$\delta_i = \frac{M_{i+1} - M_i}{6h_{i+1}}. \tag{2.86}$$

Finally we calculate M_i from the continuity of $s'(x)$. Substituting for A_i in $p_i'(x)$ we have

$$p_i'(x) = M_{i+1}\frac{(x - x_i)^2}{2h_{i+1}} - M_i\frac{(x_{i+1} - x)^2}{2h_{i+1}} + \frac{y_{i+1} - y_i}{h_{i+1}} - \frac{h_{i+1}}{6}(M_{i+1} - M_i) \tag{2.87}$$

and from $p_{i-1}'(x_i) = p_i'(x_i)$ it follows

$$M_i\frac{h_i}{2} + \frac{y_i - y_{i-1}}{h_i} - \frac{h_i}{6}(M_i - M_{i-1})$$

$$= -M_i\frac{h_{i+1}}{2} + \frac{y_{i+1} - y_i}{h_{i+1}} - \frac{h_{i+1}}{6}(M_{i+1} - M_i) \tag{2.88}$$

$$M_i\frac{h_i}{3} + M_{i-1}\frac{h_i}{6} + M_i\frac{h_{i+1}}{3} + M_{i+1}\frac{h_{i+1}}{6} = \frac{y_{i+1} - y_i}{h_{i+1}} - \frac{y_i - y_{i-1}}{h_i} \tag{2.89}$$

which is a system of linear equations for the M_i. Using the abbreviations

$$\lambda_i = \frac{h_{i+1}}{h_i + h_{i+1}} \tag{2.90}$$

$$\mu_i = 1 - \lambda_i = \frac{h_i}{h_i + h_{i+1}} \tag{2.91}$$

$$d_i = \frac{6}{h_i + h_{i+1}}\left(\frac{y_{i+1} - y_i}{h_{i+1}} - \frac{y_i - y_{i-1}}{h_i}\right) \tag{2.92}$$

we have

$$\mu_i M_{i-1} + 2M_i + \lambda_i M_{i+1} = d_i \quad i = 1 \cdots n - 1. \tag{2.93}$$

We define for natural boundary conditions

$$\lambda_0 = 0 \quad \mu_n = 0 \quad d_0 = 0 \quad d_n = 0 \tag{2.94}$$

and in case of given derivative values

$$\lambda_0 = 1 \quad \mu_n = 1 \quad d_0 = \frac{6}{h_1}\left(\frac{y_1 - y_0}{h_1} - y_0'\right) \quad d_n = \frac{6}{h_n}\left(y_n' - \frac{y_n - y_{n-1}}{h_n}\right). \tag{2.95}$$

The system of equation has the form

$$\begin{bmatrix} 2 & \lambda_0 & & & & \\ \mu_1 & 2 & \lambda_1 & & & \\ & \mu_2 & 2 & \lambda_2 & & \\ & & \ddots & \ddots & \ddots & \\ & & & \mu_{n-1} & 2 & \lambda_{n-1} \\ & & & & \mu_n & 2 \end{bmatrix} \begin{bmatrix} M_0 \\ M_1 \\ M_2 \\ \vdots \\ M_{n-1} \\ M_n \end{bmatrix} = \begin{bmatrix} d_0 \\ d_1 \\ d_2 \\ \vdots \\ d_{n-1} \\ d_n \end{bmatrix}. \tag{2.96}$$

For periodic boundary conditions we define

$$\lambda_n = \frac{h_1}{h_1 + h_n} \quad \mu_n = 1 - \lambda_n \quad d_n = \frac{6}{h_1 + h_n}\left(\frac{y_1 - y_n}{h_1} - \frac{y_n - y_{n-1}}{h_n}\right) \tag{2.97}$$

and the system of equations is (with $M_n = M_0$)

$$\begin{bmatrix} 2 & \lambda_1 & & & & \mu_1 \\ \mu_2 & 2 & \lambda_2 & & & \\ & \mu_3 & 2 & \lambda_3 & & \\ & & \ddots & \ddots & \ddots & \\ & & & \mu_{n-1} & 2 & \lambda_{n-1} \\ \lambda_n & & & & \mu_n & 2 \end{bmatrix} \begin{bmatrix} M_1 \\ M_2 \\ M_3 \\ \vdots \\ M_{n-1} \\ M_n \end{bmatrix} = \begin{bmatrix} d_1 \\ d_2 \\ d_3 \\ \vdots \\ d_{n-1} \\ d_n \end{bmatrix}. \tag{2.98}$$

All this tridiagonal systems can be easily solved with a special Gaussian elimination method (Sects. 5.3, 5.4).

2.4 Rational Interpolation

The use of rational approximants allows to interpolate functions with poles, where polynomial interpolation can give poor results [2]. Rational approximants without poles [22] are also well suited for the case of equidistant x_i, where higher order polynomials tend to become unstable. The main disadvantages are additional poles which are difficult to control and the appearance of unattainable points. Recent developments using the barycentric form of the interpolating function [22–24] helped to overcome these difficulties.

2.4.1 Pade Approximant

The Pade approximant [25] of order $[M/N]$ to a function $f(x)$ is the rational function

$$R_{M/N}(x) = \frac{P_M(x)}{Q_N(x)} = \frac{p_0 + p_1 x + \ldots p_M x^M}{q_0 + q_1 x + \ldots q_N x^N} \tag{2.99}$$

which reproduces the McLaurin series (the Taylor series at $x = 0$) of

$$f(x) = a_0 + a_1 x + a_2 x^2 + \ldots \tag{2.100}$$

up to order $M + N$, i.e.

$$f(0) = R(0)$$

$$\frac{\mathrm{d}}{\mathrm{d}x} f(0) = \frac{\mathrm{d}}{\mathrm{d}x} R(0)$$

$$\vdots$$

$$\frac{\mathrm{d}^{(M+N)}}{\mathrm{d}x^{(M+N)}} f(0) = \frac{\mathrm{d}^{(M+N)}}{\mathrm{d}x^{(M+N)}} R(0). \tag{2.101}$$

Multiplication gives

$$p_0 + p_1 x + \cdots + p_M x^M = (q_0 + q_1 x + \cdots + q_N x^N)(a_0 + a_1 x + \cdots) \tag{2.102}$$

and collecting powers of x we find the system of equations

$$p_0 = q_0 a_0$$

$$p_1 = q_0 a_1 + q_1 a_0$$

$$p_2 = q_0 a_2 + a_1 q_1 + a_0 q_2$$

$$\vdots$$

$$p_M = q_0 a_M + a_{M-1} q_1 + \cdots + a_0 q_M$$

$$0 = q_0 a_{M+1} + q_1 a_M + \cdots + q_N a_{M-N+1}$$

$$\vdots$$

$$0 = q_0 a_{M+N} + q_1 a_{M+N-1} + \cdots + q_N a_M \qquad (2.103)$$

where

$$a_n = 0 \quad \text{for } n < 0 \qquad (2.104)$$

$$q_j = 0 \quad \text{for } j > N. \qquad (2.105)$$

Example: Calculate the [3, 3] approximant to $\tan(x)$
The Laurent series of the tangent is

$$\tan(x) = x + \frac{1}{3}x^3 + \frac{2}{15}x^5 + \ldots. \qquad (2.106)$$

We set $q_0 = 1$. Comparison of the coefficients of the polynomial

$$p_0 + p_1 x + p_2 x^2 + p_3 x^3 = (1 + q_1 x + q_2 x^2 + q_3 x^3)(x + \frac{1}{3}x^3 + \frac{2}{15}x^5) \qquad (2.107)$$

gives the equations

$$\begin{aligned} x^0 &: p_0 = 0 \\ x^{1:} & p_1 = 1 \\ x^2 &: p_2 = q_1 \\ x^3 &: p_3 = q_2 + \frac{1}{3} \\ x^4 &: 0 = q_3 + \frac{1}{3}q_1 \\ x^5 &: 0 = \frac{2}{15} + \frac{1}{3}q_2 \\ x^6 &: 0 = \frac{2}{15}q_1 + \frac{1}{3}q_3. \end{aligned} \qquad (2.108)$$

We easily find

$$p_2 = q_1 = q_3 = 0 \quad q_2 = -\frac{2}{5} \quad p_3 = -\frac{1}{15} \qquad (2.109)$$

and the approximant of order [3, 3] is

$$R_{3,3} = \frac{x - \frac{1}{15}x^3}{1 - \frac{2}{5}x^2}. \qquad (2.110)$$

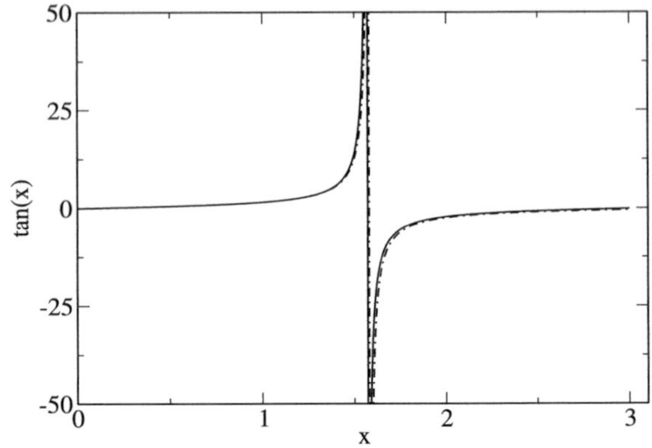

Fig. 2.6 Pade approximation to $\tan(x)$. The Pade approximant (2.110, dash dotted curve) reproduces the tangent (full curve) quite well

This expression reproduces the tangent quite well (Fig. 2.6). Its pole at $\sqrt{10}/2 \approx 1.581$ is close to the pole of the tangent function at $\pi/2 \approx 1.571$.

2.4.2 Barycentric Rational Interpolation

If the weights of the barycentric form of the interpolating polynomial (2.21) are taken as general parameters $u_i \neq 0$ it becomes a rational function

$$R(x) = \frac{\sum_{i=0}^{n} f_i \frac{u_i}{x-x_i}}{\sum_{i=0}^{n} \frac{u_i}{x-x_i}} \tag{2.111}$$

which obviously interpolates the data points since

$$\lim_{x \to x_i} R(x) = f_i. \tag{2.112}$$

With the polynomials[2]

$$P(x) = \sum_{i=0}^{n} u_i f_i \prod_{j=0;\, j \neq i}^{n} (x - x_j) = \sum_{i=0}^{n} u_i f_i \frac{\omega(x)}{x - x_i}$$

$$Q(x) = \sum_{i=0}^{n} u_i \prod_{j=0;\, j \neq i}^{n} (x - x_j) = \sum_{i=0}^{n} u_i \frac{\omega(x)}{x - x_i}$$

---

[2] $\omega(x) = \prod_{i=0}^{n}(x - x_i)$ as in (2.62).

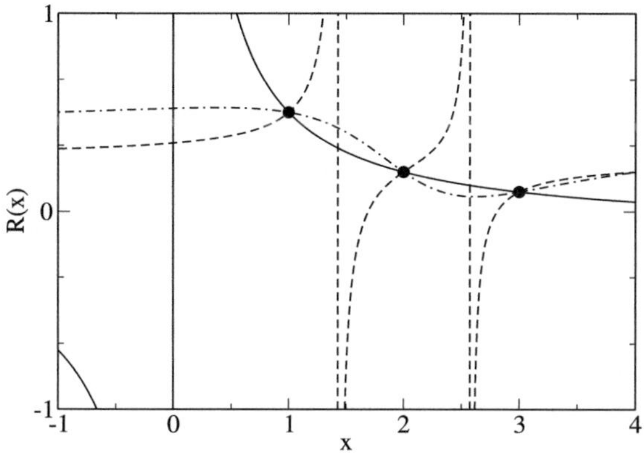

Fig. 2.7 Rational interpolation. The data points $(1, \frac{1}{2})$, $(2, \frac{1}{5})$, $(3, \frac{1}{10})$ are interpolated by several rational functions. The [1, 1] approximant (2.118) corresponding to $\mathbf{u} = (5, -20, 15)$ is shown by the solid curve, the dashed curve shows the function $R(x) = \frac{8x^2-36x+38}{10(3x^2-12x+11)}$ which is obtained for $\mathbf{u} = (1, 1, 1)$ and the dash dotted curve shows the function $R(x) = \frac{4x^2-20x+26}{10(5-4x+x^2)}$ which follows for $\mathbf{u} = (1, -1, 1)$ and has no real poles

a rational interpolating function is given by[3]

$$R(x) = \frac{P(x)}{Q(x)}.$$

Obviously there are infinitely different rational interpolating functions which differ by the weights $\mathbf{u} = (u_0, u_1 \dots u_n)$ (an example is shown in Fig. 2.7). To fix the parameters u_i, additional conditions have to be imposed.

Rational Interpolation of Order [M, N]

One possibility is to assume that $P(x)$ and $Q(x)$ are of order $\leq M$ and $\leq N$, respectively with $M + N = n$. This gives n additional equations for the $2(n + 1)$ polynomial coefficients. The number of unknown equals $n + 1$ and the rational interpolant is uniquely determined up to a common factor in numerator and denominator.

Example consider the data points $f(1) = \frac{1}{2}$, $f(2) = \frac{1}{5}$, $f(3) = \frac{1}{10}$
The polynomials are

$$P(x) = \frac{1}{2}u_0(x-2)(x-3) + \frac{1}{5}u_1(x-1)(x-3) + \frac{1}{10}u_2(x-1)(x-2)$$

[3] It can be shown that any rational interpolant can be written in this form.

$$= 3u_0 + \frac{3}{5}u_1 + \frac{1}{5}u_2 + \left[-\frac{5}{2}u_0 - \frac{4}{5}u_1 - \frac{3}{10}u_2\right]x + \left[\frac{1}{2}u_0 + \frac{1}{5}u_1 + \frac{1}{10}u_2\right]x^2 \tag{2.113}$$

$$Q(x) = u_0(x-2)(x-3) + u_1(x-1)(x-3) + u_2(x-1)(x-2)$$

$$= 6u_0 + 3u_1 + 2u_2 + [-5u_0 - 4u_1 - 3u_2]x + [u_0 + u_1 + u_2]x^2. \tag{2.114}$$

To obtain a [1, 1] approximant we have to solve the equations

$$\frac{1}{2}u_0 + \frac{1}{5}u_1 + \frac{1}{10}u_2 = 0 \tag{2.115}$$

$$u_0 + u_1 + u_2 = 0 \tag{2.116}$$

which gives

$$u_2 = 3u_0 \quad u_1 = -4u_0 \tag{2.117}$$

and thus

$$R(x) = \frac{\frac{6}{5}u_0 - \frac{1}{5}u_0 x}{2u_0 x} = \frac{6-x}{10x}. \tag{2.118}$$

General methods to obtain the coefficients u_i for a given data set are described in [23, 24]. They also allow to determine unattainable points corresponding to $u_i = 0$ and to locate the poles. Without loss of generality it can be assumed [24] that $M \geq N$[4]

Let $P(x)$ be the unique polynomial which interpolates the product $f(x)Q(x)$

$$P(x_i) = f(x_i)Q(x_i) \quad i = 0 \ldots M. \tag{2.119}$$

Then from (2.31) we have

$$f(x)Q(x) - P(x) = (fQ)[x_0 \cdots x_M, x](x - x_0) \cdots (x - x_M). \tag{2.120}$$

Setting

$$x = x_i \quad i = M+1, \ldots n \tag{2.121}$$

we have

$$f(x_i)Q(x_i) - P(x_i) = (fQ)[x_0 \ldots x_M, x_i](x_i - x_0) \ldots (x - x_M) \tag{2.122}$$

which is zero if $P(x_i)/Q(x_i) = f_i$ for $i = 0, \ldots n$. But then

$$(fQ)[x_0 \ldots x_M, x_i] = 0 \quad i = M+1, \ldots n. \tag{2.123}$$

---

[4] The opposite case can be treated by considering the reciprocal function values $1/f(x_i)$.

The polynomial $Q(x)$ can be written in Newtonian form (2.30)

$$Q(x) = \sum_{i=0}^{N} \nu_i \prod_{j=0}^{i-1}(x - x_j) = \nu_0 + \nu_1(x - x_0) + \cdots + \nu_N(x - x_0)\ldots(x - x_{N-1}). \tag{2.124}$$

With the abbreviation

$$g_j(x) = x - x_j \quad j = 0\ldots N \tag{2.125}$$

we find

$$(fg_j)[x_0\ldots x_M, x_i] = \sum_{k=0\ldots M,i} \frac{f(x_k)g(x_k)}{\prod_{r\neq k}(x_k - x_r)} = \sum_{k=0\ldots M,i,k\neq j} \frac{f(x_k)}{\prod_{r\neq k,r\neq j}(x_k - x_r)}$$

$$= f[x_0\ldots x_{j-1}, x_{j+1}\ldots x_M, x_i] \tag{2.126}$$

which we apply repeatedly to (2.123) to get the system of $n - M = N$ equations for $N + 1$ unknowns

$$\sum_{j=0}^{N} \nu_j f[x_j, x_{j+1}\ldots x_M, x_i] = 0 \quad i = M + 1\ldots n \tag{2.127}$$

from which the coefficients ν_j can be found by Gaussian elimination up to a scaling factor. The Newtonian form of $Q(x)$ can then be converted to the barycentric form as described in [6].

Rational interpolation without poles

Polynomial interpolation of larger data sets can be ill behaved, especially for the case of equidistant $x-$values. Rational interpolation without poles can be a much better choice here (Fig. 2.8).

Berrut [26] suggested to choose the following weights

$$u_k = (-1)^k.$$

With this choice $Q(x)$ has no real roots. Floater and Horman [22] used the different choice

$$u_k = (-1)^{k-1}\left(\frac{1}{x_{k+1} - x_k} + \frac{1}{x_k - x_{k-1}}\right) \quad k = 1\ldots n - 1$$

$$u_0 = -\frac{1}{x_1 - x_0} \quad u_n = (-1)^{n-1}\frac{1}{x_n - x_{n-1}} \tag{2.128}$$

which becomes very similar for equidistant x-values.

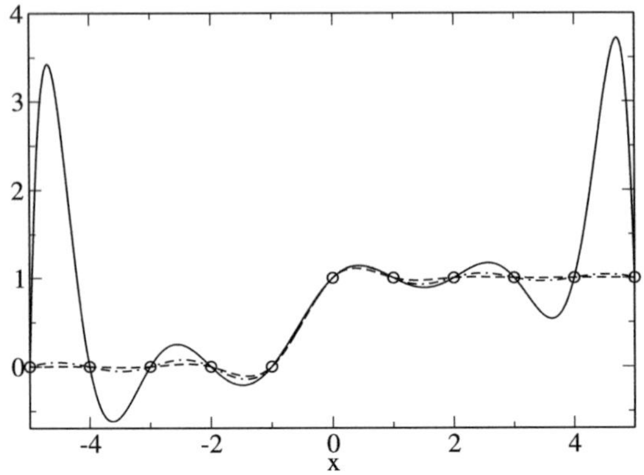

Fig. 2.8 Interpolation of a step function. A step function with uniform x-values (circles) is interpolated by a polynomial (full curve), a cubic spline (dashed curve) and with the rational Floater-Horman $d = 1$ function (2.128, dash-dotted curve). The rational function behaves similar to the spline function but provides in addition an analytical function with continuous derivatives

Table 2.1 Floater-Horman weights for uniform data

$\lvert u_k \rvert$	d
1, 1, 1 . . . , 1, 1, 1	0
1, 2, 2, 2, . . . , 2, 2, 2, 1	1
1, 3, 4, 4, 4, . . . , 4, 4, 4, 3, 1	2
1, 4, 7, 8, 8, 8, . . . , 8, 8, 8, 7, 4, 1	3
1, 5, 11, 15, 16, 16, 16, . . . , 16, 16, 16, 15, 11, 5, 1	4

Floater and Horman generalized this expression and found a class of rational interpolants without poles given by the weights

$$u_k = (-1)^{k-d} \sum_{i=\max(k-d,0)}^{\min(k,n-d)} \prod_{j=i, j\neq k}^{i+d} \frac{1}{|x_k - x_j|} \tag{2.129}$$

where $0 \leq d \leq n$ and the approximation order increases with d. In the uniform case this simplifies to (Table 2.1)

$$u_k = (-1)^{k-d} \sum_{i=\min(k-d,0)}^{\max(k,n-d)} \binom{d}{k-i}. \tag{2.130}$$

2.5 Multivariate Interpolation

The simplest two-dimensional interpolation method isbilinear interpolation.[5] It uses linear interpolation for both coordinates within the rectangle $x_i \leq x \leq x_{i+1} \quad y_i \leq y_i \leq y_{i+1}$:

$$p(x_i + h_x, y_i + h_y) = p(x_i + h_x, y_i) + h_y \frac{p(x_i + h_x, y_{i+1}) - p(x_i + h_x, y_i)}{y_{i+1} - y_i}$$

$$= f(x_i, y_i) + h_x \frac{f(x_{i+1}, y_i) - f(x_i, y_i)}{x_{i+1} - x_i} \tag{2.131}$$

$$+ h_y \frac{f(x_i, y_{i+1}) + h_x \frac{f(x_{i+1}, y_{i+1}) - f(x_i, y_{i+1})}{x_{i+1} - x_i} - f(x_i, y_i) - h_x \frac{f(x_{i+1}, y_i) - f(x_i, y_i)}{x_{i+1} - x_i}}{y_{i+1} - y_i}$$

which can be written as a two dimensional polynomial

$$p(x_i + h_x, y_i + h_y) = a_{00} + a_{10}h_x + a_{01}h_y + a_{11}h_x h_y \tag{2.132}$$

with

$$a_{00} = f(x_i, y_i)$$

$$a_{10} = \frac{f(x_{i+1}, y_i) - f(x_i, y_i)}{x_{i+1} - x_i}$$

$$a_{01} = \frac{f(x_i, y_{i+1}) - f(x_i, y_i)}{y_{i+1} - y_i}$$

$$a_{11} \frac{f(x_{i+1}, y_{i+1}) - f(x_i, y_{i+1}) - f(x_{i+1}, y_i) + f(x_i, y_i)}{(x_{i+1} - x_i)(y_{i+1} - y_i)}. \tag{2.133}$$

Application of higher order polynomials is straightforward. For image processing purposes bicubic interpolation is often used.

If high quality is needed, more sophisticated interpolation methods can be applied. Consider for instance two-dimensional spline interpolation on a rectangular mesh of data to create a new data set with finer resolution[6]

$$f_{i,j} = f(ih_x, jh_y) \text{ with } 0 \leq i < N_x \quad 0 \leq j < N_y. \tag{2.134}$$

First perform spline interpolation in x-direction for each data row j to calculate new data sets

[5] Bilinear means linear interpolation in two dimensions. Accordingly linear interpolation in three dimensions is called trilinear.

[6] A typical task of image processing.

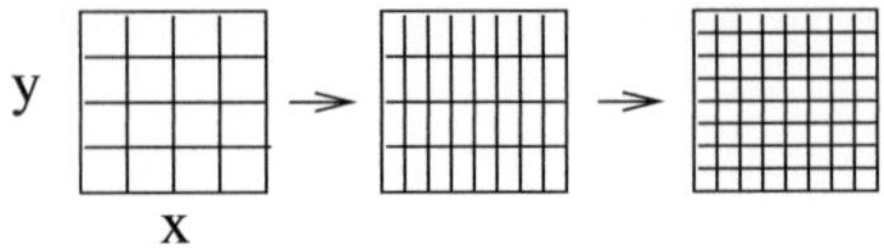

Fig. 2.9 Bispline interpolation

$$f_{i',j} = s(x_{i'}, f_{ij}, 0 \le i < N_x) \quad 0 \le j \le N_y \quad 0 \le i' < N'_x \tag{2.135}$$

and then interpolate in y direction to obtain the final high resolution data (Fig. 2.9)

$$f_{i',j'} = s(y_{j'}, f_{i'j}, 0 \le j < N_y) \quad 0 \le i' < N'_x \quad 0 \le j' < N'_y. \tag{2.136}$$

Exercises

Exercise 2.1 Polynomial Interpolation

This computer experiment interpolates a given set of n data points by

- a polynomial

$$p(x) = \sum_{i=0}^{n} f_i \prod_{k=0,k\ne i}^{n} \frac{x - x_k}{x_i - x_k}, \tag{2.137}$$

- a linear spline which connects successive points by straight lines

$$s_i(x) = a_i + b_i(x - x_i) \text{ for } x_i \le x \le x_{i+1} \tag{2.138}$$

- a cubic spline with natural boundary conditions

$$s(x) = p_i(x) = \alpha_i + \beta_i(x - x_i) + \gamma_i(x - x_i)^2 + \delta_i(x - x_i)^3 \quad x_i \le x \le x_{i+1} \tag{2.139}$$

$$s''(x_n) = s''(x_0) = 0 \tag{2.140}$$

- a rational function without poles

$$R(x) = \frac{\sum_{i=0}^{n} f_i \frac{u_i}{x-x_i}}{\sum_{i=0}^{n} \frac{u_i}{x-x_i}} \tag{2.141}$$

with weights according to Berrut

$$u_k = (-1)^k \tag{2.142}$$

or Floater-Hormann

$$u_k = (-1)^{k-1} \left(\frac{1}{x_{k+1} - x_k} + \frac{1}{x_k - x_{k-1}} \right) \quad k = 1 \ldots n-1 \tag{2.143}$$

$$u_0 = -\frac{1}{x_1 - x_0} \quad u_n = (-1)^{n-1} \frac{1}{x_n - x_{n-1}}. \tag{2.144}$$

- Interpolate the data (Table 2.2) in the range

$-1.5 < x < 0$.

- Now add some more sample points (Table 2.3) for $-1.5 < x < 4.5$.

- Interpolate the function $f(x) = \sin(x)$ at the points $x = 0, \frac{\pi}{2}, \pi, \frac{3\pi}{2}, 2\pi$. Take more sample points and check if the quality of the fit is improved.

- Investigate the oscillatory behavior for a discontinuous pulse or step function as given by the data (Table 2.4).

Table 2.2 Zener diode voltage/current data

Voltage	−1.5	−1.0	−0.5	0.0
Current	−3.375	−1.0	−0.125	0.0

Table 2.3 Additional voltage/current data

Voltage	1.0	2.0	3.0	4.0	4.1	4.2	4.5
Current	0.0	0.0	0.0	0.0	1.0	3.0	10.0

Table 2.4 Pulse and step function data

x	−3	−2	−1	0	1	2	3
y_{pulse}	0	0	0	1	0	0	0
y_{step}	0	0	0	1	1	1	1

Table 2.5 Data set for two-dimensional interpolation

x	0	1	2	0	1	2	0	1	2
y	0	0	0	1	1	1	2	2	2
f	1	0	−1	0	0	0	−1	0	1

Exercise 2.2 Two-dimensional Interpolation

This computer experiment uses bilinear interpolation or bicubic spline interpolation to interpolate the data (Table 2.5) on a finer grid $\Delta x = \Delta y = 0.1$.

Chapter 3
Differentiation

For more complex problems analytical derivatives are not always available and have to be approximated by numerical methods. Numerical differentiation is also very important for the discretization of differential equations (Sect. 13.2). The simplest approximation uses a forward difference quotient (Fig. 3.1) and is not very accurate. A symmetric difference quotient improves the quality. Even higher precision is obtained with the extrapolation method. Approximations to higher order derivatives can be obtained systematically with the help of polynomial interpolation. Apart from approximate difference quotients and symbolic differentiation by algebra programs, algorithmic differentiation, also known as automatic differentiation or autograd, is a further alternative. It has become especially important in the field of machine learning and artificial intelligence.

3.1 One-Sided Difference Quotient

The simplest approximation of a derivative is the ordinary difference quotient which can be taken forward

$$\frac{\mathrm{d}f}{\mathrm{d}x}(x) \approx \frac{\Delta f}{\Delta x} = \frac{f(x+h) - f(x)}{h} \tag{3.1}$$

or backward

$$\frac{\mathrm{d}f}{\mathrm{d}x}(x) \approx \frac{\Delta f}{\Delta x} = \frac{f(x) - f(x-h)}{h}. \tag{3.2}$$

Its truncation error can be estimated from the Taylor series expansion

---

Supplementary Information The online version contains supplementary material available at https://doi.org/10.1007/978-3-032-07856-8_3.

P. O. J. Scherer, *Computational Physics I*, Graduate Texts in Physics,
https://doi.org/10.1007/978-3-032-07856-8_3

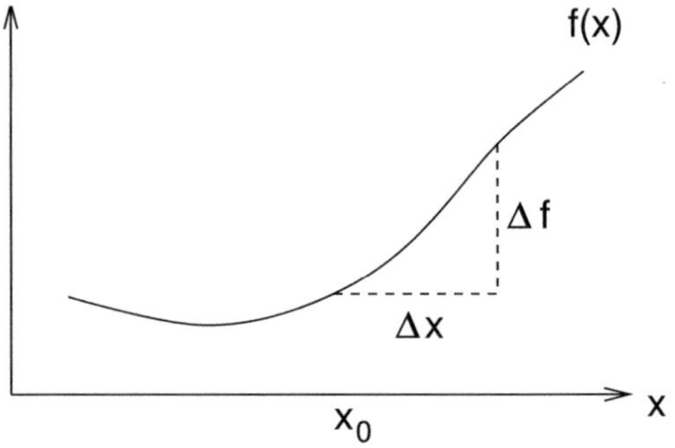

Fig. 3.1 Numerical differentiation. Numerical differentiation approximates the differential quotient by a difference quotient $\frac{df}{dx} \approx \frac{\Delta f}{\Delta x}$. However, approximation by a simple forward difference $\frac{df}{dx}(x_0) \approx \frac{f(x_0+h)-f(x_0)}{h}$, is not very accurate

$$\frac{f(x+h)-f(x)}{h} = \frac{f(x)+hf'(x)+\frac{h^2}{2}f''(x)+\cdots-f(x)}{h}$$

$$= f'(x)+\frac{h}{2}f''(x)+\cdots . \tag{3.3}$$

The error order is $O(h)$. The step width should not be too small to avoid rounding errors. Error analysis gives

$$\begin{aligned}\widetilde{\Delta f} &= fl_-(f(x+h)(1+\varepsilon_1), f(x)(1+\varepsilon_2))\\ &= (\Delta f + f(x+h)\varepsilon_1 - f(x)\varepsilon_2)(1+\varepsilon_3)\\ &= \Delta f + \Delta f\varepsilon_3 + f(x+h)\varepsilon_1 - f(x)\varepsilon_2 + \cdots\end{aligned} \tag{3.4}$$

$$\begin{aligned}fl_\div(\widetilde{\Delta f}, h(1+\varepsilon_4)) &= \frac{\Delta f + \Delta f\varepsilon_3 + f(x+h)\varepsilon_1 - f(x)\varepsilon_2}{h(1+\varepsilon_4)}(1+\varepsilon_5)\\ &= \frac{\Delta f}{h}(1+\varepsilon_5-\varepsilon_4+\varepsilon_3)+\frac{f(x+h)}{h}\varepsilon_1-\frac{f(x)}{h}\varepsilon_2.\end{aligned} \tag{3.5}$$

The errors are uncorrelated and the relative error of the result can be estimated by

$$\frac{\left|\frac{\widetilde{\Delta f}}{\Delta x}-\frac{\Delta f}{\Delta x}\right|}{\frac{\Delta f}{\Delta x}} \le 3\varepsilon_M + \left|\frac{f(x)}{\frac{\Delta f}{\Delta x}}\right| 2\frac{\varepsilon_M}{h}. \tag{3.6}$$

Numerical extinction produces large relative errors for small step width h. The optimal value of h gives comparable errors from rounding and truncation. It can be found from

$$\frac{h}{2}|f''(x)| = |f(x)|\frac{2\varepsilon_M}{h}. \tag{3.7}$$

Assuming that the magnitude of the function and the derivative are comparable, we have the rule of thumb (double precision)

$$h_o = \sqrt{\varepsilon_M} \approx 10^{-8}.$$

The corresponding relative error is of the same order.

3.2 Central Difference Quotient

Accuracy is much higher if a symmetriccentral difference quotient is used (Fig. 3.2):

$$\frac{\Delta f}{\Delta x} = \frac{f(x+\frac{h}{2}) - f(x-\frac{h}{2})}{h}$$
$$= \frac{f(x) + \frac{h}{2}f'(x) + \frac{h^2}{8}f''(x) + \cdots - \left(f(x) - \frac{h}{2}f'(x) + \frac{h^2}{8}f''(x) + \cdots\right)}{h}$$
$$= f'(x) + \frac{h^2}{24}f'''(x) + \cdots. \tag{3.8}$$

The error order is $O(h^2)$. The optimal step width is estimated from

$$\frac{h^2}{24}|f'''(x)| = |f(x)|\frac{2\varepsilon_M}{h} \tag{3.9}$$

again with the assumption that function and derivatives are of similar magnitude as

$$h_0 =^3 \sqrt{48\varepsilon_M} \approx 10^{-5}. \tag{3.10}$$

The relative error has to be expected in the order of $\frac{h_0^2}{24} \approx 10^{-11}$.

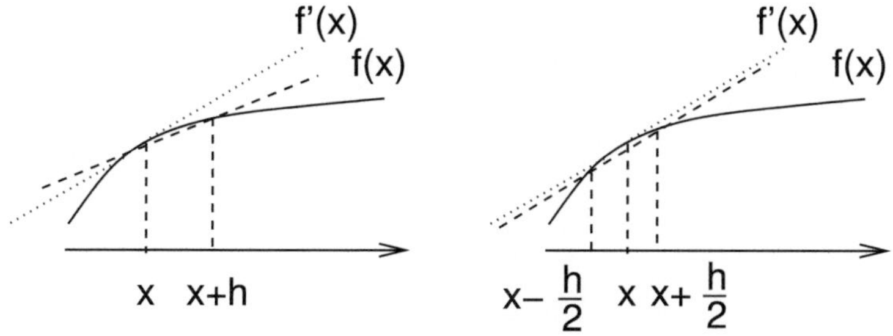

Fig. 3.2 Difference quotients. **Left:** One-sided difference quotient; **Right:** The central difference quotient approximates the derivative (dotted) much more accurately

3.3 Extrapolation Methods

The Taylor series of the symmetric difference quotient contains only even powers of h:

$$D(h) = \frac{f(x+h) - f(x-h)}{2h} = f'(x) + \frac{h^2}{3!}f'''(x) + \frac{h^4}{5!}f^{(5)}(x) + \cdots . \quad (3.11)$$

The Extrapolation method [27] uses a series of step widths, e.g.

$$h_{i+1} = \frac{h_i}{2} \quad (3.12)$$

and calculates an estimate of $D(0)$ by polynomial interpolation (Fig. 3.3). Consider $D_0 = D(h_0)$ and $D_1 = D(\frac{h_0}{2})$. The polynomial of degree 1 (with respect to h^2) $p(h) = a + bh^2$ can be found by the Lagrange method

$$p(h) = D_0 \frac{h^2 - \frac{h_0^2}{4}}{h_0^2 - \frac{h_0^2}{4}} + D_1 \frac{h^2 - h_0^2}{\frac{h_0^2}{4} - h_0^2}. \quad (3.13)$$

Extrapolation for $h = 0$ gives

$$p(0) = -\frac{1}{3}D_0 + \frac{4}{3}D_1. \quad (3.14)$$

Taylor series expansion shows

$$p(0) = -\frac{1}{3}\left(f'(x) + \frac{h_0^2}{3!}f'''(x) + \frac{h_0^4}{5!}f^{(5)}(x) + \cdots\right) +$$

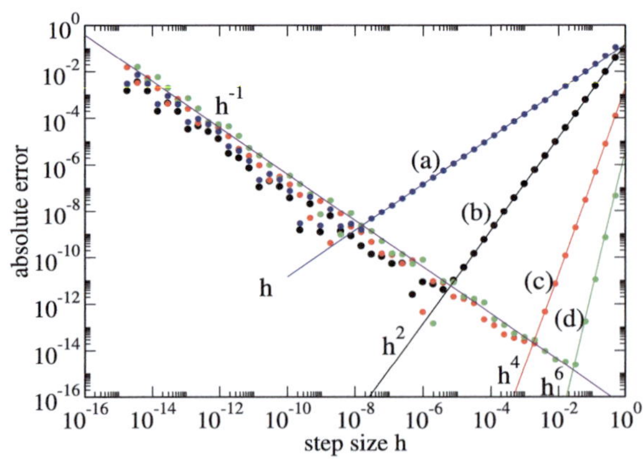

Fig. 3.3 Numerical differentiation. The derivative $\frac{d}{dx}\sin(x)$ is calculated numerically using algorithms with increasing error order (3.1, 3.8, 3.14, 3.18). For very small step sizes the error increases as h^{-1} due to rounding errors

$$+\frac{4}{3}\left(f'(x)+\frac{h_0^2}{4\cdot 3!}f'''(x)+\frac{h_0^4}{16\cdot 5!}f^{(5)}(x)+\cdots\right) \tag{3.15}$$

$$= f'(x)-\frac{1}{4}\frac{h_0^4}{5!}f^{(5)}(x)+\cdots . \tag{3.16}$$

The error is of order $O(h_0^4)$. For 3 step widths $h_0 = 2h_1 = 4h_2$ we obtain the polynomial of second order (in h^2)

$$p(h) = D_0\frac{(h^2-\frac{h_0^2}{4})(h^2-\frac{h_0^2}{16})}{(h_0^2-\frac{h_0^2}{4})(h_0^2-\frac{h_0^2}{16})}+D_1\frac{(h^2-h_0^2)(h^2-\frac{h_0^2}{16})}{(\frac{h_0^2}{4}-h_0^2)(\frac{h_0^2}{4}-\frac{h_0^2}{16})}+D_2\frac{(h^2-h_0^2)(h^2-\frac{h_0^2}{4})}{(\frac{h_0^2}{16}-h_0^2)(\frac{h_0^2}{16}-\frac{h_0^2}{4})} \tag{3.17}$$

and the improved expression

$$p(0) = D_0\frac{\frac{1}{64}}{\frac{3}{4}\cdot\frac{15}{16}}+D_1\frac{\frac{1}{16}}{\frac{-3}{4}\cdot\frac{3}{16}}+D_2\frac{\frac{1}{4}}{\frac{-15}{16}\cdot\frac{-3}{16}} =$$

$$= \frac{1}{45}D_0-\frac{4}{9}D_1+\frac{64}{45}D_2 = f'(x)+O(h_0^6). \tag{3.18}$$

Often used is the following series of step widths:

$$h_i^2 = \frac{h_0^2}{2^i}. \tag{3.19}$$

The Neville method

$$P_{i\cdots k}(h^2) = \frac{(h^2-\frac{h_0^2}{2^i})P_{i+1\cdots k}(h^2)-(h^2-\frac{h_0^2}{2^k})P_{i\cdots k-1}(h^2)}{\frac{h_0^2}{2^k}-\frac{h_0^2}{2^i}} \tag{3.20}$$

gives for $h = 0$

$$P_{i\cdots k} = \frac{P_{i\cdots k-1}-2^{k-i}P_{i+1\cdots k}}{1-2^{k-i}} \tag{3.21}$$

which can be written as

$$P_{i\cdots k} = P_{i+1\cdots k}+\frac{P_{i\cdots k-1}-P_{i+1\cdots k}}{1-2^{k-i}} \tag{3.22}$$

and can be calculated according to the following scheme:

$$\begin{array}{llll} P_0 = D(h^2) & P_{01} & P_{012} & P_{0123} \\ P_1 = D(\frac{h^2}{2}) & P_{12} & P_{123} & \\ P_2 = D(\frac{h^2}{4} & P_{23} & & \cdot \\ \vdots & \vdots & \vdots & \ddots \end{array} \tag{3.23}$$

Here the values of the polynomials are arranged in matrix form

$$P_{i\cdots k} = T_{i,k-i} = T_{i,j} \tag{3.24}$$

with the recursion formula

$$T_{i,j} = T_{i+1,j-1} + \frac{T_{i,j-1} - T_{i+1,j}}{1 - 2^j}. \tag{3.25}$$

3.4 Higher Derivatives

Difference quotients for higher derivatives can be obtained systematically using polynomial interpolation. Consider equidistant points

$$x_n = x_0 + nh = \ldots x_0 - 2h, x_0 - h, x_0, x_0 + h, x_0 + 2h, \ldots. \tag{3.26}$$

From the second order polynomial

$$p(x) = y_{-1}\frac{(x-x_0)(x-x_1)}{(x_{-1}-x_0)(x_{-1}-x_1)} + y_0\frac{(x-x_{-1})(x-x_1)}{(x_0-x_{-1})(x_0-x_1)}$$

$$+ y_1\frac{(x-x_{-1})(x-x_0)}{(x_1-x_{-1})(x_1-x_0)} =$$

$$= y_{-1}\frac{(x-x_0)(x-x_1)}{2h^2} + y_0\frac{(x-x_{-1})(x-x_1)}{-h^2}$$

$$+ y_1\frac{(x-x_{-1})(x-x_0)}{2h^2} \tag{3.27}$$

we calculate the derivatives

$$p'(x) = y_{-1}\frac{2x-x_0-x_1}{2h^2} + y_0\frac{2x-x_{-1}-x_1}{-h^2} + y_1\frac{2x-x_{-1}-x_0}{2h^2} \tag{3.28}$$

$$p''(x) = \frac{y_{-1}}{h^2} - 2\frac{y_0}{h^2} + \frac{y_1}{h^2} \tag{3.29}$$

which are evaluated at x_0

$$f'(x_0) \approx p'(x_0) = -\frac{1}{2h}y_{-1} + \frac{1}{2h}y_1 = \frac{f(x_0+h) - f(x_0-h)}{2h} \tag{3.30}$$

$$f''(x_0) \approx p''(x_0) = \frac{f(x_0-h) - 2f(x_0) + f(x_0+h)}{h^2}. \tag{3.31}$$

Higher order polynomials can be evaluated with an algebra program. For five sample points

$$x_0 - 2h, x_0 - h, x_0, x_0 + h, x_0 + 2h$$

we find

$$f'(x_0) \approx \frac{f(x_0-2h) - 8f(x_0-h) + 8f(x_0+h) - f(x_0+2h)}{12h} \tag{3.32}$$

$$f''(x_0) \approx \frac{-f(x_0-2h) + 16f(x_0-h) - 30f(x_0) + 16f(x_0+h) - f(x_0+2h)}{12h^2} \tag{3.33}$$

$$f'''(x_0) \approx \frac{-f(x_0-2h) + 2f(x_0-h) - 2f(x_0+h) + f(x_0+2h)}{2h^3} \tag{3.34}$$

$$f^{(4)}(x_0) \approx \frac{f(x_0-2h) - 4f(x_0-h) + 6f(x_0+h) - 4f(x_0+h) + f(x_0+2h)}{h^4}. \tag{3.35}$$

3.5 Partial Derivatives of Multivariate Functions

Consider polynomials of more than one variable. In two dimensions we use the Lagrange polynomials

$$L_{i,j}(x,y) = \prod_{k \neq i} \frac{(x - x_k)}{(x_i - x_k)} \prod_{j \neq l} \frac{(y - y_l)}{(y_j - y_l)}. \tag{3.36}$$

The interpolating polynomial is

$$p(x,y) = \sum_{i,j} f_{i,j} L_{i,j}(x,y). \tag{3.37}$$

For the nine sample points

$$\begin{matrix} (x_{-1}, y_1) & (x_0, y_1) & (x_1, y_1) \\ (x_{-1}, y_0) & (x_0, y_0) & (x_1, y_0) \\ (x_{-1}, y_{-1}) & (x_0, y_{-1}) & (x_1, y_{-1}) \end{matrix} \tag{3.38}$$

we obtain the polynomial

$$p(x, y) = f_{-1,-1} \frac{(x - x_0)(x - x_1)(y - y_0)(y - y_1)}{(x_{-1} - x_0)(x_{-1} - x_1)(y_{-1} - y_0)(y_{-1}) - y_1)} + \cdots \tag{3.39}$$

which gives an approximation to the gradient

$$\operatorname{grad} f(x_0 y_0) \approx \operatorname{grad} p(x_0 y_0) = \begin{pmatrix} \frac{f(x_0+h, y_0) - f(x_0-h, y_0)}{2h} \\ \frac{f(x_0, y_0+h) - f(x_0, y_0-h)}{2h} \end{pmatrix}, \tag{3.40}$$

the Laplace operator

$$\left(\frac{\partial^2}{\partial x^2} + \frac{\partial^2}{\partial y^2}\right) f(x_0, y_0) \approx \left(\frac{\partial^2}{\partial x^2} + \frac{\partial^2}{\partial y^2}\right) p(x_0, y_0)$$

$$= \frac{1}{h^2} \left(f(x_0, y_0 + h) + f(x_0, y_0 - h) + f(x_0, y_0 + h) + f(x_0, y_0 - h) - 4f(x_0, y_0)\right) \tag{3.41}$$

and the mixed second derivative

$$\frac{\partial^2}{\partial x \partial y} f(x_0, y_0) \approx \frac{\partial^2}{\partial x \partial y} p(x_0, y_0)$$

$$= \frac{1}{4h^2} \left(f(x_0 + h, y_0 + h) + f(x_0 - h, y_0 - h) - f(x_0 - h, y_0 + h) - f(x_0 + h, y_0 - h)\right). \tag{3.42}$$

3.6 Algorithmic Differentiation

Apart from approximate difference quotients and symbolic differentiation by algebra programs, algorithmic differentiation, also known as automatic differentiation or autograd, is a third alternative. It has become especially important in the field of machine learning and artificial intelligence [28] and has been incorporated into ML frameworks like TensorFlow [29], PyTorch [30] and MATLAB [31]. A comparison of several programs can be found in [32].

We consider an algorithm which calculates a function value[1] as a finite sequence of elementary operations and functions, which can be represented by a computational directed acyclic graph [33, 34].[2] The mapping of the input vector $\mathbf{x}$ onto the output vector

$$\mathbf{x} \in \Re^N \to \mathbf{f}(\mathbf{x}) \in \Re^{\mathrm{M}}$$

is split into a sequence of p elementary operations (addition, multiplication, exponential etc.) represented as arrows. The graph is partitioned into input nodes $x_n, n = 1 \cdots N$, output nodes $f_m, m = 1 \cdots M$ and intermediate nodes $a_k, k = 1 \cdots K$ and has been topologically ordered.[3]

Initially the input nodes are fed with the input data. Then following the ordered graph, a chain of operations[4]

$$\mathbf{x} \to \mathbf{a}^{(1)}(\mathbf{x}) \to \mathbf{a}^{(2)}(\mathbf{a}^{(1)}) \to \cdots \mathbf{f}(\mathbf{a}^{(p-1)})$$

produces the desired output values. We assume that during the j-th step the node a_j is assigned a value[5] and never overwritten later. Therefore we ignore all nodes $a_1 \cdots a_{j-1}$, which have not yet been assigned and use vectors of increasing length to represent all assigned values after the j-th step[6]

$$\mathbf{u}^{(j)} = \left(x_1, \cdots x_N, a_1, \cdots a_j\right)^T . \tag{3.43}$$

Following the graph in topological order, the chain rule provides the variations of intermediate and final results as a function of variations of the input values:

$$\delta\mathbf{u}^{(1)} = \frac{\partial(x_1 \cdots x_N, a_1)}{\partial(x_1, x_2, \cdots x_N)}\delta\mathbf{x}$$

$$\delta\mathbf{u}^{(2)} = \frac{\partial(x_1 \cdots x_N, a_1, a_2)}{\partial(x_1, \cdots x_N, a_1)}\delta\mathbf{u}^{(1)}$$

$$\vdots$$

$$\delta\mathbf{f} = \frac{\partial(f_1 \cdots f_M)}{\partial(x_1, \cdots x_N, a_1, \cdots a_K)}\delta\mathbf{u}^{(p-1)}. \tag{3.44}$$

---

[1] Input and output can be scalar, vectorial or even higher rank tensors.

[2] Loops and conditional statements have to be treated separately and are not under consideration in the following [35].

[3] Every directed acyclic graph has at least one linear order where for every arrow from node i to node j, i comes before j in the order [36, 37].

[4] A Wengert list [38] where any variable can be assigned a value only once.

[5] If several nodes have the same position in the order, parallel operation would be possible.

[6] We formally include also the input values which do not change but may be required in later steps. Effectively, every step adds one new row and one new column to the Jacobian.

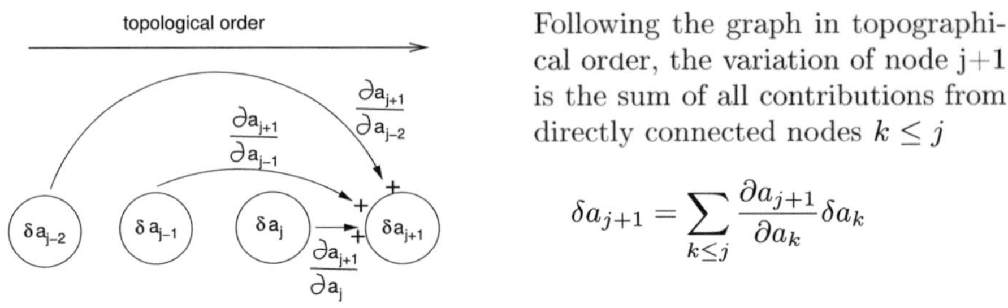

Fig. 3.4 Forward differentiation. The forward mode of algorithmic differentiation follows the nodes one by one and sums up all contributions to the gradient from previous nodes in one step

Forward Mode

The forward mode of algorithmic differentiation (Fig. 3.4) follows the same steps and calculates

$$\delta \mathbf{f} = J \delta \mathbf{x} \tag{3.45}$$

where the Jacobian matrix factorizes

$$J = \frac{\partial(f_1 \cdots f_M)}{\partial(x_1, \cdots x_N)}$$

$$= \frac{\partial(f_1 \cdots f_M)}{\partial(x_1, \cdots x_N, a_1, \cdots a_K)} \frac{\partial(x_1 \cdots x_N, a_1 \cdots a_K)}{\partial(x_1, \cdots x_N, a_1, \cdots a_{K-1})} \cdots \frac{\partial(x_1 \cdots x_N, a_1)}{\partial(x_1, \cdots x_N)}$$

$$= J^{(p)} J^{(p-1)} \cdots J^{(2)} J^{(1)}. \tag{3.46}$$

Each factor corresponds to one step. Partial derivatives[7] $\frac{\partial f_i(\mathbf{a}(\mathbf{x}))}{\partial x_j}$ can be obtained by using a standard unit vector as input, e.g. the first column of J is obtained from

$$\delta \mathbf{x} \leftarrow \begin{pmatrix} 1 \\ 0 \\ \vdots \\ 0 \end{pmatrix} \qquad \mathbf{f} = J \delta \mathbf{x} = \begin{pmatrix} \frac{\partial f_1(\mathbf{a}(\mathbf{x}))}{\partial x_1} \\ \vdots \\ \frac{\partial f_M(\mathbf{a}(\mathbf{x}))}{\partial x_1} \end{pmatrix}. \tag{3.47}$$

Forward mode is efficient if the number of input nodes is small since each element of the input vector requires a separate path. It is also efficient to calculate the product of the Jacobian matrix with a column vector from the right side

[7] As common in the literature, in the following partial derivatives $\frac{\partial a_i}{\partial a_j}$ only consider the direct connection of two nodes shown as an arrow $j \to i$ in the graph (the values of all but these two nodes are kept fixed). Partial derivatives including all possible indirect connections are denoted as $\frac{\partial f_i(\mathbf{a}(\mathbf{x}))}{\partial x_j}$. These are elements of the Jacobian matrix $J = \frac{\partial(f_1 \cdots f_M)}{\partial(x_1, \cdots x_N)}$.

$$\delta \mathbf{x} \leftarrow \begin{pmatrix} v_1 \\ v_2 \\ \vdots \\ v_N \end{pmatrix} \qquad \mathbf{f} = J\delta \mathbf{x} = \begin{pmatrix} v_1 \frac{\partial f_1(\mathbf{a}(\mathbf{x}))}{\partial x_1} + \cdots + v_N \frac{\partial f_1(\mathbf{a}(\mathbf{x}))}{\partial x_N} \\ \vdots \\ v_1 \frac{\partial f_M(\mathbf{a}(\mathbf{x}))}{\partial x_1} + \cdots + v_N \frac{\partial f_M(\mathbf{a}(\mathbf{x}))}{\partial x_N} \end{pmatrix}. \tag{3.48}$$

Reverse Mode

The reverse mode follows the graph backwards (Fig. 3.5) in reverse topological order. Essentially it applies the chain rule

$$\frac{\partial f_i(\mathbf{a}(\mathbf{x}))}{\partial x_j} = \frac{\partial f_i}{\partial x_j} + \frac{\partial f_i}{\partial a_K}\frac{\partial a_K}{\partial x_j} + \frac{\partial f_i}{\partial a_K}\frac{\partial a_K}{\partial a_{K-1}}\frac{\partial a_{K-1}}{\partial x_j} + \cdots . \tag{3.49}$$

Usually it follows a forward path which calculates all intermediate values and partial derivatives. The simplest approach mirrors the forward mode and considers the outgoing (reversed) arrows from node $j \to k, \quad k < j$. Starting from an initial row vector $\overline{f}^T$ a transpose vector of (N+K) adjoint values $\overline{v}^T = (\overline{x}_1, \cdots \overline{x}_N, \overline{a}_1, \cdots \overline{a}_K)$ is calculated according to

$$\overline{v}_k = \sum_{i=1}^{M} \overline{f}_i \frac{\partial \overline{f}_i}{\partial v_k} \quad k = 1 \cdots N + K \tag{3.50}$$

$$\left[\overline{v}_k = \overline{v}_k + \overline{a}_m \frac{\partial a_m}{\partial v_k} \quad k = 1 \cdots N + K - 1 \right] \quad m = K, K-1, \cdots 2 \tag{3.51}$$

$$\overline{x}_k = \overline{x}_k + \overline{a}_1 \frac{\partial a_1}{\partial x_k} \quad k = 1 \cdots N. \tag{3.52}$$

At the end,

$$\overline{x}_k = \sum_{i=1}^{M} \overline{f}_i \left[\frac{\partial f_i}{\partial x_k} + \frac{\partial f_i}{\partial a_K}\frac{\partial a_K}{\partial x_k} + \frac{\partial f_i}{\partial a_K}\frac{\partial a_K}{\partial a_{K-1}}\frac{\partial a_{K-1}}{\partial x_k} + \cdots \right] = \sum_{i=1}^{M} \overline{f}_i \left(\frac{\partial f_i(\mathbf{a}(\mathbf{x}))}{\partial x_j} \right) \tag{3.53}$$

i.e. the resulting row vector $(\overline{x}_1 \cdots \overline{x}_N)$ is the product of the Jacobian with the seed vector from the left side

$$(\overline{x}_1 \cdots \overline{x}_N) = \overline{f}^T J. \tag{3.54}$$

Starting from a standard unit row vector we obtain one row of the Jacobian matrix, e.g.

$$(1, 0 \cdots 0)\, J = \left(\frac{\partial f_1}{\partial x_1} \cdots \frac{\partial f_1}{\partial x_N} \right). \tag{3.55}$$

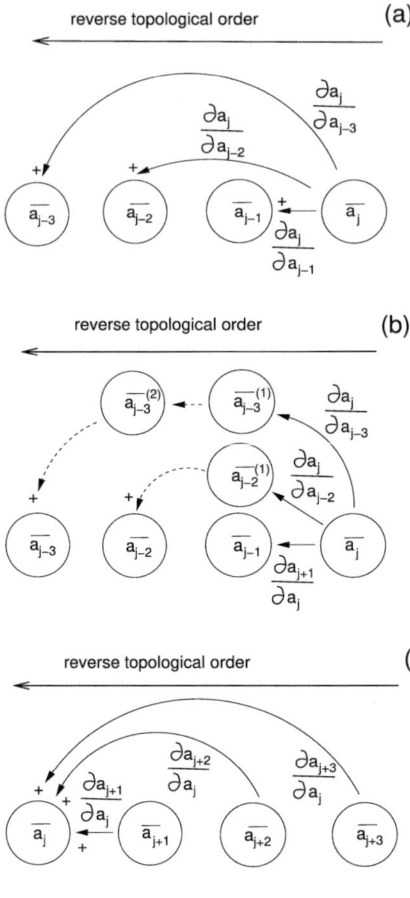

(a) Following the graph in reverse order, the outgoing arrows from node j to nodes $k < j$ are considered. Adjoint nodes may be reached several times which is not compatible with a Wengert list.

$$\bar{a}_k = \bar{a}_k + \bar{a}_j \frac{\partial a_j}{\partial a_k}$$

(b) Following the graph in reverse order, the outgoing arrows from node j to nodes $k < j$ are considered. Copies of adjoint nodes are added to provide a Wengert list

$$\bar{a}_k = \bar{a}_k + \bar{a}_j \frac{\partial a_j}{\partial a_k}$$

(c) Following the graph in reverse order, the arrows incoming to node j from nodes $k > j$ are considered. This scheme provides a Wengert list

$$\bar{a}_j = \sum_{k>j} \bar{a}_k \frac{\partial a_k}{\partial a_j}$$

Fig. 3.5 Backward differentiation. The reverse mode of algorithmic differentiation follows the nodes one by one in reverse order. (**a**) A mirror image of the forward path distributes the contributions to the gradient over all previous nodes. (**b**) A Wengert list is obtained by adding extra temporary nodes to store intermediate results until they are needed. (**c**) The message passing mode considers all contributions from later nodes in one step

The reverse mode is efficient if the number of output nodes is small (e.g. one loss function of a neural net) or if the scalar product of J with a row vector from the left side is required.

The reverse topological order does not provide a Wengert list in general, since the adjoint nodes may be reached several times. This is problematic for an efficient implementation and can be avoided by adding extra temporary nodes [35] to store intermediate values until they are required in a later step. Alternatively, a different approach can be used which considers incoming instead of outgoing arrows [39]:

$$\overline{a}_K = \sum_{i=1}^{N} \overline{f}_i \frac{\partial f_i}{\partial a_K} \tag{3.56}$$

$$\overline{a}_k = \sum_{i=1}^{M} \overline{f}_i \frac{\partial f_i}{\partial a_k} + \sum_{m=k+1}^{K} \overline{a}_m \frac{\partial a_m}{\partial a_k} \quad k = K-1, K-2, \cdots 1 \tag{3.57}$$

$$\overline{x}_j = \sum_{i=1}^{M} \overline{f}_i \frac{\partial f_i}{\partial x_j} + \sum_{m=1}^{K} \overline{a}_m \frac{\partial a_m}{\partial x_j} \quad j = 1 \cdots N. \tag{3.58}$$

The reverse mode provides also the gradients with respect to the intermediate values and therefore has become a key ingredient for the training of neural networks where it is known as back-propagation.

Example

Figure 3.6 shows as a simple example a function of three variables

$$f = \sqrt{x_1} + e^{x_3} \sin(x_1 x_2) \tag{3.59}$$

which is evaluated in three steps

$$a_1 = x_1 \times x_2 \tag{3.60}$$

$$a_2 = e^{x_3} \sin a_1 \tag{3.61}$$

$$f = \sqrt{x_1} + a_2. \tag{3.62}$$

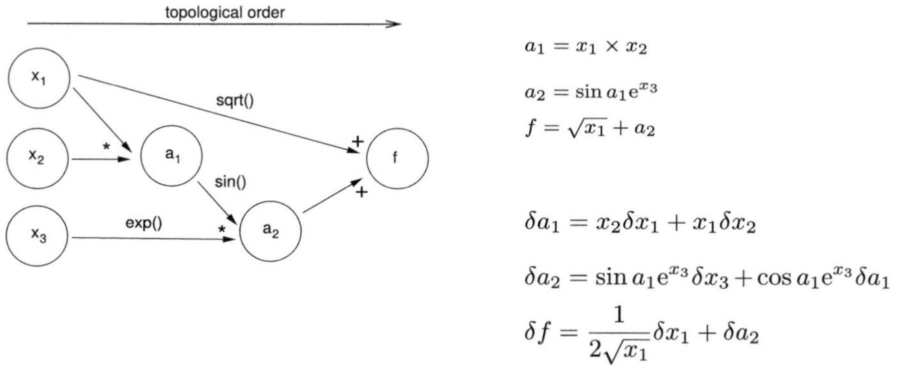

Fig. 3.6 Forward differentiation of example (3.59)

Applying the chain rule along the graph we obtain

$$\delta a_1 = \frac{\partial a_1}{\partial x_1}\delta x_1 + \frac{\partial a_1}{\partial x_2}\delta x_2 = x_2 \delta x_1 + x_1 \delta x_2 \tag{3.63}$$

$$\delta a_2 = \frac{\partial a_2}{\partial x_3}\delta x_3 + \frac{\partial a_2}{\partial a_1}\delta a_1 = \mathrm{e}^{x_3} \sin a_1 \delta x_3 + \mathrm{e}^{x_3} \cos a_1 \delta a_1 \tag{3.64}$$

$$\delta f = \left(\frac{\partial f}{\partial x_1}\right)_{a_1 a_2} \delta x_1 + \frac{\partial f}{\partial a_2}\delta a_2 = \frac{1}{2\sqrt{x_1}}\delta x_1 + \delta a_2. \tag{3.65}$$

The Jacobian matrix is the gradient

$$\frac{\partial(f)}{\partial(x_1 x_2 x_3)} = \left(\tfrac{1}{2\sqrt{x_1}}\; 0\; 0\; 0\; 1\right) \begin{pmatrix} 1 & 0 & 0 & 0 \\ 0 & 1 & 0 & 0 \\ 0 & 0 & 1 & 0 \\ 0 & 0 & 0 & 1 \\ 0 & 0 & \mathrm{e}^{x_3} \sin a_1 & \mathrm{e}^{x_3} \cos a_1 \end{pmatrix} \begin{pmatrix} 1 & 0 & 0 \\ 0 & 1 & 0 \\ 0 & 0 & 1 \\ x_2 & x_1 & 0 \end{pmatrix} \tag{3.66}$$

and the resulting output variation is

$$\delta f = (\frac{1}{2\sqrt{x_1}} + x_2 \mathrm{e}^{x_3} \cos a_1)\delta x_1 \quad + x_1 \mathrm{e}_1^{x_3} \cos a_1 \delta x_2, \quad + \mathrm{e}^{x_3} \sin a_1 \delta x_3. \tag{3.67}$$

In reverse mode we begin with

$$\overline{f} = 1 \tag{3.68}$$

and follow the graph backwards (Fig. 3.7)

$$\overline{a_2} = \overline{f}(\frac{\partial f}{\partial a_2}) = 1 \tag{3.69}$$

$$\overline{a}_1 = \overline{a}_2 \frac{\partial a_2}{\partial a_1} = \mathrm{e}^{x_3} \cos a_1 \tag{3.70}$$

$$\overline{x}_3 = \overline{a}_2 (\frac{\partial a_2}{\partial x_3}) = \mathrm{e}^{x_3} \sin a_1 \tag{3.71}$$

$$\overline{x}_2 = \overline{a}_1 \frac{\partial a_1}{\partial x_2} = x_1 \mathrm{e}^{x_3} \cos a_1 \tag{3.72}$$

$$\overline{x}_1 = \overline{f}\frac{\partial f}{\partial x_1} + \overline{a}_1 \frac{\partial a_1}{\partial x_1} = \frac{1}{2\sqrt{x_1}} + x_2 \mathrm{e}^{x_3} \cos a_1. \tag{3.73}$$

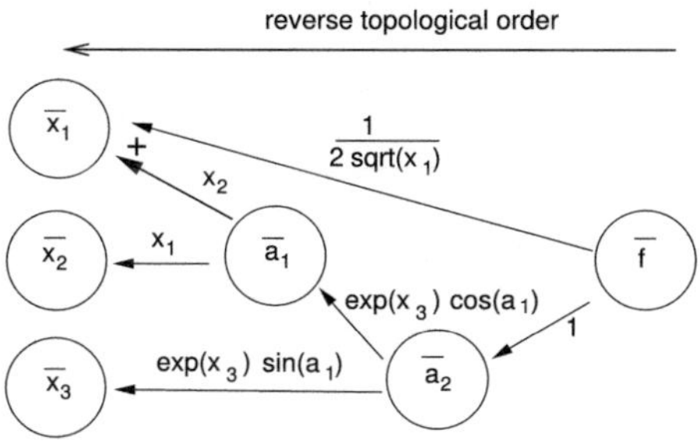

$$\overline{a}_2 = \overline{f}$$

$$\overline{a}_1 = \overline{a}_2 \cos a_1 e^{x_3}$$

$$\overline{x}_3 = \overline{a}_2 e^{x_3} \sin a_1$$

$$\overline{x}_2 = \overline{a}_1 x_1$$

$$\overline{x}_1 = \overline{f}\frac{1}{2\sqrt{x_1}} + \overline{a}_1 x_2$$

Fig. 3.7 Backward differentiation of example (3.59)

The gradient now involves also the intermediate values

$$\frac{\partial(f)}{\partial(x_1, x_2, x_3, a_1, a_2, f)}$$

$$= (1, 1)\begin{pmatrix} \cos a_1 e^{x_3} & 1 & 0 \\ 0 & 0 & 1 \end{pmatrix}\begin{pmatrix} 0 & 1 & 0 & 0 \\ e^{x_3}\sin a_1 & 0 & 1 & 0 \\ 0 & 0 & 0 & 1 \end{pmatrix}\begin{pmatrix} 0 & 1 & 0 & 0 & 0 \\ x_1 & 0 & 1 & 0 & 0 \\ 0 & 0 & 0 & 1 & 0 \\ 0 & 0 & 0 & 0 & 1 \end{pmatrix}\begin{pmatrix} 0 & 1 & 0 & 0 & 0 & 0 \\ 0 & 0 & 1 & 0 & 0 & 0 \\ x_2 & 0 & 0 & 1 & 0 & 0 \\ 0 & 0 & 0 & 0 & 1 & 0 \\ \frac{1}{2\sqrt{x_1}} & 0 & 0 & 0 & 0 & 1 \end{pmatrix}$$

$$= \left(x_2 e^{x_3}\cos a_1 + \frac{1}{2\sqrt{x_1}}, x_1 e^{x_3}\cos a_1, e^{x_3}\sin a_1, \cos a_1 e^{x_3}, 1, 1\right). \qquad (3.74)$$

All derivatives of one output function[8] with respect to input and intermediate values are obtained in one path.

Exercises

Exercise 3.1 Numerical Differentiation

In this computer experiment we calculate the derivative of $f(x) = \sin(x)$ numerically with

[8] Or a linear combination of several.

- the single sided difference quotient

$$\frac{\mathrm{d}f}{\mathrm{d}x} \approx \frac{f(x+h) - f(x)}{h}, \tag{3.75}$$

- the symmetrical difference quotient

$$\frac{\mathrm{d}f}{\mathrm{d}x} \approx D_h f(x) = \frac{f(x+h) - f(x-h)}{2h}, \tag{3.76}$$

- higher order approximations which can be derived using the extrapolation method

$$-\frac{1}{3} D_h f(x) + \frac{4}{3} D_{h/2} f(x) \tag{3.77}$$

$$\frac{1}{45} D_h f(x) - \frac{4}{9} D_{h/2} f(x) + \frac{64}{45} D_{h/4} f(x). \tag{3.78}$$

The error of the numerical approximation is shown on a log-log plot as a function of the step width h.

Chapter 4
Integration

Physical simulations often involve the calculation of definite integrals over complicated functions, for instance the Coulomb interaction between two electrons. Integration is also the elementary step in solving equations of motion.

An integral over a finite interval $[a, b]$ can always be transformed into an integral over $[0, 1]$ or $[-1, 1]$

$$\int_a^b f(x)dx = \int_0^1 f(a + (b - a)t)\,(b - a)dt$$

$$= \int_{-1}^1 f\left(\frac{a + b}{2} + \frac{b - a}{2}t\right)\frac{b - a}{2}dt. \tag{4.1}$$

An Integral over an infinite interval may have to be transformed into an integral over a finite interval by substitution of the integration variable, for example

$$\int_0^\infty f(x)dx = \int_0^1 f\left(\frac{t}{1 - t}\right)\frac{dt}{(1 - t)^2} \tag{4.2}$$

$$\int_{-\infty}^\infty f(x)dx = \int_{-1}^1 f\left(\frac{t}{1 - t^2}\right)\frac{t^2 + 1}{(t^2 - 1)^2}dt. \tag{4.3}$$

In general a definite integral can be approximated numerically as the weighted average over a finite number of function values

Supplementary Information The online version contains supplementary material available at https://doi.org/10.1007/978-3-032-07856-8_4.

P. O. J. Scherer, *Computational Physics I*, Graduate Texts in Physics,
https://doi.org/10.1007/978-3-032-07856-8_4

$$\int_a^b f(x)dx \approx \sum_{x_i} w_i f(x_i). \tag{4.4}$$

Specific sets of quadrature points x_i and quadrature weights w_i are known as "integral rules". Newton-Cotes rules like the trapezoidal rule, the midpoint rule or Simpson's rule, use equidistant points x_i and are easy to apply. Accuracy can be improved by dividing the integration range into sub-intervals and applying composite Newton-Cotes rules. Extrapolation methods reduce the error almost to machine precision but need many function evaluations. Equidistant sample points are convenient but not the best choice. Clenshaw-Curtis expressions use non uniform sample points and a rapidly converging Chebyshev expansion. Gaussian integration fully optimizes the sample points with the help of orthogonal polynomials.

4.1 Equidistant Sample Points

For equidistant points

$$x_i = a + ih \quad i = 0 \cdots N \quad h = \frac{b-a}{N} \tag{4.5}$$

the interpolating polynomial of order N with $p(x_i) = f(x_i)$ is given by the Lagrange method

$$p(x) = \sum_{i=0}^{N} f_i \prod_{k=0, k \neq i}^{N} \frac{x - x_k}{x_i - x_k}. \tag{4.6}$$

Integration of the polynomial gives

$$\int_a^b p(x)dx = \sum_{i=0}^{N} f_i \int_a^b \prod_{k=0, k \neq i}^{N} \frac{x - x_k}{x_i - x_k} dx. \tag{4.7}$$

After substituting

$$\begin{aligned} x &= a + hs \\ x - x_k &= h(s-k) \\ x_i - x_k &= (i-k)h \end{aligned} \tag{4.8}$$

we have

$$\int_a^b \prod_{k=0,k\neq i}^{N} \frac{x-x_k}{x_i-x_k}dx = \int_0^N \prod_{k=0,k\neq i}^{N} \frac{s-k}{i-k}hds = h\alpha_i \tag{4.9}$$

and hence

$$\int_a^b p(x)dx = (b-a)\sum_{i=0}^{N} f_i\alpha_i. \tag{4.10}$$

The weight factors are given by

$$w_i = (b-a)\alpha_i = Nh\alpha_i. \tag{4.11}$$

4.1.1 Closed Newton-Cotes Formulae

For $N = 1$ the polynomial is

$$p(x) = f_0\frac{x-x_1}{x_0-x_1} + f_1\frac{x-x_0}{x_1-x_0} \tag{4.12}$$

and the integral is

$$\int_a^b p(x)dx = f_0\int_0^1 \frac{s-1}{0-1}hds + f_1\int_0^1 \frac{s-0}{1-0}hds$$

$$= -f_0h(\frac{(1-1)^2}{2} - \frac{(0-1)^2}{2}) + f_1h(\frac{1^2}{2} - \frac{0^2}{2})$$

$$= h\frac{f_0+f_1}{2} \tag{4.13}$$

which is known as the trapezoidal rule (Fig. 4.1). $N = 2$ gives Simpson's rule

$$2h\,\frac{f_0+4f_1+f_2}{6}. \tag{4.14}$$

Larger N give further integration rules

$$
\begin{array}{ll}
3h\,\frac{f_0+3f_1+3f_2+f_3}{8} & \text{3/8-rule} \\
4h\,\frac{7f_0+32f_1+12f_2+32f_3+7f_4}{90} & \text{Milne-rule} \\
5h\,\frac{19f_0+75f_1+50f_2+50f_3+75f_4+19f_5}{288} & \\
6h\,\frac{41f_0+216f_1+27f_2+272f_3+27f_4+216f_5+41f_6}{840} & \text{Weddle-rule.}
\end{array}
\tag{4.15}
$$

For even larger N negative weight factors appear and the formulas are not numerically stable.

4.1.2 Open Newton-Cotes Formulae

Alternatively, the integral can be computed from only interior points

$$x_i = a + ih \quad i = 1, 2, \cdots N \quad h = \frac{b-a}{N+1}. \tag{4.16}$$

The simplest case is the midpoint rule (Fig. 4.1)

$$\int_a^b f(x)dx \approx 2hf_1 = (b-a)f(\frac{a+b}{2}). \tag{4.17}$$

The next two are

$$\frac{3h}{2}(f_1 + f_2) \tag{4.18}$$

$$\frac{4h}{3}(2f_1 - f_2 + 2f_3). \tag{4.19}$$

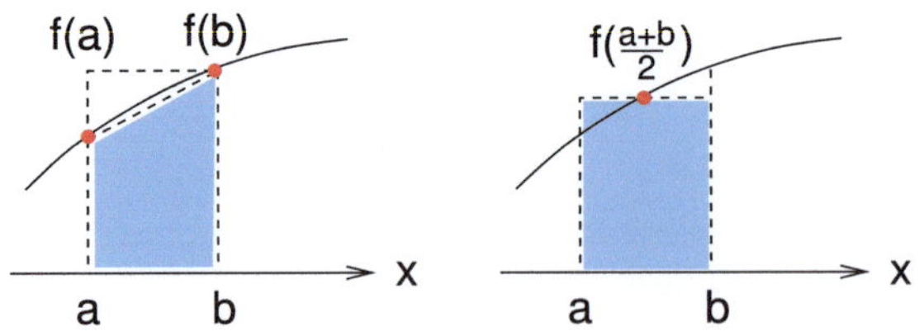

Fig. 4.1 Trapezoidal rule and midpoint rule. **Left:** The trapezoidal rule approximates the integral by the average of the function values at the boundaries. **Right:** The midpoint rule evaluates the function in the center of the interval and has the same error order

4.1.3 Composite Newton-Cotes Rules

Newton-Cotes formulas are only accurate, if the step width is small. Usually the integration range is divided into small sub-intervals

$$[x_i, x_{i+1}] \quad x_i = a + ih \quad i = 0 \cdots N \tag{4.20}$$

for which a simple quadrature formula can be used. Application of the trapezoidal rule for each interval

$$I_i = \frac{h}{2}(f(x_i) + f(x_{i+1})). \tag{4.21}$$

gives the composite trapezoidal rule

$$T = h(\frac{f(a)}{2} + f(a+h) + \cdots f(b-h) + \frac{f(b)}{2}) \tag{4.22}$$

with error order $O(h^2)$. Repeated application of Simpson's rule for $[a, a+2h]$, $[a+2h, a+4h]\ldots$ gives the composite Simpson's rule

$$S = \frac{h}{3}(f(a) + 4f(a+h) + 2f(a+2h) + 4f(a+3h) + \ldots$$

$$\cdots + 2f(b-2h) + 4f(b-h) + f(b)) \tag{4.23}$$

with error order $O(h^4)$.[1]
Repeated application of the midpoint rule gives the composite midpoint rule

$$M = 2h(f(a+h) + f(a+3h) + \cdots f(b-h)) \tag{4.24}$$

with error order $O(h^2)$.

4.1.4 Extrapolation Method (Romberg Integration)

For the trapezoidal rule the Euler-McLaurin expansion exists which for a $2m$ times differentiable function has the form

$$\int_{x_0}^{x_N} f(x)dx - T = \alpha_2 h^2 + \alpha_4 h^4 + \cdots \alpha_{2m-2} h^{2m-2} + O(h^{2m}). \tag{4.25}$$

Therefore extrapolation methods are applicable. From the composite trapezoidal rule for h and $h/2$ an approximation of error order $O(h^4)$ results:

[1] The number of sample points must be even.

$$\int_{x_0}^{x_N} f(x)dx - T(h) = \alpha_2 h^2 + \alpha_4 h^4 + \cdots \tag{4.26}$$

$$\int_{x_0}^{x_N} f(x)dx - T(h/2) = \alpha_2 \frac{h^2}{4} + \alpha_4 \frac{h^4}{16} + \cdots \tag{4.27}$$

$$\int_{x_0}^{x_N} f(x)dx - \frac{4T(h/2) - T(h)}{3} = -\alpha_4 \frac{h^4}{4} + \cdots . \tag{4.28}$$

More generally, for the series of step widths

$$h_k = \frac{h_0}{2^k} \tag{4.29}$$

the Neville method gives the recursion for the interpolating polynomial

$$P_{i\cdots k}(h^2) = \frac{(h^2 - \frac{h_0^2}{2^{2i}})P_{i+1\cdots k}(h^2) - (h^2 - \frac{h_0^2}{2^{2k}})P_{i\cdots k-1}(h^2)}{\frac{h_0^2}{2^{2k}} - \frac{h_0^2}{2^{2i}}} \tag{4.30}$$

which for $h = 0$ becomes the higher order approximation to the integral (Fig. 4.2)

$$P_{i\cdots k} = \frac{2^{-2k}P_{i\cdots k-1} - 2^{-2i}P_{i+1\cdots k}}{2^{-2k} - 2^{-2i}} = \frac{P_{i\cdots k-1} - 2^{2k-2i}P_{i+1\cdots k}}{1 - 2^{2k-2i}}$$

$$= P_{i+1\cdots k} + \frac{P_{i\cdots k-1} - P_{i+1\cdots k}}{1 - 2^{2k-2i}}. \tag{4.31}$$

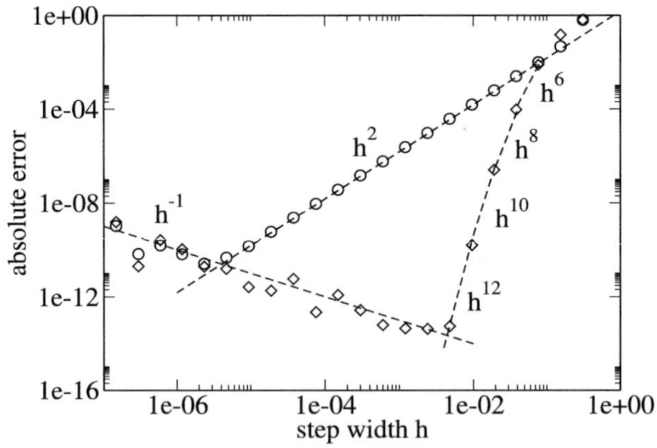

Fig. 4.2 Romberg integration. The integral $\int_0^{\pi^2} \sin(x^2)dx$ is calculated numerically. Circles show the absolute error of the composite trapezoidal rule (4.22) for the step size sequence $h_{i+1} = h_i/2$. Diamonds show the absolute error of the extrapolated value (4.31). The error order of the trapezoidal rule is $O(h^2)$ whereas the error order of the Romberg method increases by factors of h^2. For very small step sizes the rounding errors dominate which increase as h^{-1}

The polynomial values can again be arranged in matrix form

$$\begin{matrix} P_0 & P_{01} & P_{012} & \cdots \\ P_1 & P_{12} & & \\ P_2 & & & \\ \vdots & & & \end{matrix} \tag{4.32}$$

with

$$T_{i,j} = P_{i \cdots i+j} \tag{4.33}$$

and the recursion formula

$$T_{i,0} = P_i = T_s(\frac{h_0}{2^i}) \tag{4.34}$$

$$T_{i,j} = T_{i+1,j-1} + \frac{T_{i,j-1} - T_{i+1,j-1}}{1 - 2^{2j}}. \tag{4.35}$$

4.2 Optimized Sample Points

The Newton-Cotes method integrates polynomials of order up to $N - 1$ exactly, using N equidistant sample points. Unfortunately the polynomial approximation converges slowly, at least for not so well behaved integrands. The accuracy of the integration can be improved by optimizing the sample point positions. Gaussian quadrature determines the N positions and N weights such, that a polynomial of order $2N - 1$ is integrated exactly. The Clenshaw-Curtis and the related Fejer methods use the roots or the extrema of the Chebyshev polynomials as nodes and determine the weights to integrate polynomials of order N. However, since the approximation by Chebyshev polynomials usually converges very fast, the accuracy is in many cases comparable to the Gaussian method [40, 41]. In the following we restrict the integration interval to $[-1, 1]$. The general case $[a, b]$ is then given by a simple change of variables.

4.2.1 Clenshaw-Curtis Expressions

Clenshaw and Curtis [42] make the variable substitution

$$x = \cos\theta \quad dx = -\sin\theta\, d\theta \tag{4.36}$$

for the integral

$$\int_{-1}^{1} f(x)dx = \int_{0}^{\pi} f(\cos t)\, \sin t\, dt \tag{4.37}$$

and approximate the function by the trigonometric polynomial (7.21 with $N = 2M, T = 2\pi$)

$$f(\cos t) = \frac{1}{2M}c_0 + \frac{1}{M}\sum_{j=1}^{M-1} c_j \cos(j\,t) + \frac{1}{2M}c_M \cos(Mt) \tag{4.38}$$

which interpolates (Sect. 7.2.1) $f(\cos t)$ at the sample points

$$t_n = n\Delta t = n\frac{\pi}{M} \quad \text{with } n = 0, 1, \ldots M \tag{4.39}$$

$$x_n = \cos t_n = \cos\left(n\frac{\pi}{M}\right) \tag{4.40}$$

and where the Fourier coefficients are given by (7.17)

$$c_j = f_0 + 2\sum_{n=1}^{M-1} f(\cos(t_n))\cos(\frac{\pi}{M}jn) + f_M\cos(j\pi). \tag{4.41}$$

The function $\cos(j\ t)$ is related to the Chebyshev polynomials of the first kind which for $-1 \leq x \leq 1$ are given by the trigonometric definition

$$T_j(x) = \cos(j\ \arccos(x)) \tag{4.42}$$

and can be calculated recursively

$$T_0(x) = 1 \tag{4.43}$$

$$T_1(x) = x \tag{4.44}$$

$$T_{j+1}(x) = 2x\,T_j(x) - T_{j-1}(x). \tag{4.45}$$

Substituting $x = \cos t$ we find

$$T_j(\cos t) = \cos(j\ t). \tag{4.46}$$

Hence the Fourier series (4.38) corresponds to a Chebyshev approximation

$$f(x) = \sum_{j=0}^{M} a_j T_j(x) = \frac{c_0}{2M}T_0(x) + \sum_{j=1}^{M-1}\frac{c_j}{M}T_j(x) + \frac{c_M}{2M}T_M(x) \tag{4.47}$$

and can be used to approximate the integral

$$\int_{-1}^{1} f(x)dx \approx \int_{0}^{\pi} \left\{ \frac{1}{2M} c_0 + \frac{1}{M} \sum_{j=1}^{M-1} c_j \cos(j\,t) + \frac{1}{2M} c_M \cos(Mt) \right\} \sin\theta \, d\theta \tag{4.48}$$

$$= \frac{1}{M} c_0 + \frac{1}{M} \sum_{j=1}^{M-1} c_j \frac{\cos(j\pi) + 1}{1 - j^2} + \frac{1}{2M} c_M \frac{\cos(M\pi) + 1}{1 - M^2}$$

where, in fact, only the even j contribute.

Example Clenshaw Curtis quadrature for $M = 5$

The function has to be evaluated at the sample points $x_k = \cos(\frac{\pi}{5}k) = (1, 0.80902, 0.30902, -0.30902, -0.80902, -1)$. The Fourier coefficients are given by

$$\begin{pmatrix} c_0 \\ c_1 \\ c_2 \\ c_3 \\ c_4 \\ c_5 \end{pmatrix} = \begin{pmatrix} 1 & 2 & 2 & 2 & 2 & 1 \\ 1 & 1.618 & 0.618 & -0.618 & -1.618 & -1 \\ 1 & 0.618 & -1.618 & -1.618 & 0.618 & 1 \\ 1 & -0.618 & -1.618 & 1.618 & 0.618 & -1 \\ 1 & -1.618 & 0.618 & 0.618 & -1.618 & 1 \\ 1 & -2 & 2 & -2 & 2 & -1 \end{pmatrix} \begin{pmatrix} f_0 \\ f_1 \\ f_2 \\ f_3 \\ f_4 \\ f_5 \end{pmatrix} \tag{4.49}$$

and the integral is approximately

$$\int_{-1}^{1} f(x)dx \approx \left(\tfrac{1}{5}\ 0\ -\tfrac{2}{15}\ 0\ -\tfrac{2}{75}\ 0 \right) \begin{pmatrix} c_0 \\ c_1 \\ c_2 \\ c_3 \\ c_4 \\ c_5 \end{pmatrix}$$

$$= 0.0400 f_0 + 0.3607 f_1 + 0.5993 f_2 + 0.5993 f_3 + 0.3607 f_4 + 0.0400 f_5. \tag{4.50}$$

Clenshaw Curtis weights of very high order can be calculated efficiently [43, 44] using the FFT algorithm (fast Fourier transformation, Sect. 7.3.2).

4.2.2 Gaussian Integration

Now we will optimize the positions of the N quadrature points x_i to obtain the maximum possible accuracy. We approximate the integral by a sum

$$\int_{a}^{b} f(x)dx \approx \sum_{i=1}^{N} f(x_i) w_i \tag{4.51}$$

and determine the $2N$ parameters x_i and w_i such that a polynomial of order $2N-1$ is integrated exactly. This can be achieved with the help of a set of polynomials which are orthogonal with respect to the scalar product

$$< fg >= \int_a^b f(x)g(x)w(x)dx \tag{4.52}$$

where the weight function $w(x)$ and the interval $[a, b]$ determine a particular set of orthogonal polynomials.

Gauss-Legendre Integration

Again we restrict the integration interval to $[-1, 1]$ in the following. For integrals with one or two infinite boundaries see Sect. 4.2.2. The simplest choice for the weight function is

$$w(x) = 1. \tag{4.53}$$

An orthogonal system of polynomials on the interval $[-1, 1]$ can be found using the Gram-Schmidt method:

$$P_0 = 1 \tag{4.54}$$

$$P_1 = x - \frac{P_0}{< P_0 P_0 >} \int_{-1}^{1} x P_0(x)dx = x \tag{4.55}$$

$$\begin{aligned} P_2 &= x^2 - \frac{P_1}{< P_1 P_1 >} \int_{-1}^{1} x^2 P_1(x)dx - \frac{P_0}{< P_0 P_0 >} \int_{-1}^{1} x^2 P_0(x)dx \\ &= x^2 - \frac{1}{3} \end{aligned} \tag{4.56}$$

$$\begin{aligned} P_n = x^n &- \frac{P_{n-1}}{< P_{n-1} P_{n-1} >} \int_{-1}^{1} x^n P_{n-1}(x)dx \\ &- \frac{P_{n-2}}{< P_{n-2} P_{n-2} >} \int_{-1}^{1} x^n P_{n-2}(x)dx - \cdots . \end{aligned} \tag{4.57}$$

The P_n are known as Legendre-polynomials. Consider now a polynomial $p(x)$ of order $2N-1$. It can be interpolated at the N quadrature points x_i using the Lagrange method by a polynomial $\widetilde{p}(x)$ of order $N-1$:

$$\widetilde{p}(x) = \sum_{j=1}^{N} L_j(x)p(x_j). \tag{4.58}$$

Then $p(x)$ can be written as

$$p(x) = \widetilde{p}(x) + (x - x_1)(x - x_2)\cdots(x - x_N)q(x). \tag{4.59}$$

Obviously $q(x)$ is a polynomial of order $(2N - 1) - N = N - 1$. Now choose the positions x_i as the roots of the Legendre polynomial of order N

$$(x - x_1)(x - x_2)\cdots(x - x_N) = P_N(x). \tag{4.60}$$

Then we have

$$\int_{-1}^{1} (x - x_1)(x - x_2)\cdots(x - x_N)q(x)dx = 0 \tag{4.61}$$

since P_N is orthogonal to the polynomial of lower order. But now

$$\int_{-1}^{1} p(x)dx = \int_{-1}^{1} \widetilde{p}(x)dx = \int_{-1}^{1} \sum_{j=1}^{N} p(x_j)L_j(x)dx = \sum_{j=1}^{N} w_j p(x_j) \tag{4.62}$$

with the weight factors

$$w_j = \int_{-1}^{1} L_j(x)dx. \tag{4.63}$$

Example (Gauss-Legendre integration with two quadrature points) The 2nd order Legendre polynomial

$$P_2(x) = x^2 - \frac{1}{3} \tag{4.64}$$

has two roots

$$x_{1,2} = \pm\sqrt{\frac{1}{3}}. \tag{4.65}$$

The Lagrange polynomials are

$$L_1 = \frac{x - \sqrt{\frac{1}{3}}}{-\sqrt{\frac{1}{3}} - \sqrt{\frac{1}{3}}} \quad L_2 = \frac{x + \sqrt{\frac{1}{3}}}{\sqrt{\frac{1}{3}} + \sqrt{\frac{1}{3}}} \tag{4.66}$$

and the weights

$$w_1 = \int_{-1}^{1} L_1 dx = -\frac{\sqrt{3}}{2}\left(\frac{x^2}{2} - \sqrt{\frac{1}{3}}x\right)\Bigg|_{-1}^{1} = 1 \tag{4.67}$$

$$w_2 = \int_{-1}^{1} L_2 dx = \frac{\sqrt{3}}{2}\left(\frac{x^2}{2} + \sqrt{\frac{1}{3}}x\right)_{-1}^{1} = 1. \tag{4.68}$$

This gives the integral rule

$$\int_{-1}^{1} f(x)dx \approx f(-\sqrt{\frac{1}{3}}) + f(\sqrt{\frac{1}{3}}). \tag{4.69}$$

For a general integration interval we substitute

$$x = \frac{a+b}{2} + \frac{b-a}{2}u \tag{4.70}$$

and find the approximation

$$\int_{a}^{b} f(x)dx = \int_{-1}^{1} f(\frac{a+b}{2} + \frac{b-a}{2}u)\frac{b-a}{2}du$$

$$\approx \frac{b-a}{2}\left(f\left(\frac{a+b}{2} - \frac{b-a}{2}\sqrt{\frac{1}{3}}\right) + f\left(\frac{a+b}{2} + \frac{b-a}{2}\sqrt{\frac{1}{3}}\right)\right). \tag{4.71}$$

The next higher order Gaussian rule is given by

$$n = 3 : w_1 = w_3 = 5/9, w_2 = 8/9, x_3 = -x_1 = 0.77459\cdots, x_2 = 0. \tag{4.72}$$

Other Types of Gaussian Integration

Further integral rules can be obtained by using other sets of orthogonal polynomials, for instance

Chebyshev polynomials

$$w(x) = \frac{1}{\sqrt{1-x^2}} \tag{4.73}$$

$$\int_{-1}^{1} f(x)dx = \int_{-1}^{1} f(x)\sqrt{1-x^2}w(x)dx \tag{4.74}$$

$$T_{n+1}(x) = 2x\,T_n(x) - T_{n-1}(x) \tag{4.75}$$

$$x_k = \cos(\frac{2k-1}{2N}\pi) \quad w_k = \frac{\pi}{N}. \tag{4.76}$$

Hermite polynomials

$$w(x) = \mathrm{e}^{-x^2} \tag{4.77}$$

$$\int_{-\infty}^{\infty} f(x)dx = \int_{-\infty}^{\infty} f(x)\mathrm{e}^{x^2} w(x)dx \tag{4.78}$$

$$H_0(x) = 1, \quad H_1(x) = 2x, \quad H_{n+1}(x) = 2xH_n(x) - 2nH_{n-1}(x).$$

Laguerre polynomials

$$w(x) = \mathrm{e}^{-x} \tag{4.79}$$

$$\int_{0}^{\infty} f(x)dx = \int_{0}^{\infty} f(x)\mathrm{e}^{x} w(x)dx \tag{4.80}$$

$$L_0(x) = 1, \quad L_1(x) = 1 - x, \quad L_{n+1}(x) = \frac{1}{n+1}\left((2n+1-x)L_n(x) - nL_{n-1}(x)\right). \tag{4.81}$$

Connection with an Eigenvalue Problem

The determination of quadrature points and weights can be formulated as an eigenvalue problem [45, 46]. Any set of orthogonal polynomials $P_n(x)$ with

$$\int_a^b P_m(x)P_n(x)w(x)dx = \delta_{m,n} \tag{4.82}$$

satisfies a three term recurrence relation

$$P_{n+1}(x) = (a_{n+1}x + b_{n+1})\, P_n(x) - c_{n+1}P_{n-1}(x) \tag{4.83}$$

with $a_n > 0$, $c_n > 0$ which can be written in matrix form [47]

$$x \begin{pmatrix} P_0(x) \\ P_1(x) \\ \vdots \\ \vdots \\ \vdots \\ P_{N-1}(x) \end{pmatrix} = \begin{pmatrix} -\frac{b_1}{a_1} & \frac{1}{a_1} & & & & \\ \frac{c_2}{a_2} & -\frac{b_2}{a_2} & \frac{1}{a_2} & & & \\ & \ddots & \ddots & \ddots & & \\ & & \ddots & \ddots & \ddots & \\ & & & \frac{c_{N-1}}{a_{N-1}} & -\frac{b_{N-1}}{a_{N-1}} & \frac{1}{a_{N-1}} \\ & & & & \frac{c_N}{a_N} & -\frac{b_N}{a_N} \end{pmatrix} \begin{pmatrix} P_0(x) \\ P_1(x) \\ \vdots \\ \vdots \\ \vdots \\ P_{N-1}(x) \end{pmatrix} + \begin{pmatrix} 0 \\ 0 \\ \vdots \\ \vdots \\ 0 \\ \frac{1}{a_N}P_N(x) \end{pmatrix} \tag{4.84}$$

or shorter

$$x\mathbf{P}(x) = T\mathbf{P}(x) + \frac{1}{a_N} P_N(x)\mathbf{e}_{N-1} \tag{4.85}$$

with a tridiagonal matrix T. Obviously $P_N(x) = 0$ if and only if

$$x_j \mathbf{P}(x_j) = T\mathbf{P}(x_j), \tag{4.86}$$

hence the roots of $P_N(x)$ are given by the eigenvalues of T. The matrix T is symmetric if the polynomials are orthonormal, otherwise it can be transformed into a symmetric tridiagonal matrix by an orthogonal transformation [47]. Finally the quadrature weight corresponding to the eigenvalue x_j can be calculated from the first component of the corresponding eigenvector $\mathbf{u}_j$ [45] as

$$w_j = u_{j,1}^2 \times \int_a^b w(x)dx. \tag{4.87}$$

Exercises

Exercise 4.1 Romberg Integration

Use the trapezoidal rule

$$T(h) = h(\frac{1}{2}f(a) + f(a+h) + \cdots f(b-h) + \frac{1}{2}f(b)) = \int_a^b f(x)dx + \cdots \tag{4.88}$$

with the step sequence

$$h_i = \frac{h_0}{2^i} \tag{4.89}$$

and calculate the elements of the triangular matrix

$$T(i,0) = T(h_i) \tag{4.90}$$

$$T(i,k) = T(i+1,k-1) + \frac{T(i,k-1) - T(i+1,k-1)}{1 - \frac{h_i^2}{h_{i+k}^2}} \tag{4.91}$$

to obtain the approximations

$$T_{01} = P_{01}, T_{02} = P_{012}, T_{03} = P_{0123}, \cdots \tag{4.92}$$

- calculate

$$\int_0^{\pi^2} \sin(x^2)dx = 0.6773089370468890331\ldots \tag{4.93}$$

and compare the absolute error of the trapezoidal sums $T(h_i) = T_{i,0}$ and the extrapolated values $T_{0,i}$.

- calculate

$$\int_{\varepsilon}^{1} \frac{dx}{\sqrt{x}} \tag{4.94}$$

for $\varepsilon = 10^{-3}$. Compare with the composite midpoint rule

$$T(h) = h(f(a + \frac{h}{2}) + f(a + \frac{3h}{2}) + \cdots + f(b - \frac{3h}{2}) + f(b - \frac{h}{2})). \tag{4.95}$$

Chapter 5
Systems of Inhomogeneous Linear Equations

Many problems in physics and especially computational physics involve systems of linear equations

$$\begin{aligned} a_{11}x_1 + \cdots + a_{1n}x_n &= b_1 \\ \vdots \quad \vdots \quad &\vdots \\ a_{n1}x_1 + \cdots + a_{nn}x_n &= b_n \end{aligned} \tag{5.1}$$

or shortly in matrix form

$$A\mathbf{x} = \mathbf{b} \tag{5.2}$$

which arise e.g. from linearization of a general nonlinear problem like (Sect. 9.2 in Vol.2)

$$0 = \begin{pmatrix} F_1(x_1 \ldots \cdots x) \\ \vdots \\ F_n(x_1 \ldots \cdots x) \end{pmatrix} = \begin{pmatrix} F_1(x_1^{(0)} \ldots \cdots x) \\ \vdots \\ F_n(x_1^{(0)} \ldots \cdots x) \end{pmatrix} + \begin{pmatrix} \frac{\partial F_1}{\partial x_1} & \cdots & \frac{\partial F_1}{\partial x_n} \\ \vdots & \ddots & \vdots \\ \frac{\partial F_n}{\partial x_1} & \cdots & \frac{\partial F_n}{\partial x_n} \end{pmatrix} \begin{pmatrix} x_1 - x_1^{(0)} \\ \vdots \\ x_n - x_n^{(0)} \end{pmatrix} + \ldots \tag{5.3}$$

or from discretization of differential equations like

$$0 = \frac{\partial f}{\partial x} - g(x) \rightarrow \begin{pmatrix} \vdots \\ \frac{f((j+1)\Delta x) - f(j\Delta x)}{\Delta x} - g(j\Delta x) \\ \vdots \end{pmatrix}$$

Supplementary Information The online version contains supplementary material available at https://doi.org/10.1007/978-3-032-07856-8_5.

P. O. J. Scherer, *Computational Physics I*, Graduate Texts in Physics,
https://doi.org/10.1007/978-3-032-07856-8_5

$$= \begin{pmatrix} \ddots & & & \\ & -\frac{1}{\Delta x} & \frac{1}{\Delta x} & \\ & & -\frac{1}{\Delta x} & \frac{1}{\Delta x} \\ & & & \ddots \end{pmatrix} \begin{pmatrix} \vdots \\ f_j \\ f_{j+1} \\ \vdots \end{pmatrix} - \begin{pmatrix} \vdots \\ g_j \\ g_{j+1} \\ \vdots \end{pmatrix}. \tag{5.4}$$

If the matrix A is non singular and has full rank, (5.2) can be formally solved by matrix inversion

$$\mathbf{x} = A^{-1}\mathbf{b} \tag{5.5}$$

if the matrix is singular or the number of equations smaller than the number of variables, a manifold of solutions exists which can be found efficiently by singular value decomposition (Sect. 11.2). The general solution is given by a particular solution and the nullspace of A

$$\mathbf{x} = \mathbf{x}_p + \mathbf{z} \text{ with } A\mathbf{x}_p = \mathbf{b} \text{ and } A\mathbf{z} = 0. \tag{5.6}$$

If the number of equations is larger than the number of variables there exists no unique solution. The "best possible solution" can be determined by minimizing the residual

$$\min|A\mathbf{x} - \mathbf{b}| \tag{5.7}$$

which leads to a least squares problem (Sect. 11.1.1).

In the following we discuss several methods to solve non singular systems. If the dimension is not too large direct methods like Gaussian elimination or QR decomposition are sufficient. Systems with a tridiagonal matrix are important for cubic spline interpolation and numerical second derivatives. They can be solved very efficiently with a specialized Gaussian elimination method. Practical applications often involve very large dimensions and require iterative methods. Stationary methods apply a simple iteration scheme repeatedly. The slow convergence of the methods by Jacobi and Gauss-Seidel can be improved with relaxation or over-relaxation. Non-stationary methods construct a sequence of improved approximations within a series of increasing subspaces of $\mathbb{R}^N$. Modern Krylov-space methods minimize the residual $\mathbf{r} = A\mathbf{x} - \mathbf{b}$ within the sequence of Krylov-spaces $K_n(A, \mathbf{r}^{(0)}) = \text{span}(\mathbf{r}^{(0)}, A\mathbf{r}^{(0)}, \ldots A^{n-1}\mathbf{r}^{(0)})$. We discuss the conjugate gradients method (CG [48]) for symmetric positive definite matrices and the method of general minimal residuals (GMRES [49]) for non symmetric matrices. Other popular methods are the methods of bi-conjugate gradients (BiCG [50] BiCGSTAB [51]), conjugate residuals (CR [52]) , minimal residual (MINRES [53]), quasi-minimal residual (QMR [54]), the symmetric LQ-method (SYMMLQ [55]) and Lanzcos type product methods (LTPM [56–58]).

5.1 Gaussian Elimination Method

A series of linear combinations of the equations transforms the matrix A into an upper triangular matrix. Start with subtracting a_{i1}/a_{11} times the first row from rows $2 \cdots n$

$$\begin{pmatrix} \mathbf{a}_1^T \\ \mathbf{a}_2^T \\ \vdots \\ \mathbf{a}_n^T \end{pmatrix} \to \begin{pmatrix} \mathbf{a}_1^T \\ \mathbf{a^T}_2 - l_{21}\mathbf{a}_1^T \\ \vdots \\ \mathbf{a}_n^T - l_{n1}\mathbf{a}_1^T \end{pmatrix} \tag{5.8}$$

which can be written as a multiplication

$$A^{(1)} = L_1 A \tag{5.9}$$

with the Frobenius matrix

$$L_1 = \begin{pmatrix} 1 & & & \\ -l_{21} & 1 & & \\ -l_{31} & & 1 & \\ \vdots & & & \ddots & \\ -l_{n1} & & & & 1 \end{pmatrix} \qquad l_{i1} = \frac{a_{i1}}{a_{11}}. \tag{5.10}$$

The result has the form

$$A^{(1)} = \begin{pmatrix} a_{11} & a_{12} & \cdots & a_{1n-1} & a_{1n} \\ 0 & a_{22}^{(1)} & \cdots & a_{2n-1}^{(1)} & a_{2n}^{(1)} \\ 0 & a_{32}^{(1)} & \cdots & \cdots & a_{3n}^{(1)} \\ 0 & \vdots & & & \vdots \\ 0 & a_{n2}^{(1)} & \cdots & \cdots & a_{nn}^{(1)} \end{pmatrix}. \tag{5.11}$$

Now subtract $\frac{a_{i2}}{a_{22}}$ times the second row from rows $3 \cdots n$. This can be formulated as

$$A^{(2)} = L_2 A^{(1)} = L_2 L_1 A \tag{5.12}$$

with

$$L_2 = \begin{pmatrix} 1 & & & \\ 0 & 1 & & \\ 0 & -l_{32} & 1 & \\ \vdots & \vdots & & \ddots & \\ 0 & -l_{n2} & & & 1 \end{pmatrix} \qquad l_{i2} = \frac{a_{i2}^{(1)}}{a_{22}^{(1)}}. \tag{5.13}$$

The result is

$$A^{(2)} = \begin{pmatrix} a_{11}^{(2)} & a_{12}^{(2)} & a_{13}^{(2)} & \cdots & a_{1n}^{(2)} \\ 0 & a_{22}^{(2)} & a_{23}^{(2)} & \cdots & a_{2n}^{(2)} \\ 0 & 0 & a_{33}^{(2)} & \cdots & a_{3n}^{(2)} \\ \vdots & \vdots & \vdots & & \vdots \\ 0 & 0 & a_{n3}^{(2)} & \cdots & a_{nn}^{(2)} \end{pmatrix}. \tag{5.14}$$

Continue until an upper triangular matrix results after n-1 steps:

$$A^{(n-1)} = L_{n-1}A^{(n-2)} \tag{5.15}$$

$$L_{n-1} = \begin{pmatrix} 1 & & & & \\ & 1 & & & \\ & & \ddots & & \\ & & & 1 & \\ & & & -l_{n,n-1} & 1 \end{pmatrix} \qquad l_{n,n-1} = \frac{a_{n,n-1}^{(n-2)}}{a_{n-1,n-1}^{(n-2)}} \tag{5.16}$$

$$A^{(n-1)} = L_{n-1}L_{n-2}\cdots L_2L_1A = U \tag{5.17}$$

$$U = \begin{pmatrix} u_{11} & u_{12} & u_{13} & \cdots & u_{1n} \\ & u_{22} & u_{23} & \cdots & u_{2n} \\ & & u_{33} & \cdots & u_{3n} \\ & & & \ddots & \vdots \\ & & & & u_{nn} \end{pmatrix}. \tag{5.18}$$

The transformed system of equations

$$U\mathbf{x} = \mathbf{y} \qquad \mathbf{y} = L_{n-1}L_{n-1}\cdots L_2L_1\mathbf{b} \tag{5.19}$$

can be solved easily by backward substitution:

$$x_n = \frac{1}{u_{nn}}y_n \tag{5.20}$$

$$x_{n-1} = \frac{y_{n-1} - x_n u_{n-1,n}}{u_{n-1,n-1}} \tag{5.21}$$

$$\vdots \tag{5.22}$$

Alternatively the matrices L_i can be inverted:

$$L_1^{-1} = \begin{pmatrix} 1 & & & & \\ l_{21} & 1 & & & \\ l_{31} & & 1 & & \\ \vdots & & & \ddots & \\ l_{n1} & & & & 1 \end{pmatrix} \cdots L = \begin{pmatrix} 1 & & & & \\ & 1 & & & \\ & & \ddots & & \\ & & & 1 & \\ & & & l_{n,n-1} & 1 \end{pmatrix}. \tag{5.23}$$

This gives

$$A = L_1^{-1} L_2^{-1} \cdots L_{n-1}^{-1} U. \tag{5.24}$$

The product of the inverted matrices is a lower triangular matrix:

$$L_1^{-1} L_2^{-1} = \begin{pmatrix} 1 & & & & \\ l_{21} & 1 & & & \\ l_{31} & l_{32} & 1 & & \\ \vdots & \vdots & & \ddots & \\ l_{n1} & l_{n2} & & & 1 \end{pmatrix}$$

$$\vdots$$

$$L = L_1^{-1} L_2^{-1} \cdots L_{n-1}^{-1} = \begin{pmatrix} 1 & & & & \\ l_{21} & 1 & & & \\ \vdots & \vdots & \ddots & & \\ l_{n-1,1} & l_{n-1,2} & \cdots & 1 & \\ l_{n1} & l_{n2} & \cdots & l_{n,n-1} & 1 \end{pmatrix}. \tag{5.25}$$

Hence the matrix A becomes decomposed into a product of a lower and an upper triangular matrix

$$A = LU \tag{5.26}$$

which can be used to solve the system of equations (5.2).

$$A\mathbf{x} = LU\mathbf{x} = \mathbf{b} \tag{5.27}$$

in two steps:

$$L\mathbf{y} = \mathbf{b} \tag{5.28}$$

which can be solved from the top

$$y_1 = b_1 \tag{5.29}$$

$$y_2 = b_2 - l_{21} y_1 \tag{5.30}$$

$$\vdots \tag{5.31}$$

and

$$U\mathbf{x} = \mathbf{y} \tag{5.32}$$

which can be solved from the bottom

$$x_n = \frac{1}{u_{nn}} y_n \tag{5.33}$$

$$x_{n-1} = \frac{y_{n-1} - x_n u_{n-1,n}}{u_{n-1,n-1}}. \tag{5.34}$$

$$\vdots \tag{5.35}$$

5.1.1 Pivoting

To improve numerical stability and to avoid division by zero pivoting is used. Most common is partial pivoting. In every step the order of the equations is changed in order to maximize the pivoting element $a_{k,k}$ in the denominator. This gives LU decomposition of the matrix PA where P is a permutation matrix. P is not needed explicitly. Instead an index vector is used which stores the new order of the equations

$$P \begin{pmatrix} 1 \\ \vdots \\ N \end{pmatrix} = \begin{pmatrix} i_1 \\ \vdots \\ i_N \end{pmatrix}. \tag{5.36}$$

Total pivoting exchanges rows and columns of A. This can be time consuming for larger matrices.

If the elements of the matrix are of different orders of magnitude it can be necessary to balance the matrix, for instance by normalizing all rows of A. This can be also achieved by selecting the maximum of

$$\frac{a_{ik}}{\sum_j |a_{ij}|} \tag{5.37}$$

as the pivoting element.

5.1.2 Direct LU Decomposition

LU decomposition can be also performed in a different order [59]. For symmetric positive definite matrices there exists the simpler and more efficient Cholesky method decomposes the matrix into the product LL^T of a lower triangular matrix and its transpose [60].

5.2 QR Decomposition

The Gaussian elimination method can become numerically unstable [61]. An alternative method to solve a system of linear equations uses the decomposition [62]

$$A = QR \tag{5.38}$$

with a unitary matrix $Q^\dagger Q = 1$ (an orthogonal matrix $Q^T Q = 1$ if A is real) and an upper right triangular matrix R. The system of linear equations (5.2) is simplified by multiplication with $Q^\dagger = Q^{-1}$

$$QR\mathbf{x} = A\mathbf{x} = \mathbf{b} \tag{5.39}$$

$$R\mathbf{x} = Q^\dagger \mathbf{b}. \tag{5.40}$$

Such a system with upper triangular matrix is easily solved (see 5.32).

5.2.1 *QR Decomposition by Orthogonalization*

Gram-Schmidt orthogonalization [2, 62] provides a simple way to perform a QR decomposition. It is used for symbolic calculations and also for least square fitting (11.1.2) but can become numerically unstable.

From the decomposition $A = QR$ we have

$$a_{ik} = \sum_{j=1}^{k} q_{ij} r_{jk} \tag{5.41}$$

$$\mathbf{a}_k = \sum_{j=1}^{k} r_{jk} \mathbf{q}_j \tag{5.42}$$

which gives the k-th column vector $\mathbf{a}_k$ of A as a linear combination of the orthonormal vectors $\mathbf{q}_1 \cdots \mathbf{q}_k$. Similarly $\mathbf{q}_k$ is a linear combination of the first k columns of A. With the help of the Gram-Schmidt method r_{jk} and $\mathbf{q}_j$ are calculated as follows:

$$r_{11} \leftarrow |a_1| \tag{5.43}$$

$$\mathbf{q}_1 \leftarrow \frac{\mathbf{a}_1}{r_{11}}. \tag{5.44}$$

For $k = 2, \cdots n$:

$$r_{ik} \leftarrow \mathbf{q}_i \mathbf{a}_k \; i = 1 \cdots k-1 \tag{5.45}$$

$$\mathbf{b}_k \leftarrow \mathbf{a}_k - r_{1k}\mathbf{q}_1 - \cdots r_{k-1,k}\mathbf{q}_{k-1} \tag{5.46}$$

$$r_{kk} \leftarrow |\mathbf{b}_k| \tag{5.47}$$

$$\mathbf{q}_k \leftarrow \frac{\mathbf{b}_k}{r_{kk}}. \tag{5.48}$$

Obviously now

$$\mathbf{a}_k = r_{kk}\mathbf{q}_k + r_{k-1,k}\mathbf{q}_{k-1} + \cdots r_{1k}\mathbf{q}_1 \tag{5.49}$$

since per definition

$$\mathbf{q}_i \mathbf{a}_k = r_{ik} \quad i = 1 \cdots k \tag{5.50}$$

and

$$r_{kk}^2 = |\mathbf{b}_k|^2 = |\mathbf{a}_k|^2 + r_{1k}^2 + \cdots r_{k-1,k}^2 - 2r_{1k}^2 - \cdots - 2r_{k-1,k}^2. \tag{5.51}$$

Hence

$$\mathbf{q}_k \mathbf{a}_k = \frac{1}{r_{kk}}(\mathbf{a}_k - r_{1k}\mathbf{q}_1 \cdots r_{k-1,k}\mathbf{q}_{k-1})\mathbf{a}_k = \frac{1}{r_{kk}}(|a_k|^2 - r_{1k}^2 \cdots - r_{k-1,k}^2) = r_{kk}. \tag{5.52}$$

Orthogonality gives

$$\mathbf{q}_i \mathbf{a}_k = 0 \quad i = k+1 \cdots n. \tag{5.53}$$

In matrix notation we have finally

$$A = (\mathbf{a}_1 \cdots \mathbf{a}_n) = (\mathbf{q}_1 \cdots \mathbf{q}_n) \begin{pmatrix} r_{11} & r_{12} & \cdots & r_{1n} \\ & r_{22} & \cdots & r_{2n} \\ & & \ddots & \vdots \\ & & & r_{nn} \end{pmatrix}. \tag{5.54}$$

If the columns of A are almost linearly dependent, numerical stability can be improved by an additional orthogonalization step

$$\mathbf{b}_k \leftarrow \mathbf{b}_k - (\mathbf{q}_1\mathbf{b}_k)\mathbf{q}_1 - \cdots (\mathbf{q}_{k-1}\mathbf{b}_k)\mathbf{q}_{k-1} \tag{5.55}$$

after (5.46) which can be iterated several times to improve the results [2, 63].

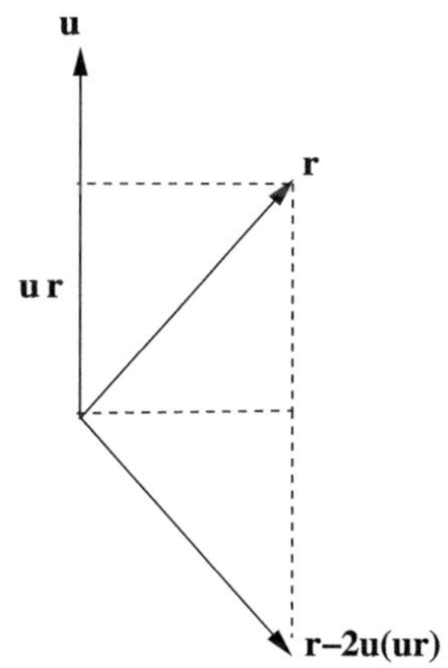

Fig. 5.1 Householder transformation. Geometrically the Householder transformation (5.56) is a mirror operation with respect to a plane with normal vector **u**

5.2.2 *QR Decomposition by Householder Reflections*

Numerically stable algorithms use a series of transformations with unitary matrices, mostly Householder reflections (Fig. 5.1) [2][1] which have the form

$$P = P^T = 1 - 2\mathbf{u}\mathbf{u}^T \tag{5.56}$$

with a unit vector

$$|\mathbf{u}| = 1. \tag{5.57}$$

Obviously P is an orthogonal matrix since

$$P^T P = (1 - 2\mathbf{u}\mathbf{u}^T)(1 - 2\mathbf{u}\mathbf{u}^T) = 1 - 4\mathbf{u}\mathbf{u}^T + 4\mathbf{u}\mathbf{u}^T\mathbf{u}\mathbf{u}^T = 1. \tag{5.58}$$

In the first step we try to find a vector $\mathbf{u}$ such that the first column vector of A

$$\mathbf{a}_1 = \begin{pmatrix} a_{11} \\ \vdots \\ a_{n1} \end{pmatrix} \tag{5.59}$$

is transformed into a vector along the 1-axis

$$P\mathbf{a}_1 = \left(1 - 2\mathbf{u}\mathbf{u}^T\right)\mathbf{a}_1 = k\mathbf{e}_1 = \begin{pmatrix} k \\ 0 \\ \vdots \\ 0 \end{pmatrix}. \tag{5.60}$$

[1] Alternatively Givens rotations [62] can be employed which need slightly more floating point operations.

Multiplication with the transpose vector gives

$$k^2 = (P\mathbf{a}_1)^T \, P\mathbf{a}_1 = \mathbf{a}_1^T P^T P\mathbf{a}_1 = |\mathbf{a}_1|^2 \tag{5.61}$$

and

$$k = \pm|\mathbf{a}_1| \tag{5.62}$$

can have both signs. From (5.60) we have

$$\mathbf{a}_1 - 2\mathbf{u}(\mathbf{u}\mathbf{a}_1) = k\mathbf{e}_1. \tag{5.63}$$

Multiplication with $\mathbf{a}_1^T$ gives

$$2(\mathbf{u}\mathbf{a}_1)^2 = |\mathbf{a}_1|^2 - k(\mathbf{a}_1\mathbf{e}_1) \tag{5.64}$$

and since

$$|\mathbf{a}_1 - k\mathbf{e}_1|^2 = |\mathbf{a}_1|^2 + k^2 - 2k(\mathbf{a}_1\mathbf{e}_1) = 2|\mathbf{a}_1|^2 - 2k(\mathbf{a}_1\mathbf{e}_1) \tag{5.65}$$

we have

$$2(\mathbf{u}\mathbf{a}_1)^2 = \frac{1}{2}|\mathbf{a}_1 - k\mathbf{e}_1|^2 \tag{5.66}$$

and from (5.63) we find

$$\mathbf{u} = \frac{\mathbf{a}_1 - k\mathbf{e}_1}{2\mathbf{u}\mathbf{a}_1} = \frac{\mathbf{a}_1 - k\mathbf{e}_1}{|\mathbf{a}_1 - k\mathbf{e}_1|}. \tag{5.67}$$

To avoid numerical extinction the sign of k is chosen such that

$$\sigma = \text{sign}(k) = -\text{sign}(a_{11}). \tag{5.68}$$

Then,

$$\mathbf{u} = \frac{1}{\sqrt{2(a_{11}^2 + \cdots + a_{n1}^2) + 2|a_{11}|\sqrt{a_{11}^2 + \cdots + a_{n1}^2}}} \begin{pmatrix} \text{sign}(a_{11}) \left(|a_{11}| + \sqrt{a_{11}^2 + a_{21}^2 + \cdots a} \right) \\ a_{21} \\ \vdots \\ a_{n1} \end{pmatrix} \tag{5.69}$$

$$2\mathbf{u}\mathbf{u}^T\mathbf{a}_1 = \begin{pmatrix} \mathrm{sign}(a_{11})\left(|a_{11}| + \sqrt{a_{11}^2 + a_{21}^2 + \ldots a_{n1}^2}\right) \\ a_{21} \\ \vdots \\ a_{n1} \end{pmatrix}$$

$$\times \frac{1}{(a_{11}^2 + \cdots + a_{n1}^2) + |a_{11}|\sqrt{a_{11}^2 + \cdots + a_{n1}^2}} \left(a_{11}^2 + |a_{11}|\sqrt{a_{11}^2 + \cdots a_{n1}^2} + a_{21}^2 + \cdots + a_{n1}^2\right) \tag{5.70}$$

and the Householder transformation of the first column vector of A gives

$$(1 - 2\mathbf{u}\mathbf{u}^T)\mathbf{a}_1 = \begin{pmatrix} -\mathrm{sign}(a_{11})\sqrt{a_{11}^2 \cdots a_{n1}^2} \\ 0 \\ \vdots \\ 0 \end{pmatrix}. \tag{5.71}$$

Thus after the first step a matrix results of the form

$$A^{(1)} = P_1 A = \left(\cdots a_{nn}^{(1)}\right).$$

In the following (n-2) steps further Householder reflections are applied in the subspace $k \le i, j \le n$ to eliminate the elements

$$a_{k+1,k} \ldots a_{n,k}$$

of the $k - th$ row vector below the diagonal of the matrix

$$A^{(k-1)} = P_{k-1} \cdots PA = \begin{pmatrix} a_{11}^{(1)} & \cdots & a_{1,k-1}^{(1)} & a_{1,k}^{(1)} & \cdots & a_{1,n}^{(1)} \\ 0 & \ddots & \vdots & \vdots & & \vdots \\ \vdots & \ddots & a_{k-1,k-1}^{(k-1)} & a_{k-1,k}^{(k-1)} & & a_{k-1,n}^{(k-1)} \\ \vdots & \vdots & 0 & a_{k,k}^{(k-1)} & & a_{k,n}^{(k-1)} \\ \vdots & \vdots & \vdots & a_{k+1,k}^{(k-1)} & & a_{k+1,n}^{(k-1)} \\ \vdots & \vdots & \vdots & \vdots & & \vdots \\ 0 & \ldots & 0 & a_{n,k}^{(k-1)} & \cdots & a_{n,n}^{(k-1)} \end{pmatrix}. \tag{5.72}$$

$$P_k = \begin{pmatrix} 1_1 & & & \\ & \ddots & & \\ & & 1_{k-1} & \\ & & & 1 - 2\mathbf{u}\mathbf{u}^T \end{pmatrix}$$

Finally an upper triangular matrix results

$$A^{(n-1)} = (P_{n-1} \cdot \ldots \cdot P)A = R = \begin{pmatrix} a_{11}^{(1)} & a_{12}^{(1)} & \cdots & a_{1,n-1}^{(1)} & a_{1,n}^{(1)} \\ 0 & a_{22}^{(2)} & \cdots & a_{2,n-1}^{(2)} & a_{2,n}^{(2)} \\ \vdots & 0 & \ddots & \vdots & \vdots \\ \vdots & \vdots & \vdots & a_{n-1,n-1}^{(n-1)} & a_{n-1,n}^{(n-1)} \\ 0 & 0 & \cdots & 0 & a_{n,n}^{(n-1)} \end{pmatrix} . \tag{5.73}$$

If the orthogonal matrix Q is needed explicitly additional numerical operations are necessary to form the product

$$Q = (P_{n-1} \ldots P_1)^T . \tag{5.74}$$

5.3 Linear Equations with Tridiagonal Matrix

Linear equations with the form

$$b_1 x_1 + c_1 x_2 = r_1 \tag{5.75}$$

$$a_i x_{i-1} + b_i x_i + c_i x_{i+1} = r_i \quad i = 2 \cdots (n-1) \tag{5.76}$$

$$a_n x_{n-1} + b_n x_n = r_n \tag{5.77}$$

can be solved very efficiently with a specialized Gaussian elimination method.[2] They are important for cubic spline interpolation or second derivatives. We begin by eliminating a_2. To that end we multiply the first line with a_2/b_1 and subtract it from the first line. The result is the equation

$$\beta_2 x_2 + c_2 x_3 = \rho_2 \tag{5.78}$$

[2] This algorithm is only well behaved if the matrix is diagonal dominant $|b_i| > |a_i| + |c_i|$.

with the abbreviations

$$\beta_2 = b_2 - \frac{c_1 a_2}{b_1} \quad \rho_2 = r_2 - \frac{r_1 a_2}{b_1}. \tag{5.79}$$

We iterate this procedure

$$\beta_i x_i + c_i x_{i+1} = \rho_i \tag{5.80}$$

$$\beta_i = b_i - \frac{c_{i-1} a_i}{\beta_{i-1}} \quad \rho_i = r_i - \frac{\rho_{i-1} a_i}{\beta_{i-1}} \tag{5.81}$$

until we reach the n-th equation, which becomes simply

$$\beta_n x_n = \rho_n \tag{5.82}$$

$$\beta_n = b_n - \frac{c_{n-1} a_n}{\beta_{n-1}} \quad \rho_n = r_n - \frac{\rho_{n-1} a_n}{\beta_{n-1}}. \tag{5.83}$$

Now we immediately have

$$x_n = \frac{\rho_n}{\beta_n} \tag{5.84}$$

and backward substitution gives

$$x_{i-1} = \frac{\rho_{i-1} - c_{i-1} x_i}{\beta_{i-1}} \tag{5.85}$$

and finally

$$x_1 = \frac{r_1 - c_1 x_2}{\beta_2}. \tag{5.86}$$

This algorithm can be formulated as LU decomposition: Multiplication of the matrices

$$L = \begin{pmatrix} 1 & & & & \\ l_2 & 1 & & & \\ & l_3 & 1 & & \\ & & \ddots & \ddots & \\ & & & l_n & 1 \end{pmatrix} \quad U = \begin{pmatrix} \beta_1 & c_1 & & & \\ & \beta_2 & c_2 & & \\ & & \beta_3 & c_3 & \\ & & & \ddots & \\ & & & & \beta_n \end{pmatrix} \tag{5.87}$$

gives

$$LU = \begin{pmatrix} \beta_1 & c_1 & & & & \\ & \ddots & \ddots & & & \\ & & \ddots & \ddots & & \\ & & l_i\beta_{i-1} & (l_ic_{i-1}+\beta_i) & c_i & \\ & & & \ddots & \ddots & \ddots \\ & & & & l_n\beta_{n-1} & (l_nc_{n-1}+\beta_n) \end{pmatrix} \tag{5.88}$$

which coincides with the matrix

$$A = \begin{pmatrix} b_1 & c_1 & & & & \\ a_2 & \ddots & \ddots & & & \\ & \ddots & \ddots & \ddots & & \\ & & a_i & b_i & c_i & \\ & & & \ddots & \ddots & \ddots \\ & & & a_{n-1} & b_{n-1} & c_{n-1} \\ & & & & a_n & b_n \end{pmatrix} \tag{5.89}$$

if we choose

$$l_i = \frac{a_i}{\beta_{i-1}} \tag{5.90}$$

since then from (5.81)

$$b_i = \beta_i + l_ic_{i-1} \tag{5.91}$$

and

$$l_i\beta_{i-1} = a_i. \tag{5.92}$$

5.4 Cyclic Tridiagonal Systems

Periodic boundary conditions lead to a small perturbation of the tridiagonal matrix

$$A = \begin{pmatrix} b_1 & c_1 & & & & a_1 \\ a_2 & \ddots & \ddots & & & \\ & \ddots & \ddots & \ddots & & \\ & & a_i & b_i & c_i & \\ & & & \ddots & \ddots & \ddots \\ & & & a_{n-1} & b_{n-1} & c_{n-1} \\ c_n & & & & a_n & b_n \end{pmatrix}. \tag{5.93}$$

The system of equations

$$A\mathbf{x} = \mathbf{r} \tag{5.94}$$

can be reduced to a tridiagonal system [64, 65] with the help of the Sherman–Morrison formula [66], which states that if A_0 is an invertible matrix and $\mathbf{u}$, $\mathbf{v}$ are vectors and

$$1 + \mathbf{v}^T A_0^{-1}\mathbf{u} \neq 0 \tag{5.95}$$

then the inverse of the matrix[3]

$$A = A_0 + \mathbf{u}\mathbf{v}^T \tag{5.96}$$

is given by

$$A^{-1} = A_0^{-1} - \frac{A_0^{-1}\mathbf{u}\mathbf{v}^T A_0^{-1}}{1 + \mathbf{v}^T A_0^{-1}\mathbf{u}}. \tag{5.97}$$

We choose

$$\mathbf{u}\mathbf{v}^T = \begin{pmatrix} \alpha \\ 0 \\ \vdots \\ 0 \\ c_n \end{pmatrix} \begin{pmatrix} 1 \; 0 \cdots 0 \; \frac{a_1}{\alpha} \end{pmatrix} = \begin{pmatrix} \alpha & & a_1 \\ & & \\ c_n & & \frac{a_1 c_n}{\alpha} \end{pmatrix}. \tag{5.98}$$

Then

$$A_0 = A - \mathbf{u}\mathbf{v}^T = \begin{pmatrix} (b_1-\alpha) & c_1 & & & & & 0 \\ a_2 & \ddots & \ddots & & & & \\ & \ddots & \ddots & \ddots & & & \\ & & a_i & b_i & c_i & & \\ & & & \ddots & \ddots & \ddots & \\ & & & & a_{n-1} & b_{n-1} & c_{n-1} \\ 0 & & & & & a_n & (b_n - \frac{a_1 c_n}{\alpha}) \end{pmatrix} \tag{5.99}$$

is tridiagonal. The free parameter α has to be chosen such that the diagonal elements do not become too small. We solve the system (5.94) by solving the two tridiagonal systems

$$\begin{aligned} A_0\mathbf{x}_0 &= \mathbf{r} \\ A_0\mathbf{q} &= \mathbf{u} \end{aligned} \tag{5.100}$$

[3] Here $\mathbf{u}\mathbf{v}^T$ is the outer or matrix product of the two vectors. For complex valued matrices the Hermitian transpose $\mathbf{v}^H$ has to be used.

and compute $\mathbf{x}$ from

$$\mathbf{x} = A^{-1}\mathbf{r} = A_0^{-1}\mathbf{r} - \frac{(A_0^{-1}\mathbf{u})\mathbf{v}^T(A_0^{-1}\mathbf{r})}{1+\mathbf{v}^T(A_0^{-1}\mathbf{u})} = \mathbf{x}_0 - \mathbf{q}\frac{\mathbf{v}^T\mathbf{x}_0}{1+\mathbf{v}^\mathbf{T}\mathbf{q}}. \tag{5.101}$$

5.5 Linear Stationary Iteration

Discretized differential equations often lead to systems of equations with large sparse matrices, which have to be solved by iterative methods which, starting from an initial guess $\mathbf{x}_0$ (often simply $\mathbf{x}_0 = 0$ or $\mathbf{x}_0 = \mathbf{b}$) construct a sequence of improved vectors by the iteration

$$\mathbf{x}^{(n+1)} = \Phi(\mathbf{x}^{(n)}, \mathbf{b}). \tag{5.102}$$

which under certain conditions converges to the fixed point

$$\mathbf{x}_{FP} = \Phi(\mathbf{x}_{FP}, \mathbf{b}) \tag{5.103}$$

which solves the system of equations

$$A\mathbf{x} = \mathbf{b}. \tag{5.104}$$

A linear iterative method is called stationary, if it has the form

$$\mathbf{x}^{(n+1)} = B\mathbf{x}^{(n)} + C\mathbf{b} \tag{5.105}$$

where the matrices B (the so called iteration matrix) and C are constant and do not depend on the iteration count n. A fixed point of (5.105) solves (5.104) and hence the method is consistent, if

$$\mathbf{x} = B\mathbf{x} + C\mathbf{b} = B\mathbf{x} + CA\mathbf{x} \tag{5.106}$$

and hence

$$B = I - CA \tag{5.107}$$

which brings the iteration to the general form

$$\mathbf{x}^{(n+1)} = (I - CA)\mathbf{x}^{(n)} + C\mathbf{b} = \mathbf{x}^{(n)} - C(A\mathbf{x}^{(n)} - \mathbf{b}) \tag{5.108}$$

$$\mathbf{r}^{(n+1)} = (1 - AC)\mathbf{r}^{(n)}. \tag{5.109}$$

5.5.1 Richardson-Iteration

The simplest of these methods uses $C = \omega I$ with a damping parameter ω. It iterates according to

$$\mathbf{x}^{(n+1)} = \mathbf{x}^{(n)} - \omega(A\mathbf{x}^{(n)} - \mathbf{b}) \tag{5.110}$$

$$\mathbf{r}^{(n+1)} = (1 - \omega A)\mathbf{r}^{(n)}. \tag{5.111}$$

The Richardson iteration is not of much practical use. It serves as the prototype of a linear stationary method. To improve convergence (5.104) usually has to be preconditioned by multiplication with a suitable matrix P

$$PA\mathbf{x} = P\mathbf{b} \tag{5.112}$$

for which the Richardson iteration

$$\mathbf{x}^{(n+1)} = \mathbf{x}^{(n)} - \omega P(A\mathbf{x}^{(n)} - \mathbf{b}) \tag{5.113}$$

is of the general form (5.108).

5.5.2 Matrix Splitting Methods

let us divide the matrix A into two (non singular) parts [2]

$$A = A_1 + A_2 \tag{5.114}$$

where A_1 can be easily inverted and rewrite (5.104) as

$$A_1\mathbf{x} = \mathbf{b} - A_2\mathbf{x} \tag{5.115}$$

which defines the iteration

$$\Phi(\mathbf{x}) = -A_1^{-1}A_2\mathbf{x} + A_1^{-1}\mathbf{b} \tag{5.116}$$

$$= -A_1^{-1}(A - A_1)\mathbf{x} + A_1^{-1}\mathbf{b} = \mathbf{x} - A_1^{-1}(A\mathbf{x} - \mathbf{b}). \tag{5.117}$$

5.5.3 Jacobi Method

Jacobi divides the matrix A into its diagonal and two triangular matrices [67]:

$$A = L + U + D. \tag{5.118}$$

For A_1 the diagonal part is chosen

$$A_1 = D \tag{5.119}$$

giving

$$\mathbf{x}^{(n+1)} = -D^{-1}(A - D)\mathbf{x}^{(n)} + D^{-1}\mathbf{b} \tag{5.120}$$

which reads explicitly

$$x_i^{(n+1)} = -\frac{1}{a_{ii}} \sum_{j \neq i} a_{ij} x_j^{(n)} + \frac{1}{a_{ii}} b_i. \tag{5.121}$$

This method is stable but converges rather slowly. Reduction of the error by a factor of 10^{-p} needs about $\frac{pN}{2}$ iterations. N grid points have to be evaluated in each iteration and the method scales with $O(N^2)$ [68].

5.5.4 *Gauss-Seidel Method*

With

$$A_1 = D + L \tag{5.122}$$

the iteration becomes

$$(D + L)\mathbf{x}^{(n+1)} = -U\mathbf{x}^{(n)} + \mathbf{b} \tag{5.123}$$

which has the form of a system of equations with triangular matrix [69]. It reads explicitly

$$\sum_{j \leq i} a_{ij} x_j^{(n+1)} = -\sum_{j > i} a_{ij} x_j^{(n)} + b_i. \tag{5.124}$$

Forward substitution starting from x_1 gives

$$i = 1: \qquad x_1^{(n+1)} = \frac{1}{a_{11}} \left(-\sum_{j \geq 2} a_{ij} x_j^{(n)} + b_1 \right)$$

$$i = 2: \qquad x_2^{(n+1)} = \frac{1}{a_{22}} \left(-a_{21} x_1^{(n+1)} - \sum_{j \geq 3} a_{ij} x_j^{(n)} + b_2 \right)$$

$$i = 3: \quad x_3^{(n+1)} = \frac{1}{a_{33}} \left(-a_{31}x_1^{(n+1)} - a_{32}x_2^{(n+1)} - \sum_{j \geq 4} a_{ij}x_j^{(n)} + b_3 \right)$$

$$\vdots$$

$$x_i^{(n+1)} = \frac{1}{a_{ii}} \left(-\sum_{j<i} a_{ij}x_j^{(n+1)} - \sum_{j>i} a_{ij}x_j^{(n)} + b_i \right). \tag{5.125}$$

This looks very similar to the Jacobi method. But here all changes are made immediately. Convergence is slightly better (roughly a factor of 2) and the numerical effort is reduced [68].

5.5.5 Damping and Successive Over-relaxation

Convergence can be improved [68] by combining old and new values. Starting from the iteration

$$A_1 \mathbf{x}^{(n+1)} = (A_1 - A)\mathbf{x}^{(n)} + \mathbf{b} \tag{5.126}$$

we form a linear combination with

$$D\mathbf{x}^{(n+1)} = D\mathbf{x}^{(n)} \tag{5.127}$$

giving the new iteration equation

$$((1-\omega)D + \omega A_1)\mathbf{x}^{(n+1)} = ((1-\omega)D + \omega A_1 - \omega A)\mathbf{x}^{(n)} + \omega \mathbf{b}. \tag{5.128}$$

In case of the Jacobi method with $D = A_1$ we have

$$D\mathbf{x}^{(n+1)} = (D - \omega A)\mathbf{x}^{(n)} + \omega \mathbf{b} \tag{5.129}$$

$$\mathbf{x}^{(n+1)} = \mathbf{x}^{(n)} - \omega D^{-1}(A\mathbf{x} - \mathbf{b}) \tag{5.130}$$

which can be also obtained directly from (5.113).

Explicitly,

$$x_i^{(n+1)} = (1-\omega)x_i^{(n)} + \frac{\omega}{a_{ii}} \left(-\sum_{j \neq i} a_{ij}x_j^{(n)} + b_i \right). \tag{5.131}$$

The changes are damped ($0 < \omega < 1$) or exaggerated ($1 < \omega < 2$).

In case of the Gauss-Seidel method with $A_1 = D + L$ the new iteration[4] (5.128) is

$$(D+\omega L)\mathbf{x}^{(n+1)} = (D+\omega L - \omega A)\mathbf{x}^{(n)} + \omega\mathbf{b} = (1-\omega)D\mathbf{x}^{(n)} - \omega U\mathbf{x}^{(n)} + \omega\mathbf{b} \tag{5.132}$$

$$\mathbf{x}^{(n+1)} = \mathbf{x}^{(n)} - (\frac{1}{\omega}D + L)^{-1}(A\mathbf{x}^{(n)} - \mathbf{b}) \tag{5.133}$$

or explicitly

$$x_i^{(n+1)} = (1-\omega)x_i^{(n)} + \frac{\omega}{a_{ii}}\left(-\sum_{j<i} a_{ij}x_j^{(n+1)} - \sum_{j>i} a_{ij}x_j^{(n)} + \mathbf{b}\right). \tag{5.134}$$

It can be shown that the successive over-relaxation method converges only for $0 < \omega < 2$. For optimal choice of ω about $\frac{1}{3}p\sqrt{N}$ iterations are needed to reduce the error by a factor of 10^{-p}. The order of the method is $O(N^{\frac{3}{2}})$ which is comparable to the most efficient matrix inversion methods [68].

5.6 Non Stationary Iterative Methods

Non stationary methods use a more general iteration

$$\mathbf{x}^{(n+1)} = \Phi_n(\mathbf{x}_n) \tag{5.135}$$

where the iteration function differs from step to step. The method can be formulated as a direction search

$$\mathbf{x}^{(n+1)} = \mathbf{x}^{(n)} + \lambda_n \mathbf{s}_n \tag{5.136}$$

with direction vectors $\mathbf{s}_n$and step lengths λ_n. The residual

$$\mathbf{r}^{(n+1)} = A\mathbf{x}^{(n+1)} - \mathbf{b} = \mathbf{r}^{(n)} + \lambda_n A\mathbf{s}_n \tag{5.137}$$

is used as a measure of the remaining error since the exact solution $\mathbf{x}_{FP}$ together with the error vector

$$\mathbf{d}^{(n)} = \mathbf{x}^{(n)} - \mathbf{x}_{FP} \tag{5.138}$$

are unknown. Both are, however, related by

[4] This is also known as the method of successive over-relaxation (SOR) and differs from the damped Gauss-Seidel method which follows from (5.113).

$$A\mathbf{d}^{(n)} = A\mathbf{x}^{(n)} - A\mathbf{x}_{FP} = A\mathbf{x}^{(n)} - \mathbf{b} = \mathbf{r}^{(n)}. \tag{5.139}$$

5.6.1 Krylov Space Methods

Solution of the linear system

$$A\mathbf{x} = \mathbf{b} \tag{5.140}$$

can be formulated as a search for the minimum

$$\min_{\mathbf{x}\in\mathbb{R}^{\mathbb{N}}} ||A\mathbf{x} - \mathbf{b}||. \tag{5.141}$$

General iterative methods look for the minimum residual in a subspace of $\mathbb{R}^{\mathbb{N}}$ which is increased at each step. The Richardson iteration, e.g. iterates

$$\mathbf{x}^{(n+1)} = (1 - \omega A)\mathbf{x}^{(n)} + \omega\mathbf{b} = \mathbf{x}^{(n)} - \omega\mathbf{r}^{(n)} \tag{5.142}$$

$$\mathbf{r}^{(n+1)} = A(\mathbf{x}^{(n)} - \omega\mathbf{r}^{(n)}) - \mathbf{b} = (1 - \omega A)\mathbf{r}^{(n)}$$

$$\mathbf{r}_0 = A\mathbf{x}_0 - \mathbf{b}$$

$$\mathbf{x}^{(1)} = \mathbf{x}^{(0)} - \omega\mathbf{r}^{(0)}$$

$$\mathbf{r}^{(1)} = \mathbf{r}^{(0)} - \omega A\mathbf{r}^{(0)}$$

$$\mathbf{x}^{(2)} = \mathbf{x}^{(1)} - \omega\mathbf{r}^{(1)} = \mathbf{x}^{(0)} - 2\omega\mathbf{r}^{(0)} + \omega^2 A\mathbf{r}^{(0)}$$

$$\mathbf{r}^{(2)} = (1 - \omega A)\mathbf{r}^{(1)} = \mathbf{r}^{(0)} - 2\omega A\mathbf{r}^{(0)} + \omega^2 A^2\mathbf{r}^{(0)}$$

$$\vdots$$

Obviously,

$$\mathbf{x}^{(n)} - \mathbf{x}_0 \in K_n(A, \mathbf{r}^{(0)}) \quad \mathbf{r}^{(n)} \in K_{n+1}(A, \mathbf{r}^{(0)})$$

with the definition of the n-th Krylov subspace[5]

$$K_n(A, \mathbf{r}^{(0)}) = \text{span}\{\mathbf{r}^{(0)}, A\mathbf{r}^{(0)}, A^2\mathbf{r}^{(0)}, \ldots A^{n-1}\mathbf{r}^{(0)}\}. \tag{5.143}$$

[5] For the most common choice $\mathbf{x}_0 = 0$ we have $\mathbf{r}^{(0)} = -\mathbf{b}$ and $\mathbf{x}_0 + K_n(A, \mathbf{r}^{(0)}) = K_n(A, \mathbf{r}^{(0)}) = K_n(A, \mathbf{b})$.

5.6.2 Minimization Principle for Symmetric Positive Definite Systems

If the matrix A is symmetric and positive definite, we consider the quadratic form defined by

$$h(\mathbf{x}) = h_0 - \mathbf{x}^T\mathbf{b} + \frac{1}{2}\mathbf{x}^T A\mathbf{x}. \tag{5.144}$$

At a local minimum the gradient

$$\nabla h(\mathbf{x}) = A\mathbf{x} - \mathbf{b} = \mathbf{r} \tag{5.145}$$

is zero and therefore the minimum of h is also a solution of the linear system of equations

$$A\mathbf{x} = \mathbf{b}. \tag{5.146}$$

5.6.3 Gradient Method

The simple Richardson iteration (Sect. 5.5.1) uses

$$\mathbf{x}^{(n+1)} = \mathbf{x}^{(n)} - \omega\mathbf{r}^{(n)} \tag{5.147}$$

with a constant value of ω.

$$\mathbf{r}^{(n+1)} = \mathbf{r}^{(n)} - \omega A\mathbf{r}^{(n)}. \tag{5.148}$$

Let us now optimize the step width along the direction of the gradient vector. From

$$0 = \frac{\mathrm{d}}{\mathrm{d}\omega}|\mathbf{r}^{(n+1)}|^2 = \mathbf{r}^{(n)T}(1-\omega A)(1-\omega A)\mathbf{r}^{(n)}$$

$$= \mathbf{r}^{(n)T}(-2A + 2\omega A^2)\mathbf{r}^{(n)} = -2\mathbf{r}^{(n)T}A\mathbf{r}^{(n)} + 2\omega|A\mathbf{r}^{(n)}|^2 \tag{5.149}$$

$$\mathbf{r} = 2\mathbf{r}^{(n)T}(-1+\omega A)A\mathbf{r}^{(n)}$$

we find the optimum value

$$\omega^{(n)} = \frac{\mathbf{r}^{(n)T} A\mathbf{r}^{(n)}}{|A\mathbf{r}^{(n)}|^2}. \tag{5.150}$$

The residuals[6]

$$\mathbf{r}_0 = -\mathbf{b} \tag{5.151}$$

$$\mathbf{r}^{(1)} = -\mathbf{b} + \omega^{(1)} A\mathbf{b} \tag{5.152}$$

$$\mathbf{r}^{(2)} = -\mathbf{b} + (\omega^{(1)} + \omega^{(2)}) A\mathbf{b} - \omega^{(2)}\omega^{(1)} A^2\mathbf{b} \tag{5.153}$$

$$\vdots$$

etc. obviously are in the Krylov subspace

$$\mathbf{r}^{(n)} \in K_{n+1}(A, \mathbf{b}) \tag{5.154}$$

and so are the approximate solutions

$$\mathbf{x}^{(1)} = \omega^{(1)}\mathbf{b} \tag{5.155}$$

$$\mathbf{x}^{(2)} = (\omega^{(1)} + \omega^{(2)})\mathbf{b} - \omega^{(2)}\omega^{(1)} A\mathbf{b} \tag{5.156}$$

$$\vdots$$

$$\mathbf{x}^{(n)} \in K_n(A, \mathbf{b}). \tag{5.157}$$

5.6.4 Conjugate Gradients Method

The gradient method gives an approximate solution

$$\mathbf{x}^{(n+1)} = \mathbf{x}^{(0)} - \omega^{(0)}\mathbf{r}^{(0)} - \omega^{(1)}\mathbf{r}^{(1)} \cdots - \omega^{(n)}\mathbf{r}^{(n)} \tag{5.158}$$

but the previously chosen $\omega^{(0)} \ldots \omega^{(n-1)}$are not optimal since the gradient vectors $\mathbf{r}^{(0)} \ldots \mathbf{r}^{(n)}$are not orthogonal. We want to optimize the solution within the space spanned by the gradients for which a new basis $\mathbf{s}^{(0)} \ldots \mathbf{s}^{(n)}$is introduced which will be determined later

$$K_{n+1} = \text{span}(\mathbf{r}^{(0)} \ldots \mathbf{r}^{(n)}) = \text{span}(\mathbf{s}^{(0)} \ldots \mathbf{s}^{(n)}). \tag{5.159}$$

[6] We assume $\mathbf{x}_0 = 0$.

Using $\mathbf{s}^{(n)}$as search direction the iteration becomes

$$\mathbf{x}^{(n+1)} = \mathbf{x}^{(n)} + \lambda^{(n)}\mathbf{s}^{(n)} \tag{5.160}$$

$$\mathbf{r}^{(n+1)} = \mathbf{r}^{(n)} + \lambda^{(n)} A\mathbf{s}^{(n)}. \tag{5.161}$$

After n+1 steps

$$\mathbf{r}^{(n+1)} = \mathbf{r}^{(0)} + \sum \lambda^{(j)} A\mathbf{s}^{(\mathbf{j})} = A(\mathbf{d}^{(0)} + \sum \lambda^{(j)}\mathbf{s}^{(j)}). \tag{5.162}$$

Multiplication with $\mathbf{s}^{(m)}$ gives

$$\mathbf{s}^{(m)T}\mathbf{r}^{(n+1)} = \mathbf{s}^{(m)T} A\mathbf{d}^{(0)} + \sum \lambda^{(j)}\mathbf{s}^{(m)T} A\mathbf{s}^{(j)} \tag{5.163}$$

which after introduction of the $A-$scalar product which is defined for a symmetric and positive definite matrix A as

$$(\mathbf{x}, \mathbf{y})_A = \mathbf{x}^T A\mathbf{y} \tag{5.164}$$

becomes

$$\mathbf{s}^{(m)T}\mathbf{r}^{(n+1)} = (\mathbf{s}^{(m)}, \mathbf{d}^{(0)})_A + \sum_{j=0}^{n} \lambda^{(j)}(\mathbf{s}^{(m)}, \mathbf{s}^{(j)})_A$$

which simplifies considerably if we assume A-orthogonality of the search directions

$$(\mathbf{s}^{(m)}, \mathbf{s}^{(j)}) = 0 \quad \text{for } m \neq j \tag{5.165}$$

because then

$$\mathbf{s}^{(m)T}\mathbf{r}^{(n+1)} = (\mathbf{s}^{(m)}, \mathbf{d}^{(0)})_A + \lambda^{(m)}(\mathbf{s}^{(m)}, \mathbf{s}^{(m)})_A = \mathbf{s}^{(m)T}\mathbf{r}^{(0)} + \lambda^{(m)}\mathbf{s}^{(m)T} A\mathbf{s}^{(m)}. \tag{5.166}$$

If we choose

$$\lambda^{(m)} = -\frac{\mathbf{s}^{(m)T}\mathbf{r}^{(0)}}{\mathbf{s}^{(m)T} A\mathbf{s}^{(m)}} \tag{5.167}$$

the projection of the new residual $\mathbf{r}^{(n+1)}$onto K_{n+1} vanishes, i.e. this is the optimal choice of the parameters $\lambda^{(0)} \dots \lambda^{(n)}$.

Obviously the first search vector must have the direction of $\mathbf{r}^{(0)}$ to span the same one-dimensional space. Therefore we begin the iteration with

$$\mathbf{s}^{(0)} = \mathbf{r}^{(0)} \tag{5.168}$$

$$\lambda^{(0)} = -\frac{|\mathbf{r}^{(0)}|^2}{\mathbf{r}^{(0)T} A\mathbf{r}^{(0)}} \tag{5.169}$$

$$\mathbf{x}^{(1)} = \mathbf{x}^{(0)} + \lambda^{(0)}\mathbf{s}^{(0)}. \tag{5.170}$$

For the next step we apply Gram-Schmidt orthogonalization

$$\mathbf{s}^{(1)} = \mathbf{r}^{(1)} - \mathbf{s}^{(0)}\frac{(\mathbf{r}^{(1)}, \mathbf{s}^{(0)})_A}{(\mathbf{s}^{(0)}, \mathbf{s}^{(0)})_A}. \tag{5.171}$$

For all further steps we have to orthogonalize $\mathbf{s}^{(n+1)}$ with respect to all of $\mathbf{s}^{(n)} \dots \mathbf{s}^{(0)}$. But, fortunately, the residual $\mathbf{r}^{(n+1)}$ is already $A-$orthogonal to $\mathbf{s}^{(n-1)} \dots \mathbf{s}^{(0)}$. This can be seen from (5.161)

$$\mathbf{r}^{(j+1)} - \mathbf{r}^{(j)} = \lambda^{(j)} A\mathbf{s}^{(j)} \tag{5.172}$$

$$(\mathbf{r}^{(n+1)}, \mathbf{s}^{(j)})_A = \mathbf{r}^{(n+1)T} A\mathbf{s}^{(j)} = \frac{1}{\lambda^{(j)}}\mathbf{r}^{(n+1)T}(\mathbf{r}^{(j+1)} - \mathbf{r}^{(j)}). \tag{5.173}$$

We already found, that $\mathbf{r}^{(n+1)}$ is orthogonal to K_{n+1} , hence to all $\mathbf{r}^{(n)}, \dots \mathbf{r}^{(0)}$. Therefore we conclude

$$(\mathbf{r}^{(n+1)}, \mathbf{s}^{(j)})_A = 0 \quad \text{for } j + 1 \le n. \tag{5.174}$$

Finally we end up with the following procedure

$$\mathbf{x}^{(n+1)} = \mathbf{x}^{(n)} + \lambda^{(n)}\mathbf{s}^{(n)} \quad \text{with } \lambda^{(n)} = -\frac{\mathbf{s}^{(n)T}\mathbf{r}^{(0)}}{\mathbf{s}^{(n)T} A\mathbf{s}^{(n)}} \tag{5.175}$$

$$\mathbf{r}^{(n+1)} = \mathbf{r}^{(n)} + \lambda^{(n)} A\mathbf{s}^{(n)} \tag{5.176}$$

$$\mathbf{s}^{(n+1)} = \mathbf{r}^{(n+1)} - \beta^{(n)}\mathbf{s}^{(n)} \quad \text{with } \beta^{(n)} = \frac{\mathbf{r}^{(n+1)T} A\mathbf{s}^{(n)}}{\mathbf{s}^{(n)T} A\mathbf{s}^{(n)}.} \tag{5.177}$$

This method [48] solves a linear system without storing the matrix A itself. Only the product $A\mathbf{s}$ is needed. In principle the solution is reached after $N = \dim(A)$ steps, but due to rounding errors more steps can be necessary and the method has to be considered as an iterative one.

The expressions for λ and β can be brought to numerically more efficient form. From (5.162) we find

$$\mathbf{s}^{(n)T}\mathbf{r}^{(0)} = \mathbf{s}^{(n)T}\left(\mathbf{r}^{(n)} - \sum_{j=0}^{n-1}\lambda^{(j)} A\mathbf{s}^{(\mathbf{j})}\right). \tag{5.178}$$

But due to A-orthogonality of the search directions

$$\mathbf{s}^{(n)T}\mathbf{r}^{(0)} = \mathbf{s}^{(n)T}\mathbf{r}^{(n)} = \mathbf{r}^{(n)T}\mathbf{r}^{(n)} \tag{5.179}$$

which simplifies

$$\lambda^{(n)} = -\frac{\mathbf{r}^{(n)T}\mathbf{r}^{(n)}}{\mathbf{s}^{(n)T}A\mathbf{s}^{(n)}}. \tag{5.180}$$

Furthermore, from (5.176) and orthogonality of the residual vectors

$$\mathbf{r}^{(n+1)T}\mathbf{r}^{(n+1)} = \lambda^{(n)}\mathbf{r}^{(n+1)T}A\mathbf{s}^{(n)} \tag{5.181}$$

from which

$$\beta^{(n)} = \frac{\mathbf{r}^{(n+1)T}A\mathbf{s}^{(n)}}{\mathbf{s}^{(n)T}A\mathbf{s}^{(n)}} = \frac{-\frac{1}{\lambda^{(n)}}\mathbf{r}^{(n+1)T}\mathbf{r}^{(n+1)}}{\mathbf{s}^{(n)T}A\mathbf{s}^{(n)}}$$

$$= -\frac{\mathbf{r}^{(n+1)T}\mathbf{r}^{(n+1)}}{\mathbf{r}^{(n)T}\mathbf{r}^{(n)}}. \tag{5.182}$$

The conjugate gradients method is not useful for non symmetric systems. It can be applied to the normal equations (11.32)

$$A^T A\mathbf{x} = A^T\mathbf{b} \tag{5.183}$$

which, for a full-rank non singular matrix have the same solution. The condition number (Sect. 5.7), however is

$$\mathrm{cond}(A^T A) = (\mathrm{cond}A)^2 \tag{5.184}$$

and the problem may be ill conditioned.

5.6.5 Non Symmetric Systems

The general minimum residual method (GMRES) searches directly for the minimum of $||A\mathbf{x} - \mathbf{b}||$ in the Krylov spaces of increasing order $K_n(A, \mathbf{r}^{(0)})$. To avoid problems with linear dependency, first an orthogonal basis

$$K_n(A, \mathbf{r}^{(0)}) = \mathrm{span}(\mathbf{q}_1, \mathbf{q}_2, \ldots \mathbf{q}_n) \tag{5.185}$$

is constructed with Arnoldi's method, a variant of Gram-Schmidt orthogonalization. Starting from the normalized vector

$$\mathbf{q}_1 = \frac{\mathbf{r}^{(0)}}{|\mathbf{r}^{(0)}|} \tag{5.186}$$

the dimension is iteratively increased by orthogonalizing the vector $A\mathbf{q}_n$ with respect to $K_n(A, \mathbf{r}^{(0)})$[7]

$$\tilde{\mathbf{q}}_{n+1} = A\mathbf{q}_n - \sum_{j=1}^{n} (\mathbf{q}_j, \mathbf{q}_n)_A \mathbf{q}_j = A\mathbf{q}_n - \sum_{j=1}^{n} (\mathbf{q}_j^T A\mathbf{q}_n)\mathbf{q}_j = A\mathbf{q}_n - \sum_{j=1}^{n} h_{jn}\mathbf{q}_j \tag{5.187}$$

and normalizing this vector

$$h_{n+1,n} = |\tilde{\mathbf{q}}_{n+1}| \quad \mathbf{q}_{n+1} = \frac{\tilde{\mathbf{q}}_{n+1}}{h_{n+1,n}}. \tag{5.188}$$

Then

$$A\mathbf{q}_n = h_{n+1,n}\mathbf{q}_{n+1} + \sum_{j=1}^{n} h_{jn}\mathbf{q}_j \tag{5.189}$$

which explicitly reads

$$A\mathbf{q}_1 = h_{21}\mathbf{q}_2 + h_{11}\mathbf{q}_1 = (\mathbf{q}_1, \mathbf{q}_2)\begin{pmatrix} h_{11} \\ h_{21} \end{pmatrix} \tag{5.190}$$

$$A\mathbf{q}_2 = h_{32}\mathbf{q}_3 + h_{12}\mathbf{q}_1 + h_{22}\mathbf{q}_2 = (\mathbf{q}_1, \mathbf{q}_2, \mathbf{q}_3)\begin{pmatrix} h_{12} \\ h_{22} \\ h_{32} \end{pmatrix} \tag{5.191}$$

$$\vdots$$

$$A\mathbf{q}_n = h_{n+1,n}\mathbf{q}_{n+1} + h_{1n}\mathbf{q}_1 + \dots h_{nn}\mathbf{q}_n. \tag{5.192}$$

The new basis vectors are orthogonal[8] since

$$\mathbf{q}_1^T\tilde{\mathbf{q}}_2 = \mathbf{q}_1^T\left[A\mathbf{q}_1 - (\mathbf{q}_1^T A\mathbf{q}_1)\mathbf{q}_1\right] = (\mathbf{q}_1^T A\mathbf{q}_1)(1 - |\mathbf{q}_1|^2) = 0 \tag{5.193}$$

and induction shows for $k = 1 \dots n$

[7] $\mathbf{q}_2$ is a linear combination of $\mathbf{q}_1$ and $A\mathbf{q}_1$, $\mathbf{q}_3$ one of $\mathbf{q}_1, \mathbf{q}_2$ and $A\mathbf{q}_2$ hence also of $\mathbf{q}_1, A\mathbf{q}_1$, $A^2\mathbf{q}_1$etc. which proves the validity of (5.185).

[8] If $\mathbf{q}_{n+1} = 0$ the algorithm has to stop and the Krylov space has the full dimension of the matrix.

$$\mathbf{q}_k^T \tilde{\mathbf{q}}_{n+1} = \mathbf{q}_k^T A\mathbf{q}_n - \sum_{j=1}^{n} (\mathbf{q}_j^T A\mathbf{q}_n)(\mathbf{q}_k^T \mathbf{q}_j) = \mathbf{q}_k^T A\mathbf{q}_n - \sum_{j=1}^{n} (\mathbf{q}_j^T A\mathbf{q}_n)\delta_{k,j} = 0. \tag{5.194}$$

We collect the new basis vectors $\mathbf{q}_1 \dots \mathbf{q}_n$ into a matrix

$$U_n = (\mathbf{q}_1, \dots \mathbf{q}_n) \tag{5.195}$$

and obtain from (5.190 to 5.192)

$$AU_n = U_{n+1} H \tag{5.196}$$

with the $(n+1) \times n$ upper Hessenberg matrix

$$H = \begin{pmatrix} h_{11} & h_{12} & \cdots & h_{1n} \\ h_{21} & h_{22} & & h_{2n} \\ & h_{32} & \ddots & \vdots \\ & & \ddots & h_{nn} \\ & & & h_{n+1,n} \end{pmatrix}. \tag{5.197}$$

Since

$$\mathbf{x}^{(n)} - \mathbf{x}_0 \in K_n(A, \mathbf{r}^{(0)}) \tag{5.198}$$

it can be written as a linear combination of $\mathbf{q}_1 \dots \mathbf{q}_n$

$$\mathbf{x}^{(n)} - \mathbf{x}_0 = (\mathbf{q}_1 \dots \mathbf{q}_n)\mathbf{v}. \tag{5.199}$$

The residual becomes

$$\begin{aligned} \mathbf{r}^{(n)} &= A(\mathbf{q}_1 \dots \mathbf{q}_n)\mathbf{v} + A\mathbf{x}_0 - \mathbf{b} = A(\mathbf{q}_1 \dots \mathbf{q}_n)\mathbf{v} + \mathbf{r}^{(0)} \\ &= U_{n+1} H\mathbf{v} + |\mathbf{r}^{(0)}|\mathbf{q}_1 \\ &= U_{n+1} \left[H\mathbf{v} + |\mathbf{r}^{(0)}| \begin{pmatrix} 1 \\ 0 \\ \vdots \\ 0 \end{pmatrix} \right] \end{aligned} \tag{5.200}$$

hence

$$|\mathbf{r}^{(n)}|^2 = \left[H\mathbf{v} + |\mathbf{r}^{(0)}| \begin{pmatrix} 1 \\ 0 \\ \vdots \\ 0 \end{pmatrix} \right]^T U_{n+1}^T U_{n+1} \left[H\mathbf{v} + |\mathbf{r}^{(0)}| \begin{pmatrix} 1 \\ 0 \\ \vdots \\ 0 \end{pmatrix} \right]. \tag{5.201}$$

But since

$$U_{n+1}^T U_{n+1} = \begin{pmatrix} \mathbf{q}_1{}^T \\ \vdots \\ \mathbf{q}_n^T \end{pmatrix} (\mathbf{q}_1 \dots \mathbf{q}_n) = \begin{pmatrix} 1 & & \\ & \ddots & \\ & & 1 \end{pmatrix} \tag{5.202}$$

is a $n \times n$ unit matrix, we have to minimize

$$|\mathbf{r}^{(n)}| = \left| H\mathbf{v} + |\mathbf{r}^{(0)}| \begin{pmatrix} 1 \\ 0 \\ \vdots \\ 0 \end{pmatrix} \right| \tag{5.203}$$

which is a least squares problem, since there are $n+1$ equations for the n unknown components of $\mathbf{v}$. It can be solved efficiently with the help of QR decomposition (11.36)

$$H = Q \begin{pmatrix} R \\ 0 \end{pmatrix} \quad R = \begin{pmatrix} r_{11} & \cdots & r_{1n} \\ & \ddots & \vdots \\ & & r_{nn} \end{pmatrix} \tag{5.204}$$

with a $(n+1) \times (n+1)$ orthogonal matrix Q. The norm of the residual vector becomes

$$|\mathbf{r}^{(n)}| = \left| Q \begin{pmatrix} R \\ 0 \end{pmatrix} \mathbf{v} + |\mathbf{r}^{(0)}| \begin{pmatrix} 1 \\ 0 \\ \vdots \\ 0 \end{pmatrix} \right| = \left| \begin{pmatrix} R\mathbf{v} \\ 0 \end{pmatrix} + |\mathbf{r}^{(0)}| Q^T \begin{pmatrix} 1 \\ 0 \\ \vdots \\ 0 \end{pmatrix} \right|. \tag{5.205}$$

Substituting

$$|\mathbf{r}^{(0)}| Q^T \begin{pmatrix} 1 \\ 0 \\ \vdots \\ 0 \end{pmatrix} = \begin{pmatrix} y_1 \\ \vdots \\ y_n \\ y_{n+1} \end{pmatrix} = \begin{pmatrix} \mathbf{y} \\ y_{n+1} \end{pmatrix} \tag{5.206}$$

we have

$$|\mathbf{r}^{(n)}| = \sqrt{y_{n+1}^2 + |R\mathbf{v} + \mathbf{y}|^2} \tag{5.207}$$

which is obviously minimized by solving the triangular system

$$R\mathbf{v} + \mathbf{y} = 0. \tag{5.208}$$

The GMRES method usually has to be preconditioned (cf. 5.112) to improve convergence. Often it is restarted after a small number (e.g.20) of iterations which avoids the necessity to store a large orthogonal basis.

5.7 Matrix Inversion

LU and QR decomposition can be also used to calculate the inverse of a non singular matrix

$$AA^{-1} = 1. \tag{5.209}$$

The decomposition is performed once and then the column vectors of A^{-1} are calculated similar to (5.27)

$$L\left(UA^{-1}\right) = 1 \tag{5.210}$$

or (5.40)

$$RA^{-1} = Q^{\dagger}. \tag{5.211}$$

Consider now a small variation of the right hand side of (5.2)

$$\mathbf{b} + \Delta\mathbf{b}. \tag{5.212}$$

Instead of

$$A^{-1}\mathbf{b} = \mathbf{x} \tag{5.213}$$

the resulting vector is

$$A^{-1}(\mathbf{b} + \Delta\mathbf{b}) = \mathbf{x} + \Delta\mathbf{x} \tag{5.214}$$

and the deviation can be measured by[9]

$$||\Delta\mathbf{x}|| = ||A^{-1}||\,||\Delta\mathbf{b}|| \tag{5.215}$$

and since

$$||A||\,||\mathbf{x}|| = ||\mathbf{b}|| \tag{5.216}$$

the relative error becomes

[9] The vector norm used here is not necessarily the Euclidean norm.

$$\frac{||\Delta\mathbf{x}||}{||\mathbf{x}||} = ||A||\,||A^{-1}||\,\frac{||\Delta\mathbf{b}||}{||\mathbf{b}||}. \tag{5.217}$$

The relative error of **b** is multiplied by the condition number for inversion

$$\text{cond}(A) = ||A||\,||A^{-1}||. \tag{5.218}$$

Exercises

Exercise 5.1 Comparison of Different Direct Solvers

In this computer experiment we solve the system of equations

$$A\mathbf{x} = \mathbf{b} \tag{5.219}$$

with several methods (Fig. 5.2):

- Gaussian elimination without pivoting (Sect. 5.1),
- Gaussian elimination with partial pivoting (Sect. 5.1.1),
- QR decomposition with Householder reflections (Sect. 5.2.2),
- QR decomposition with Gram-Schmidt orthogonalization (Sect. 5.2.1),
- QR decomposition with Gram-Schmidt orthogonalization with extra orthogonalization step (5.55).

The right hand side is chosen as

$$\mathbf{b} = A\begin{pmatrix}1\\2\\\vdots\\n\end{pmatrix} \tag{5.220}$$

hence the exact solution is

$$\mathbf{x} = \begin{pmatrix}1\\2\\\vdots\\n\end{pmatrix}. \tag{5.221}$$

Several test matrices can be chosen:

- Gaussian elimination is theoretically unstable but is stable in many practical cases. The instability can be demonstrated with the example [61]

$$A = \begin{pmatrix} 1 & & & & 1 \\ -1 & 1 & & & 1 \\ & -1 & 1 & & 1 \\ & \vdots & \vdots & \ddots & \vdots \\ & -1 & -1 & -1 & 1 \end{pmatrix}. \tag{5.222}$$

No pivoting takes place in the LU decomposition of this matrix and the entries in the last column double in each step:

$$A^{(1)} = \begin{pmatrix} 1 & & & & 1 \\ & 1 & & & 2 \\ & -1 & 1 & & 2 \\ & \vdots & \vdots & \ddots & \vdots \\ & -1 & -1 & -1 & 2 \end{pmatrix} \; A^{(2)} = \begin{pmatrix} 1 & & & & 1 \\ & 1 & & & 2 \\ & & 1 & & 4 \\ & & \vdots & \ddots & \vdots \\ & & -1 & -1 & 4 \end{pmatrix} \ldots A^{(n-1)} = \begin{pmatrix} 1 & & & 1 \\ & 1 & & 2 \\ & & \ddots & 4 \\ & & & \vdots \\ & & & 2^{n-1} \end{pmatrix}. \tag{5.223}$$

Since the machine precision is $\epsilon_M = 2^{-53}$ for double precision calculations we have to expect numerical inaccuracy for dimension $n > 53$.

- Especially well conditioned are matrices [70] which are symmetric

$$A_{ij} = A_{ji} \tag{5.224}$$

and also diagonal dominant

$$\sum_{j \neq i} |A_{ij}| < |A_{ii}|. \tag{5.225}$$

We use the matrix

$$A = \begin{pmatrix} n & 1 & \cdots & 1 & 1 \\ 1 & n & \cdots & 1 & 1 \\ \vdots & & \ddots & & \vdots \\ 1 & 1 & 1 & n & 1 \\ 1 & 1 & 1 & 1 & n \end{pmatrix} \tag{5.226}$$

which can be inverted explicitly by

$$A^{-1} = \begin{pmatrix} a & b & \cdots & b & b \\ b & a & & b & b \\ \vdots & & \ddots & & \\ b & b & b & a & b \\ b & b & b & b & a \end{pmatrix} \quad a = \frac{1}{n - \frac{1}{2}}, b = -\frac{1}{2n^2 - 3n + 1} \tag{5.227}$$

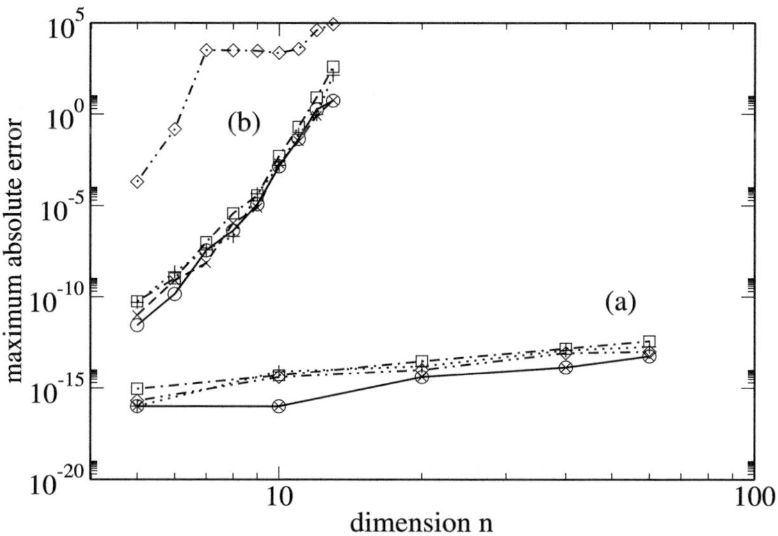

Fig. 5.2 Comparison of different direct solvers. Gaussian elimination without (circles) and with (x) pivoting, QR decomposition with Householder reflections (squares), with Gram–Schmidt orthogonalization (diamonds) and including extra orthogonalization (+) are compared. The maximum difference $\max_{i=1\dots n}(|x_i - x_i^{exact}|)$ increases only slightly with the dimension n for the well behaved matrix (5.226a) but quite dramatically for the ill conditioned Hilbert matrix (5.228b)

and has a condition number[10] which is proportional to the dimension n (Fig. 5.3)

- The Hilbert matrix [71, 72]

$$A = \begin{pmatrix} 1 & \frac{1}{2} & \frac{1}{3} & \cdots & \frac{1}{n} \\ 12 & \frac{1}{3} & \frac{1}{4} & & \frac{1}{n+1} \\ 13 & \frac{1}{4} & \frac{1}{5} & & \frac{1}{n+2} \\ & & & \ddots & \vdots \\ 1n & \frac{1}{n+1} & \frac{1}{n+2} & \cdots & \frac{1}{2n-1} \end{pmatrix} \tag{5.228}$$

is especially ill conditioned [73] even for moderate dimension. It is positive definite and therefore the inverse matrix exists and even can be written down explicitly [74]. Its column vectors are very close to linearly dependent and the condition number grows exponentially with its dimension (Fig. 5.3). Numerical errors are large for all methods compared (Fig. 5.2).

[10] Using the Frobenius norm $||A|| = \sqrt{\sum_{ij} A_{ij}^2}$.

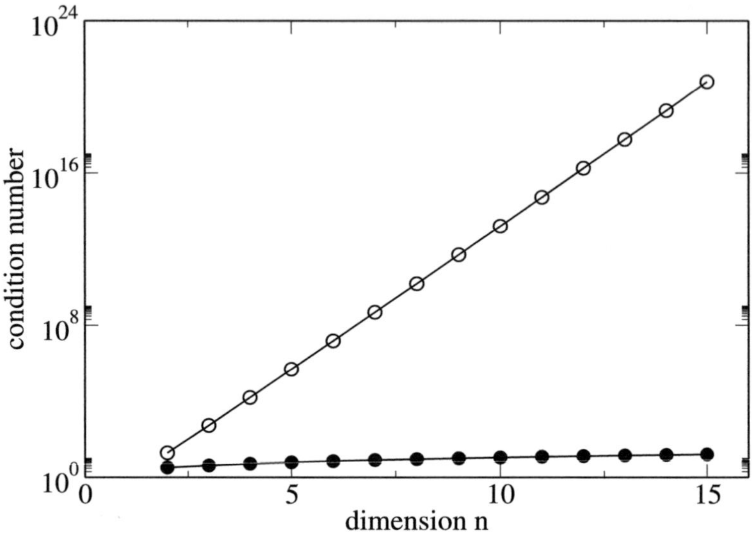

Fig. 5.3 Condition numbers. The condition number cond(A) increases only linearly with the dimension n for the well behaved matrix (5.226, full circles) but exponentially for the ill conditioned Hilbert matrix (5.228, open circles)

- random matrices

$$A_{ij} = \xi \in [-1, 1]. \tag{5.229}$$

Chapter 6
Roots and Extremal Points

In computational physics very often roots of a function, i.e solutions of an equation like

$$f(x_1 \cdots x_N) = 0 \tag{6.1}$$

have to be determined. A related problem is the search for local extrema (Fig. 6.1)

$$\max f(x_1 \cdots x_N) \quad \min f(x_1 \cdots x_N) \tag{6.2}$$

which for a smooth function are solutions of the equations

$$\frac{\partial f(x_1 \cdots x_N)}{\partial x_i} = 0, \quad i = 1 \ldots N. \tag{6.3}$$

In one dimension bisection is a very robust but rather inefficient root finding method. If a good starting point close to the root is available and the function smooth enough, the Newton-Raphson method converges much faster. Special strategies are necessary to find roots of not so well behaved functions or higher order roots. The combination of bisection and interpolation like in Dekker's and Brent's methods provides generally applicable algorithms. In multidimensions calculation of the Jacobian matrix is not always possible and Quasi-Newton methods are a good choice. Whereas local extrema can be found as the roots of the gradient, at least in principle, direct optimization can be more efficient. In one dimension the ternary search method or Brent's more efficient golden section search method can be used. In multidimensions the

Supplementary Information The online version contains supplementary material available at https://doi.org/10.1007/978-3-032-07856-8_6.

P. O. J. Scherer, *Computational Physics I*, Graduate Texts in Physics,
https://doi.org/10.1007/978-3-032-07856-8_6

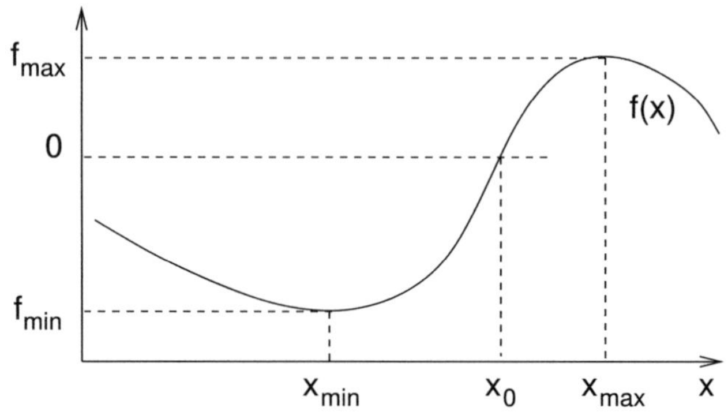

Fig. 6.1 Roots and local extrema of a function

class of direction set search methods is very popular. The basic steepest descent method can be improved by introducing momentum and step size adjustment. Very large dimensions as they appear in machine learning, can be handled with stochastic gradient descent methods like the famous ADAM. More traditional methods are conjugate gradients, the Newton-Raphson method and, if calculation of the full Hessian matrix is too expensive, the Quasi-Newton methods.

6.1 Root Finding

If there is exactly one root in the interval $a_0 < x < b_0$ then one of the following methods can be used to locate the position with sufficient accuracy. If there are multiple roots, these methods will find one of them and special care has to be taken to locate the other roots.

6.1.1 Bisection

Bisection (Algorithm 5) is the simplest method [75] to solve

$$f(x) = 0. \tag{6.4}$$

Algorithm 5: Root finding by Bisection

Determine an interval $[a_0, b_0]$, which contains a sign change of $f(x)$. If no such interval can be found then $f(x)$ does not have any zero crossings;

repeat

  Divide the interval into $[a_0, a_0 + \frac{b_0-a_0}{2}]$, $[a_0 + \frac{b_0-a_0}{2}, b_0]$ and choose that interval $[a_1, b_1]$, where $f(x)$ changes its sign

until $b_n - a_n < \varepsilon$ *is small enough*;

```
/* Usually a combination like ε = 2ε_M + |b_n|ε_r of an
   absolute and a relative tolerance is taken          */
```

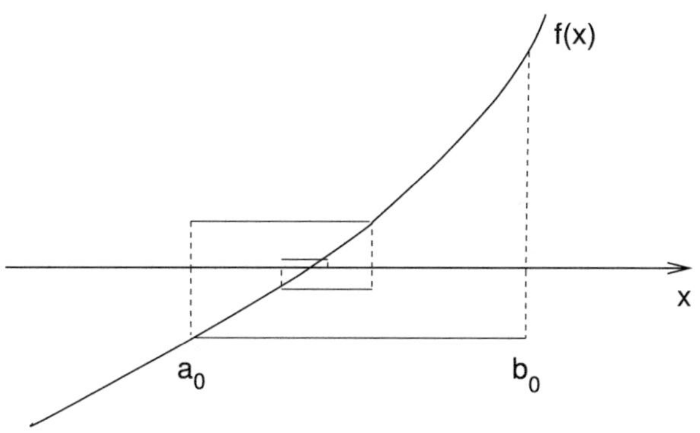

Fig. 6.2 Root finding by bisection

The bisection method needs two starting points which bracket a sign change of the function. It converges but only slowly since each step reduces the uncertainty by a factor of 2 (Fig. 6.2).

6.1.2 Regula Falsi (False Position) Method

The regula falsi [76] method (Fig. 6.3) is similar to the bisection method [75]. However, polynomial interpolation is used to divide the interval $[x_r, a_r]$ with $f(x_r)f(a_r) < 0$. The root of the linear polynomial

$$p(x) = f(x_r) + (x - x_r)\frac{f(a_r) - f(x_r)}{a_r - x_r} \tag{6.5}$$

is given by

$$\xi_r = x_r - f(x_r)\frac{a_r - x_r}{f(a_r) - f(x_r)} = \frac{a_r f(x_r) - x_r f(a_r)}{f(x_r) - f(a_r)} \tag{6.6}$$

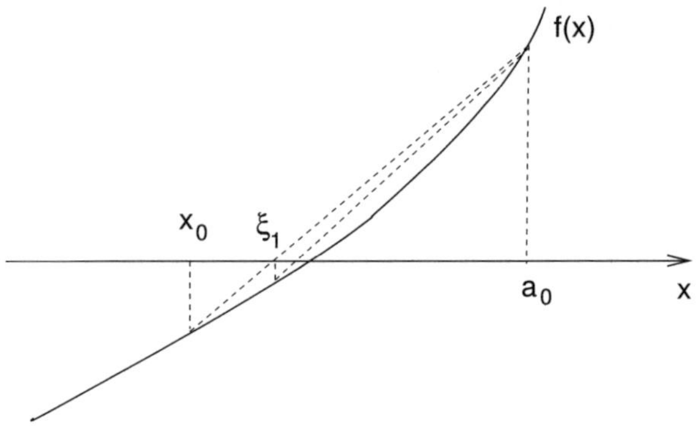

Fig. 6.3 Regula falsi method

which is inside the interval $[x_r, a_r]$. Choose the sub-interval which contains the sign change:

$$f(x_r)f(\xi_r) < 0 \rightarrow [x_{r+1}, a_{r+1}] = [x_r, \xi_r]$$
$$f(x_r)f(\xi_r) > 0 \rightarrow [x_{r+1}, a_{r+1}] = [\xi_r, a_r]. \tag{6.7}$$

Then ξ_r provides a series of approximations with increasing precision to the root of $f(x) = 0$.

6.1.3 Newton-Raphson Method

Consider a function which is differentiable at least two times around the root ξ. Taylor series expansion around a point x_0 in the vicinity

$$f(x) = f(x_0) + (x - x_0)f'(x_0) + \frac{1}{2}(x - x_0)^2 f''(x_0) + \cdots \tag{6.8}$$

gives for $x = \xi$

$$0 = f(x_0) + (\xi - x_0)f'(x_0) + \frac{1}{2}(\xi - x_0)^2 f''(x_0) + \cdots . \tag{6.9}$$

Truncation of the series and solving for ξ gives the first order Newton-Raphson [75, 77] method (Fig. 6.4)

$$x_{r+1} = x_r - \frac{f(x_r)}{f'(x_r)} \tag{6.10}$$

and the second order Newton-Raphson method (Fig. 6.4)

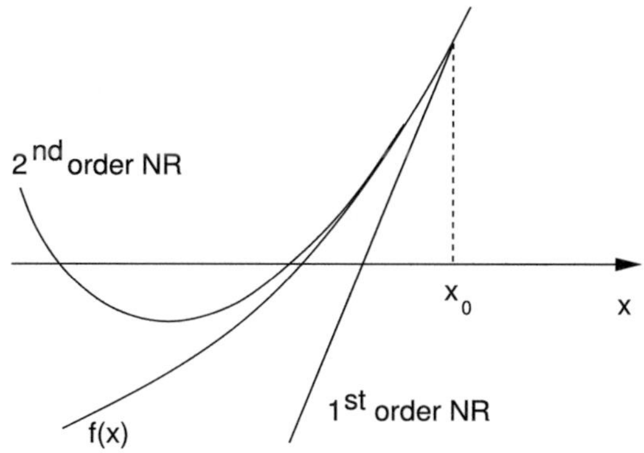

Fig. 6.4 Newton-Raphson method

$$x_{r+1} = x_r - \frac{f'(x_r) \pm \sqrt{f'(x_r)^2 - 2f(x_r)f''(x_r)}}{f''(x_r)}. \tag{6.11}$$

The Newton-Raphson method converges fast if the starting point is close enough to the root. Analytic derivatives are needed. It may fail if two or more roots are close by.

6.1.4 Secant Method

Replacing the derivative in the first order Newton Raphson method by a finite difference quotient gives the secant method [75] (Fig. 6.5) which has been known for thousands of years before [78]

$$x_{r+1} = x_r - f(x_r)\frac{x_r - x_{r-1}}{f(x_r) - f(x_{r-1})}. \tag{6.12}$$

Round-off errors can become important as $|f(x_r) - f(x_{r-1})|$ gets small. At the beginning choose a starting point x_0 and determine

$$x_1 = x_0 - f(x_0)\frac{2h}{f(x_0 + h) - f(x_0 - h)} \tag{6.13}$$

using a symmetrical difference quotient.

6.1.5 Interpolation

The secant method is also obtained by linear interpolation

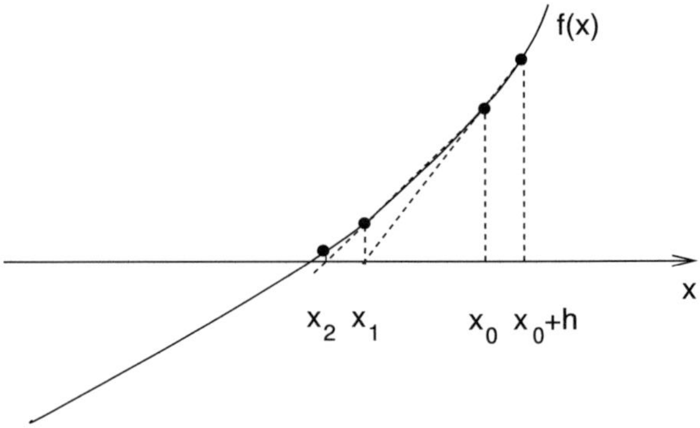

Fig. 6.5 Secant method

$$p(x) = \frac{x - x_r}{x_{r-1} - x_r} f_{r-1} + \frac{x - x_{r-1}}{x_r - x_{r-1}} f_r. \tag{6.14}$$

The root of the polynomial $p(x_{r+1}) = 0$ determines the next iterate x_{r+1}

$$x_{r+1} = \frac{1}{f_{r-1} - f_r}(x_r f_{r-1} - x_{r-1} f_r) = x_r - f_r \frac{x_r - x_{r-1}}{f_r - f_{r-1}}. \tag{6.15}$$

Quadratic interpolation of three function values is known as Muller's method [79]. Newton's form of the interpolating polynomial is

$$p(x) = f_r + (x - x_r) f[x_r, x_{r-1}] + (x - x_r)(x - x_{r-1}) f[x_r, x_{r-1}, x_{r-2}] \tag{6.16}$$

which can be rewritten as

$$\begin{aligned} p(x) &= f_r + (x - x_r) f[x_r, x_{r-1}] + (x - x_r)^2 f[x_r, x_{r-1}, x_{r-2}] \\ &\quad + (x_r - x_{r-1})(x - x_r) f[x_r, x_{r-1}, x_{r-2}] \\ &= f_r + (x - x_r)^2 f[x_r, x_{r-1}, x_{r-2}] + (x - x_r)\left(f[x_r, x_{r-1}] + f[x_r, x_{r-2}]\right. \\ &\quad \left. - f[x_{r-1}, x_{r-2}]\right) \\ &= f_r + A(x - x_r) + B(x - x_r)^2 \end{aligned} \tag{6.17}$$

and has the roots

$$x_{r+1} = x_r - \frac{A}{2B} \pm \sqrt{\frac{A^2}{4B^2} - \frac{f_r}{B}}. \tag{6.18}$$

To avoid numerical cancellation, this is rewritten

$$
\begin{aligned}
x_{r+1} &= x_r + \frac{1}{2B}\left(-A \pm \sqrt{A^2 - 4Bf_r}\right) \\
&= x_r + \frac{-2f_r}{A^2 - (A^2 - 4Bf_r)}\left(A \mp \sqrt{A^2 - 4Bf_r}\right) \\
&= x_r + \frac{-2f_r}{A \pm \sqrt{A^2 - 4Bf_r}}.
\end{aligned} \tag{6.19}
$$

The sign in the denominator is chosen such that x_{r+1} is the root closer to x_r. The roots of the polynomial can become complex valued and therefore this method is useful to find complex roots.

6.1.6 Inverse Interpolation

Complex values of x_r can be avoided by interpolation of the inverse function instead

$$
x = f^{-1}(y). \tag{6.20}
$$

Using the two points x_r, x_{r-1} the Lagrange method gives

$$
p(y) = x_{r-1}\frac{y - f_r}{f_{r-1} - f_r} + x_r \frac{y - f_{r-1}}{f_r - f_{r-1}} \tag{6.21}
$$

and the next approximation of the root corresponds again to the secant method (6.12)

$$
x_{r+1} = p(0) = \frac{x_{r-1}f_r - x_r f_{r-1}}{f_r - f_{r-1}} = x_r + \frac{(x_{r-1} - x_r)}{f_r - f_{r-1}} f_r. \tag{6.22}
$$

Inverse quadratic interpolation needs three starting points x_r, x_{r-1}, x_{r-2} together with the function values f_r, f_{r-1}, f_{r-2} (Fig. 6.6). The inverse function $x = f^{-1}(y)$ is interpolated with the Lagrange method

$$
\begin{aligned}
p(y) = &\frac{(y - f_{r-1})(y - f_r)}{(f_{r-2} - f_{r-1})(f_{r-2} - f_r)} x_{r-2} + \frac{(y - f_{r-2})(y - f_r)}{(f_{r-1} - f_{r-2})(f_{r-1} - f_r)} x_{r-1} \\
&+ \frac{(y - f_{r-1})(y - f_{r-2})}{(f_r - f_{r-1})(f_r - f_{r-2})} x_r.
\end{aligned} \tag{6.23}
$$

For $y = 0$ we find the next iterate

$$
\begin{aligned}
x_{r+1} = p(0) = &\frac{f_{r-1}f_r}{(f_{r-2} - f_{r-1})(f_{r-2} - f_r)} x_{r-2} + \frac{f_{r-2}f_r}{(f_{r-1} - f_{r-2})(f_{r-1} - f_r)} x_{r-1} \\
&+ \frac{f_{r-1}f_{r-2}}{(f_r - f_{r-1})(f_r - f_{r-2})} x_r.
\end{aligned} \tag{6.24}
$$

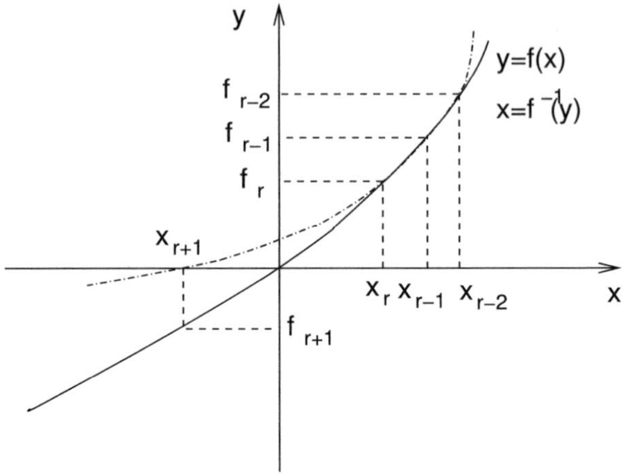

Fig. 6.6 Root finding by interpolation of the inverse function

Inverse quadratic interpolation is only a good approximation if the interpolating parabola is single valued and hence if it is a monotonous function in the range of f_r, f_{r-1}, f_{r-2}. For the following discussion we assume that the three values of x are renamed such that $x_1 < x_2 < x_3$.

Consider the limiting case (a) in Fig. 6.7 where the polynomial has a horizontal tangent at $y = f_3$ and can be written as

$$p(y) = x_3 + (x_1 - x_3)\frac{(y - f_3)^2}{(f_1 - f_3)^2}. \tag{6.25}$$

Its value at $y = 0$ is

$$p(0) = x_3 + (x_1 - x_3)\frac{f_3^2}{(f_1 - f_3)^2} = x_1 + (x_3 - x_1)\left(1 - \frac{f_3^2}{(f_1 - f_3)^2}\right). \tag{6.26}$$

If f_1 and f_3 have different sign and $|f_1| < |f_3|$ (6.1.7) we find

$$1 - \frac{f_3^2}{(f_1 - f_3)^2} < \frac{3}{4}. \tag{6.27}$$

Brent [80] used this as a criterion for the applicability of the inverse quadratic interpolation. However, this does not include all possible cases where interpolation is applicable. Chandrupatla [81] gave a more general discussion. The limiting condition is that the polynomial $p(y)$ has a horizontal tangent at one of the boundaries $x_{1,3}$. The derivative values are

$$\frac{dp}{dy}(y = f_1) = \frac{x_2(f_1 - f_3)}{(f_2 - f_1)(f_2 - f_3)} + \frac{x_3(f_1 - f_2)}{(f_3 - f_1)(f_3 - f_2)} + \frac{x_1}{f_1 - f_2} + \frac{x_1}{f_1 - f_3} \tag{6.28}$$

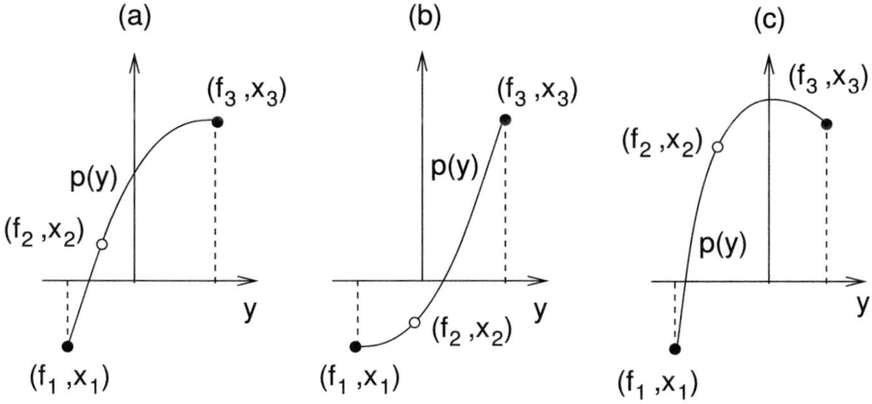

Fig. 6.7 Validity of inverse quadratic interpolation. Inverse quadratic interpolation is only applicable if the interpolating polynomial $p(y)$ is monotonous in the range of the interpolated function values $f_1 \dots f_3$. (**a**) and (**b**) show the limiting cases where the polynomial has a horizontal tangent at f_1 or f_3. (**c**) shows the case where the extremum of the parabola is inside the interpolation range and interpolation is not feasible

$$= \frac{(f_2 - f_1)}{(f_3 - f_1)(f_3 - f_2)} \left[\frac{x_2(f_3 - f_1)^2}{(f_2 - f_1)^2} - x_3 - \frac{x_1(f_3 - f_1)^2 - x_1(f_2 - f_1)^2}{(f_2 - f_1)^2} \right]$$

$$= \frac{(f_2 - f_1)(x_2 - x_1)}{(f_3 - f_1)(f_3 - f_2)} \left[\Phi^{-2} - \xi^{-1} \right]$$

$$\frac{dp}{dy}(y = f_3) = \frac{x_2(f_3 - f_1)}{(f_2 - f_1)(f_2 - f_3)} + \frac{x_1(f_3 - f_2)}{(f_1 - f_2)(f_1 - f_3)} + \frac{x_3}{f_3 - f_2} + \frac{x_3}{f_3 - f_1} \tag{6.29}$$

$$= \frac{(f_3 - f_2)}{(f_2 - f_1)(f_3 - f_1)} \left[-\frac{x_2(f_3 - f_1)^2}{(f_3 - f_2)^2} + x_3 \frac{(f_3 - f_1)^2}{(f_3 - f_2)^2} - x_3 \frac{(f_3 - f_2)^2}{(f_3 - f_2)^2} + x_1 \right]$$

$$= \frac{(f_3 - f_2)(x_3 - x_2)}{(f_2 - f_1)(f_3 - f_1)} \left[\left(\frac{1}{\Phi - 1} \right)^2 - \frac{1}{1 - \xi} \right]$$

with [81]

$$\xi = \frac{x_2 - x_1}{x_3 - x_1} \quad \Phi = \frac{f_2 - f_1}{f_3 - f_1} \tag{6.30}$$

$$\xi - 1 = \frac{x_2 - x_3}{x_3 - x_1} \quad \Phi - 1 = \frac{f_2 - f_3}{f_3 - f_1}. \tag{6.31}$$

Since for a parabola either $f_1 < f_2 < f_3$ or $f_1 > f_2 > f_3$ the conditions for applicability of inverse interpolation finally become

$$\Phi^2 < \xi \tag{6.32}$$

$$1 - \xi > (1 - \Phi)^2 \tag{6.33}$$

which can be combined into

$$1 - \sqrt{1 - \xi} < |\Phi| < \sqrt{\xi}. \tag{6.34}$$

This method is usually used in combination with other methods (Sect. 6.1.7).

6.1.7 Combined Methods

Bisection converges slowly. The interpolation methods converge faster but are less reliable. The combination of both gives methods which are reliable and converge faster than pure bisection.

Dekker's Method

Dekker's method [82, 83] combines bisection and secant method (Algorithm 6). The root is bracketed by intervals $[c_r, b_r]$ with decreasing width where b_r is the best approximation to the root found and c_r is an earlier guess for which $f(c_r)$ and $f(b_r)$ have different sign. First an attempt is made to use linear interpolation between the points $(b_r, f(b_r))$ and $(a_r, f(a_r))$ where a_r is usually the preceding approximation $a_r = b_{r-1}$ and is set to the second interval boundary $a_r = c_{r-1}$ if the last iteration did not lower the function value (Fig. 6.8).

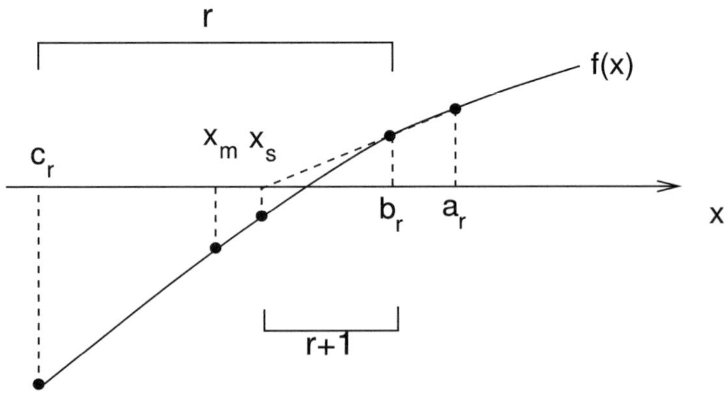

Fig. 6.8 Dekker's method

Algorithm 6: Dekker'sroot finding method [83]

/* Start with an initial interval $[x_0, x_1]$ with $f(x_0)f(x_1) \leq 0$ */

$f_1 \leftarrow f(x_1)$;

$f_0 \leftarrow f(x_0)$;

if $|f_1| < |f_0|$ **then**

  $b \leftarrow x_1$;

  $c \leftarrow x_0 \quad a \leftarrow x_0$;

  $f_b \leftarrow f_1$;

  $f_c \leftarrow f_0 \quad f_a \leftarrow f_0$;

else

  $b \leftarrow x_0$;

  $c \leftarrow x_1 \quad a \leftarrow x_1$;

  $f_b \leftarrow f_0$;

  $f_c \leftarrow f_1 \quad f_a \leftarrow f_1$;

end

/* iteration */

repeat

  $x_s = b - f_b \frac{b-a}{f_b - f_a}$;

  $x_m = \frac{c+b}{2}$;

  /* If x_s is very close to the last b then increase the distance to avoid too small steps else choose x_s if it is between b and x_m, otherwise choose x_m (thus choosing the smaller interval) */

$$x_r = \begin{cases} b + \delta \text{sign}(c-b) & \text{if abs}(x_s - b) < \delta \\ x_s \text{ if } b + \delta < x_s < x_m \text{ or } b - \delta > x_s > x_m \\ x_m \text{else} \end{cases} ;$$

  /* Determine x_k as the latest of the previous iterates $x_0 \ldots x_{r-1}$ for which $\text{sign}(f(x_k)) \neq \text{sign}(f(x_r))$. If the new function value is lower update the approximation to the root, otherwise keep the old approximation and update the second interval boundary */

  $f_r \leftarrow (x_r)$;

  if $|f_r| < |f_k|$ **then**

    $a \leftarrow b \quad b \leftarrow x_r \quad c \leftarrow x_k$;

    $f_a \leftarrow f_b \quad f_b \leftarrow f_r \quad f_c \leftarrow f_k$;

  else

    $b \leftarrow x_k \quad a \leftarrow c \leftarrow x_r$;

    $f_b \leftarrow f_k \quad f_a \leftarrow f_r \quad f_c \leftarrow f_r$;

  end

until $|c - b| < \epsilon \quad or \quad f_r = 0$;

Brent's Method

In certain cases Dekker's method converges very slowly making many small steps of the order ϵ. Brent [80, 83, 84] introduced some modifications to reduce such problems and tried to speed up convergence by the use of inverse quadratic interpolation (Sect. 6.1.6). To avoid numerical problems the iterate (6.24) is written with the help of a quotient

$$x_{r+1} = \frac{f_b f_c}{(f_a - f_b)(f_a - f_c)} a + \frac{f_a f_c}{(f_b - f_a)(f_b - f_c)} b + \frac{f_b f_a}{(f_c - f_b)(f_c - f_a)} c = b + \frac{p}{q} \tag{6.35}$$

with

$$p = \frac{f_b}{f_a}\left((c-b)\frac{f_a}{f_c}\left(\frac{f_a}{f_c} - \frac{f_b}{f_c}\right) - (b-a)\left(\frac{f_b}{f_c} - 1\right)\right)$$
$$= (c-b)\frac{f_b(f_a - f_b)}{f_c^2} - (b-a)\frac{f_b(f_b - f_c)}{f_a f_c}$$
$$= \frac{a f_b f_c(f_b - f_c) + b\left[f_a f_b(f_b - f_a) + f_b f_c(f_c - f_b)\right] + c f_a f_b(f_a - f_b)}{f_a f_c^2} \tag{6.36}$$

$$q = -\left(\frac{f_a}{f_c} - 1\right)\left(\frac{f_b}{f_c} - 1\right)\left(\frac{f_b}{f_a} - 1\right) = -\frac{(f_a - f_c)(f_b - f_c)(f_b - f_a)}{f_a f_c^2}. \tag{6.37}$$

If only two points are available linear interpolation is used. The iterate (6.22) then is written as

$$x_{r+1} = b + \frac{(a-b)}{f_b - f_a} f_b = b + \frac{p}{q} \tag{6.38}$$

with

$$p = (a-b)\frac{f_b}{f_a} \quad q = \left(\frac{f_b}{f_a} - 1\right). \tag{6.39}$$

The division is only performed if interpolation is appropriate and division by zero cannot happen. Brent's method (Algorithm 7) is fast and robust at the same time. It is often recommended by text books.

Algorithm 7: Brent's root finding method [85]

```
/* Start with an initial interval [x0,x1] with f(x0)f(x1) ≤ 0                 */
a ← x0  b ← x1  c ← x0 ;
fa ← f(a)  fb ← f(b)  fc ← f(a) ;
d ← (b − a)  e ← (b − a);
Loop
    /* If c is a better approximation than b exchange values                  */
    if |fc| < |fb| then
        a ← b  b ← c  c ← a;
        fa ← fb  fb ← fc  fc ← fa;
    end
    xm = 0.5(c − b) ;                      /* calculate midpoint relative to b */
    ;
    if |xm| < ε or fb = 0 then
        ExitLoop ;                          /* stop if accuracy is sufficient */
    end
    /* use bisection if the previous step width e was too small or the last
       step did not improve, otherwise try interpolation                      */
    if |e| < ε or |fa| ≤ |fb| then
        d =← xm  e ← xm ;
    else
        if a = c then
            p ← 2xm fb/fa   q ← (fb−fa)/fa;
        else
            p ← 2xm fb(fa−fb)/fc^2 − (b − a) fb(fb−fc)/(fa fc);
            q ← (fa/fc − 1)(fb/fc − 1)(fb/fa − 1);
        end
        if p > q then                             /* make p a positive quantity */
            q ← −q;
        else
            p ← −p;
        end
        s ← e  e ← d ;                            /* update previous step width */
        if 2p < 3xm q − |εq| and p < |0.5 sq| then          /* interpolate      */
            d ← p/q
        else                                                 /* use bisection   */
            e ← xm  d ← xm
        end
    end
    a ← b  fa ← fb if d| > ε then
        b ← b + d
    else
        b ← b + ε sign (xm)
    end
    fb ← f(b)                               /* calculate new function value */
    if sign(fb) = sign(fc) then             /* be sure to bracket the root */
        c ← a  fc ← fa;
        e ← b − a  d ← b − a;
    end
EndLoop
```

Chandrupatla's Method

In 1997 Chandrupatla [81] published a method which tries to use inverse quadratic interpolation whenever possible according to 6.34. He calculates the relative position of the new iterate as (Fig. 6.9)

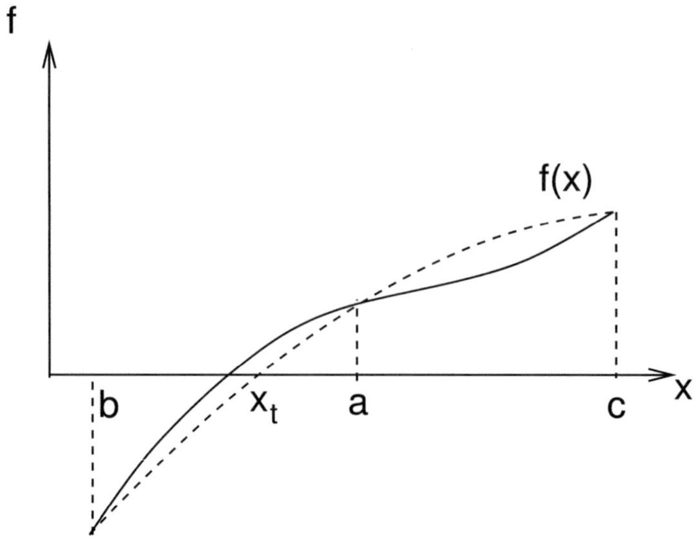

Fig. 6.9 Chandrupatla's method

$$t = \frac{x-c}{b-c}$$

$$= \frac{1}{b-c}\left[\frac{f_c f_b}{(f_a-f_b)(f_a-f_c)}a + \frac{f_a f_c}{(f_b-f_a)(f_b-f_c)}b + \frac{f_b f_a}{(f_c-f_b)(f_c-f_a)}c - c\right]$$

$$= \frac{a-c}{b-c}\frac{f_c}{f_c-f_a}\frac{f_b}{f_b-f_a} + \frac{f_a f_c}{(f_b-f_a)(f_b-f_c)}. \tag{6.40}$$

Algorithm 8 summarizes the method.

Chandrupatla's method is more efficient than Dekker's and Brent's, especially for higher order roots (Figs. 6.10, 6.11, 6.12).

6.1.8 Multidimensional Root Finding

The Newton-Raphson method can be easily generalized for functions of more than one variable. We search for the solution of a system of n nonlinear equations in n variables x_i

Algorithm 8: Chandrupatla's root finding method

```
/* Start with an initial interval [x_0, x_1] with
   f(x_0) f(x_1) ≤ 0                                        */
b ← x_0   a ← x_1   c ← x_1 ;
f_a ← f(c)   f_b ← f(b)   f_c ← f_a ;
t ← 0.5;
Loop
    x_t = a + t (b − a);
    f_t = f(x_t);
    if sign(f_t) = sign(f_a) then
        c ← a   f_c ← f_a;
        a ← x_t   f_a ← f_t;
    else
        c ← b   b ← a   a ← x_t;
        f_c ← f_b   f_b ← f_a   f_a ← f_t;
    end
    x_m ← a   f_m ← f_a;
    if abs(f_b) < abs(f_a);
    then
        x_m = b   f_m = f_b;
        tol = 2ε_M|x_m| + ε_a;
        t_l = tol / |b−c|;
        if t_l > 0.5 or f_m = 0 then ExitLoop;
        ξ ← (a−b)/(c−b)   Φ ← (f_a−f_b)/(f_c−f_b);
        if 1 − √(1 − ξ) < Φ < √ξ then
            t = f_a/(f_b−f_a) · f_c/(f_b−f_c) + (c−a)/(b−a) · f_a/(f_c−f_a) · f_b/(f_c−f_b);
        end
    else
        t=0.5;
    end
    if t < t_l then t = t_l if t > (1 − t_l) then t = 1 − t_l
EndLoop
```

$$\mathbf{f}(\mathbf{x}) = \begin{pmatrix} f_1(x_1 \cdots x_n) \\ \vdots \\ f_n(x_1 \cdots x_n) \end{pmatrix} = 0. \tag{6.41}$$

The first order Newton-Raphson method results from linearization of

$$0 = \mathbf{f}(\mathbf{x}) = \mathbf{f}(\mathbf{x}^0) + J(\mathbf{x}^0)(\mathbf{x} - \mathbf{x}^0) + \cdots \tag{6.42}$$

with the Jacobian matrix

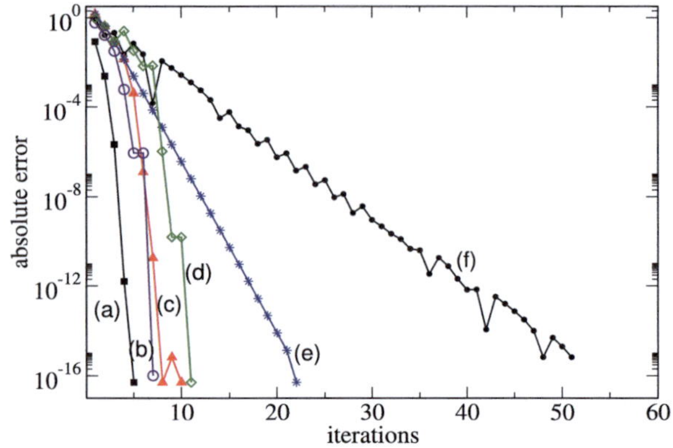

Fig. 6.10 Comparison of different solvers. The root of the equation $f(x) = x^2 - 2$ is determined with different methods: Newton-Raphson (**a**, black squares), Chandrupatla (**b**, indigo circles), Brent (**c**, red triangles up), Dekker (**d**, green diamonds), regula falsi (**e**, blue stars), pure bisection (**f**, black dots). Starting values are $x_1 = -1$, $x_2 = 2$. The absolute error is shown as a function of the number of iterations. For $x_1 = -1$, the Newton-Raphson method converges against $-\sqrt{2}$

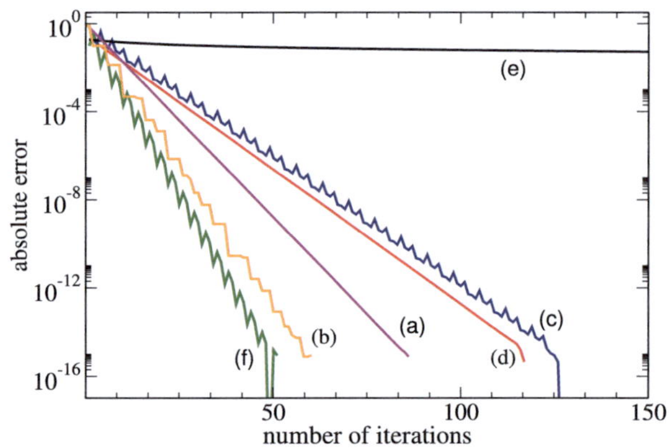

Fig. 6.11 Comparison of different solvers for a third order root. The root of the equation $f(x) = (x-1)^3$ is determined with different methods: Newton-Raphson (**a**, magenta), Chandrupatla (**b**, orange), Brent (**c**, blue), Dekker (**d**, red), regula falsi (**e**, black), pure bisection (**f**, green). Starting values are $x_1 = 0$, $x_2 = 1.8$. The absolute error is shown as function of the number of iterations

$$J = \begin{pmatrix} \frac{\partial f_1}{\partial x_1} & \cdots & \frac{\partial f_1}{\partial x_n} \\ \vdots & \ddots & \vdots \\ \frac{\partial f_n}{\partial x_1} & \cdots & \frac{\partial f_n}{\partial x_n} \end{pmatrix}. \tag{6.43}$$

If the Jacobian matrix is not singular the equation

$$0 = \mathbf{f}(\mathbf{x}^0) + J(\mathbf{x}^0)(\mathbf{x} - \mathbf{x}^0) \tag{6.44}$$

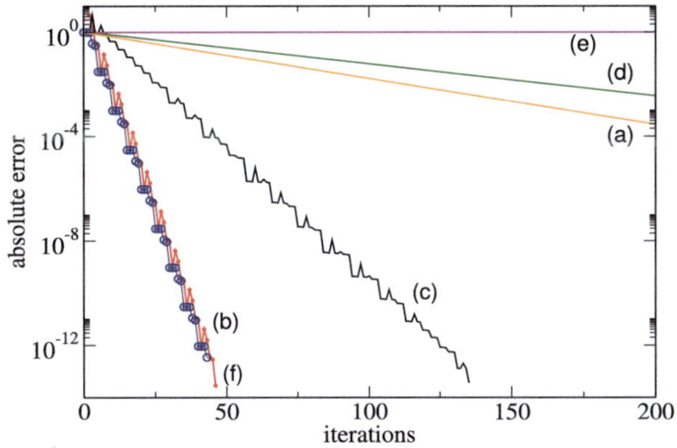

Fig. 6.12 Comparison of different solvers for a high order root. The root of the equation $f(x) = x^{25}$ is determined with different methods: Newton-Raphson (**a**, orange), Chandrupatla (**b**, blue circles), Brent (**c**, black), Dekker (**d**, green), regula falsi (**e**, magenta), pure bisection (**f**, red dots). Starting values are $x_1 = -1, x_2 = 2$. The absolute error is shown as a function of the number of iterations

can be solved by

$$\mathbf{x} = \mathbf{x}^0 - \left(J(\mathbf{x}^0)\right)^{-1} \mathbf{f}(\mathbf{x}^0). \tag{6.45}$$

This can be repeated iteratively

$$\mathbf{x}^{(r+1)} = \mathbf{x}^{(r)} - (J(\mathbf{x}^{(r)}))^{-1}\mathbf{f}(\mathbf{x}^{(r)}). \tag{6.46}$$

6.1.9 Quasi-Newton Methods

Calculation of the Jacobian matrix can be very time consuming. Quasi-Newton methods use instead an approximation to the Jacobian which is updated during each iteration. Defining the differences

$$\mathbf{d}^{(r)} = \mathbf{x}^{(r+1)} - \mathbf{x}^{(r)} \tag{6.47}$$

$$\mathbf{y}^{(r)} = \mathbf{f}(\mathbf{x}^{(r+1)}) - \mathbf{f}(\mathbf{x}^{(r)}) \tag{6.48}$$

we obtain from the truncated Taylor series

$$\mathbf{f}(\mathbf{x}^{(r+1)}) = \mathbf{f}(\mathbf{x}^{(r)}) + J(\mathbf{x}^{(r)})(\mathbf{x}^{(r+1)} - \mathbf{x}^{(r)}) \tag{6.49}$$

the so called Quasi-Newton or secant condition

$$\mathbf{y}^{(r)} = J(\mathbf{x}^{(r)})\mathbf{d}^{(r)}. \tag{6.50}$$

We attempt to construct a family of successive approximation matrices J_r so that, if J were a constant, the procedure would become consistent with the quasi-Newton condition. Then for the new update J_{r+1} we have

$$J_{r+1}\mathbf{d}^{(r)} = \mathbf{y}^{(r)}. \tag{6.51}$$

Since $\mathbf{d}^{(r)}$, $\mathbf{y}^{(r)}$ are already known, these are only n equations for the n^2 elements of J_{r+1}. To specify J_{r+1} uniquely, additional conditions are required. For instance, it is reasonable to assume, that

$$J_{r+1}\mathbf{u} = J_r\mathbf{u} \quad \text{for all } \mathbf{u} \perp \mathbf{d}^{(r)}. \tag{6.52}$$

Then J_{r+1} differs from J_r only by a rank one updating matrix

$$J_{r+1} = J_r + \mathbf{u}\,\mathbf{d}^{(r)T}. \tag{6.53}$$

From the secant condition we obtain

$$J_{r+1}\mathbf{d}^{(r)} = J_r\mathbf{d}^{(r)} + \mathbf{u}(\mathbf{d}^{(r)}\mathbf{d}^{(r)T}) = \mathbf{y}^{(r)} \tag{6.54}$$

hence

$$\mathbf{u} = \frac{1}{|d^{(r)}|^2}\left(\mathbf{y}^{(r)} - J_r\mathbf{d}^{(r)}\right). \tag{6.55}$$

This gives Broyden's update formula [86]

$$J_{r+1} = J_r + \frac{1}{|d^{(r)}|^2}\left(\mathbf{y}^{(r)} - J_r\mathbf{d}^{(r)}\right)\mathbf{d}^{(r)T}. \tag{6.56}$$

To update the inverse Jacobian matrix, the Sherman-Morrison formula [66]

$$(A + \mathbf{u}\mathbf{v}^T)^{-1} = A^{-1} - \frac{A^{-1}\mathbf{u}\mathbf{v}^T A^{-1}}{1 + \mathbf{v}^T A^{-1}\mathbf{u}} \tag{6.57}$$

can be applied to have

$$\begin{aligned} J_{r+1}^{-1} &= J_r^{-1} - \frac{J_r^{-1}\frac{1}{|d^{(r)}|^2}\left(\mathbf{y}^{(r)} - J_r\mathbf{d}^{(r)}\right)\mathbf{d}^{(r)T}J_r^{-1}}{1 + \frac{1}{|d^{(r)}|^2}\mathbf{d}^{(r)T}J_r^{-1}\left(\mathbf{y}^{(r)} - J_r\mathbf{d}^{(r)}\right)} \\ &= J_r^{-1} - \frac{\left(J_r^{-1}\mathbf{y}^{(r)} - \mathbf{d}^{(r)}\right)\mathbf{d}^{(r)T}J_r^{-1}}{\mathbf{d}^{(r)T}J_r^{-1}\mathbf{y}^{(r)}}. \end{aligned} \tag{6.58}$$

6.2 Function Minimization

Minimization or maximization of a function[1] is a fundamental task in numerical mathematics and closely related to root finding. If the function $f(x)$ is continuously differentiable then at the extremal points the derivative is zero

$$\frac{\mathrm{d}f}{\mathrm{d}x} = 0. \tag{6.59}$$

Hence, in principle root finding methods can be applied to locate local extrema of a function. However, in some cases the derivative cannot be easily calculated or the function even is not differentiable. Then derivative free methods similar to bisection for root finding have to be used.

6.2.1 *The Ternary Search Method*

Ternary search is a simple method to determine the minimum of a unimodal function $f(x)$. Initially we have to find an interval $[a_0, b_0]$ which is certain to contain the minimum. Then the interval is divided into three equal parts $[a_0, c_0]$, $[c_0, d_0]$, $[d_0, b_0]$ and either the first or the last of the three intervals is excluded (Fig. 6.13). The procedure is repeated with the remaining interval $[a_1, b_1] = [a_0, d_0]$ or $[a_1, b_1] = [c_0, b_0]$.

Each step needs two function evaluations and reduces the interval width by a factor of $2/3$ until the maximum possible precision is obtained. It can be determined by considering a differentiable function which can be expanded around the minimum x_0 as

$$f(x) = f(x_0) + \frac{(x - x_0)^2}{2} f''(x_0) + \cdots . \tag{6.60}$$

Numerically calculated function values $f(x)$ and $f(x_0)$ only differ, if

$$\frac{(x - x_0)^2}{2} f''(x_0) > \varepsilon_M f(x_0) \tag{6.61}$$

which limits the possible numerical accuracy to

$$\varepsilon(x_0) = \min|x - x_0| = \sqrt{\frac{2f(x_0)}{f''(x_0)}\varepsilon_M} \tag{6.62}$$

and for reasonably well behaved functions (Fig. 6.14) we have the rule of thumb [87]

[1] In the following we consider only a minimum since a maximum could be found as the minimum of $-f(x)$.

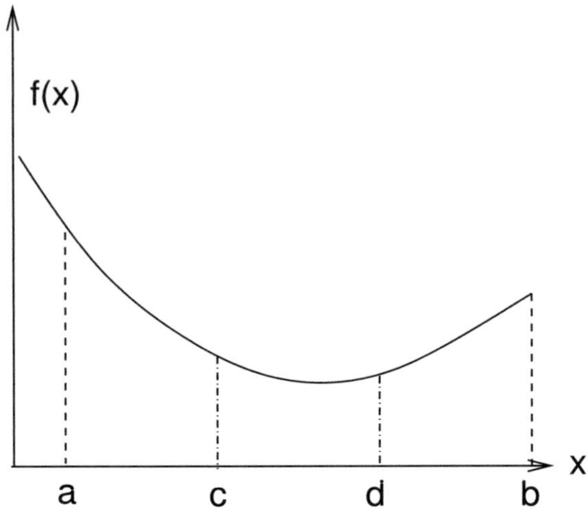

Fig. 6.13 Ternary search method

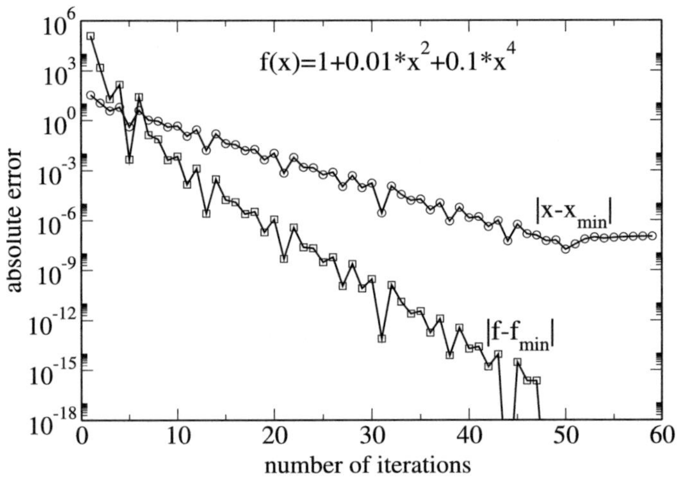

Fig. 6.14 Ternary search method. The minimum of the function $f(x) = 1 + 0.01\,x^2 + 0.1\,x^4$ is determined with the ternary search method. Each iteration needs two function evaluations. After 50 iterations the function minimum $f_{\text{min}} = 1$ is reached to machine precision $\varepsilon_M \approx 10^{-16}$. The position of the minimum x_{min} cannot be determined with higher precision than $\sqrt{\varepsilon_M} \approx 10^{-8}$ (6.63)

$$\varepsilon(x_0) \approx \sqrt{\varepsilon_M}. \tag{6.63}$$

However, it may be impossible to reach even this precision, if the quadratic term of the Taylor series vanishes (Fig. 6.15).

Algorithm 9 summarizes the method.

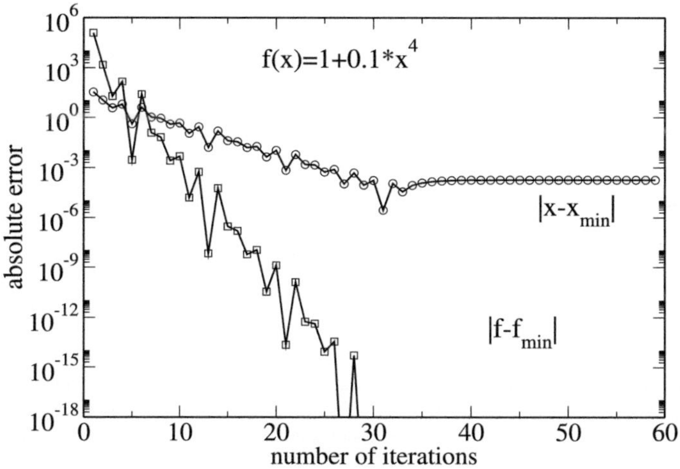

Fig. 6.15 Ternary search method for a higher order minimum. The minimum of the function $f(x) = 1 + 0.1\,x^4$ is determined with the ternary search method. Each iteration needs two function evaluations. After 30 iterations the function minimum $f_{\min} = 1$ is reached to machine precision $\varepsilon_M \approx 10^{-16}$. The position of the minimum $x_{\min}$ cannot be determined with higher precision than $\sqrt[4]{\varepsilon_M} \approx 10^{-4}$

Algorithm 9: Ternarysearch method

Loop
  if $b - a < \delta$ **then ExitLoop**;
  $c \leftarrow a + \frac{1}{3}(b-a) \quad d \leftarrow a + \frac{2}{3}(b-a)$;
  $f_c \leftarrow f(c) \quad f_d \leftarrow f(d)$;
  if $f_c < f_d$ **then**
    $b \leftarrow d$
  else
    $a \leftarrow c$
  end
EndLoop

6.2.2 *The Golden Section Search Method (Brent's Method)*

To bracket a local minimum of a unimodal function $f(x)$ three points a, b, c are necessary (Fig. 6.16) with

$$f(a) > f(b) \quad f(c) > f(b). \tag{6.64}$$

The position of the minimum can be determined iteratively by choosing a new value ξ in the interval $a < \xi < c$ and dropping either a or c, depending on the ratio

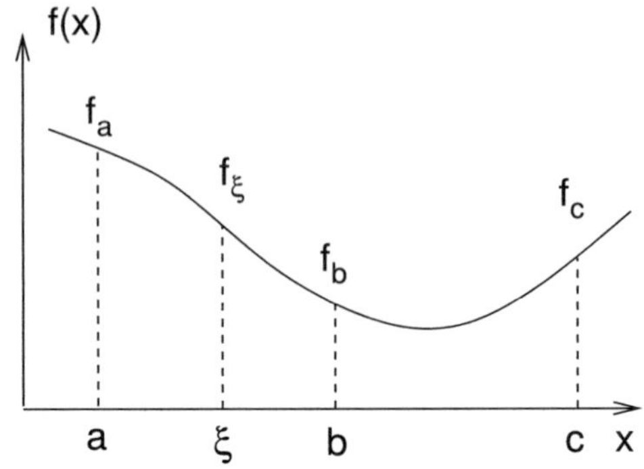

Fig. 6.16 Golden section search method. A local minimum of the function $f(x)$ is bracketed by three points a, b, c. To reduce the uncertainty of the minimum position a new point ξ is chosen in the interval $a < \xi < c$ and either a or c is dropped according to the relation of the function values. For the example shown a has to be replaced by ξ

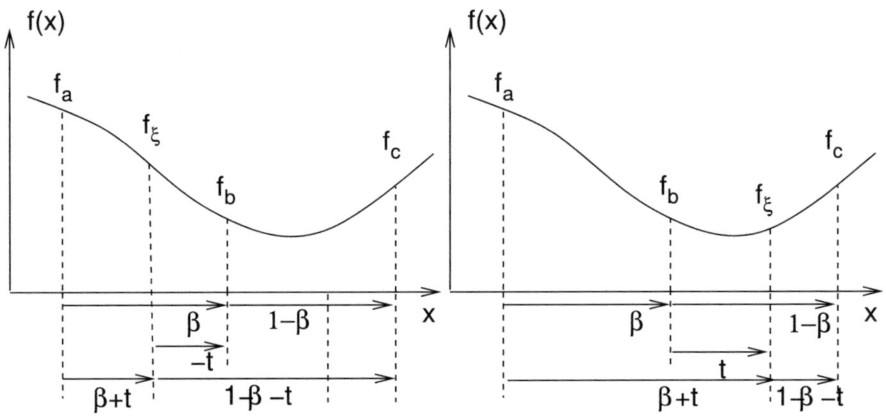

Fig. 6.17 Golden section search method

of the function values. A reasonable choice for ξ can be found as follows (Fig. 6.17) [87, 88]. Let us denote the relative positions of the middle point and the trial point as

$$\frac{b-a}{c-a}=\beta \quad \frac{c-b}{c-a}=1-\beta \quad \frac{b-a}{c-b}=\frac{\beta}{1-\beta} \quad \frac{\xi-b}{c-a}=t \tag{6.65}$$

$$\frac{\xi-a}{c-a}=\frac{\xi-b+b-a}{c-a}=t+\beta. \tag{6.66}$$

The relative width of the new interval will be

$$\frac{c-\xi}{c-a}=(1-\beta-t) \quad \text{or} \quad \frac{b-a}{c-a}=\beta \quad \text{if } a<\xi<b \tag{6.67}$$

$$\frac{\xi - a}{c - a} = (t + \beta) \quad \text{or} \quad \frac{c - b}{c - a} = (1 - \beta) \quad \text{if } b < \xi < c. \tag{6.68}$$

The golden search method requires that

$$t = 1 - 2\beta = \frac{c + a - 2b}{c - a} = \frac{(c - b) - (b - a)}{c - a}. \tag{6.69}$$

Otherwise it would be possible that the larger interval width is selected many times slowing down the convergence. The value of t is positive if $c - b > b - a$ and negative if $c - b < b - a$, hence the trial point always is in the larger of the two intervals. In addition the golden search method requires that the ratio of the spacing remains constant. Therefore we set

$$\frac{\beta}{1 - \beta} = -\frac{t + \beta}{t} = -\frac{1 - \beta}{t} \quad \text{if } a < \xi < b \tag{6.70}$$

$$\frac{\beta}{1 - \beta} = \frac{t}{1 - \beta - t} = \frac{t}{\beta} \quad \text{if } b < \xi < c. \tag{6.71}$$

Eliminating t we obtain for $a < \xi < b$ the equation

$$\frac{(\beta - 1)}{\beta}(\beta^2 + \beta - 1) = 0. \tag{6.72}$$

Besides the trivial solution $\beta = 1$ there is only one positive solution

$$\beta = \frac{\sqrt{5} - 1}{2} \approx 0.618. \tag{6.73}$$

For $b < \xi < c$ we end up with

$$\frac{\beta}{\beta - 1}(\beta^2 - 3\beta + 1) = 0 \tag{6.74}$$

which has the nontrivial positive solution

$$\beta = \frac{3 - \sqrt{5}}{2} \approx 0.382. \tag{6.75}$$

Hence the lengths of the two intervals $[a, b]$, $[b, c]$ have to be in the golden ratio $\varphi = \frac{1+\sqrt{5}}{2} \approx 1.618$ which gives the method its name. Using the golden ratio the width of the interval bracketing the minimum reduces by a factor of 0.618 (Figs. 6.18 and 6.19).

Algorithm 10 summarizes the method.

Algorithm 10: Golden section search method

```
Loop
  if c − a < δ then ExitLoop;
  if (b − a) ≥ (c − b) then
    x ← 0.618 b + 0.382 a;
    f_x = f(x);
    if f_x < f_b then
      c ← b   b ← x   f_c ← f_b   f_b ← f_x;
    else
      a = x   f_a = f_x;
    end
  end
  if (b − a) < (c − b) then
    x ← 0.618 b + 0.382 c;
    f_x ← f(x);
    if f_x < f_b then
      a ← b   b ← x   f_a ← f_b   f_b ← f_x;
    else
      c ← x   f_c ← f_x;
    end
  end
EndLoop
```

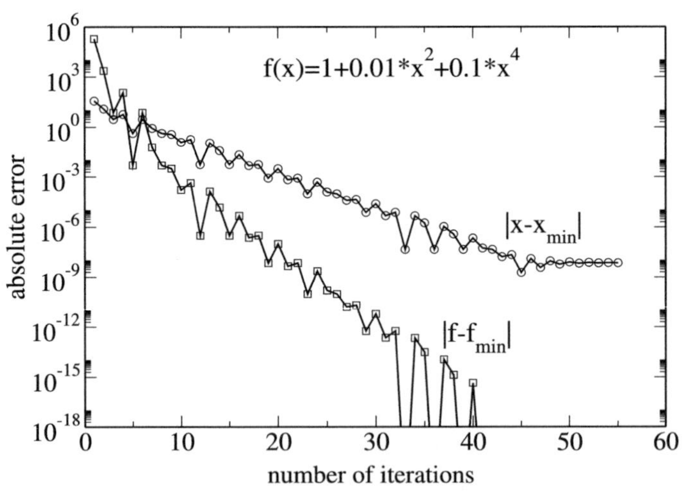

Fig. 6.18 The minimum of the function $f(x) = 1 + 0.01\,x^2 + 0.1\,x^4$ is determined with the golden section search method. Each iteration needs only one function evaluation. After 40 iterations the function minimum $f_{\min} = 1$ is reached to machine precision $\varepsilon_M \approx 10^{-16}$. The position of the minimum $x_{\min}$ cannot be determined to higher precision than $\sqrt{\varepsilon_M} \approx 10^{-8}$ (6.63)

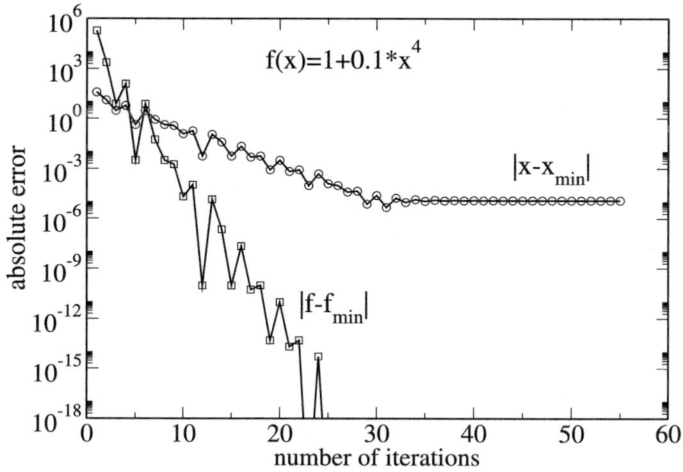

Fig. 6.19 Golden section search for a higher order minimum. The minimum of the function $f(x) = 1 + 0.1\,x^4$ is determined with the golden section search method. Each iteration needs only one function evaluation. After 28 iterations the function minimum $f_{\min} = 1$ is reached to machine precision $\varepsilon_M \approx 10^{-16}$. The position of the minimum $x_{\min}$ cannot be determined to higher precision than $\sqrt[4]{\varepsilon_M} \approx 10^{-4}$

To start the method we need three initial points which can be found by Brent's exponential search method (Fig. 6.20). Begin with three points

$$a_0, b_0 = a_0 + h, c_0 = b_0 + 1.618\,h \tag{6.76}$$

where h_0 is a suitable initial step width which depends on the problem. If the minimum is not already bracketed then if necessary exchange a_0 and b_0 to have

$$f(a_0) > f(b_0) > f(c_0). \tag{6.77}$$

Then replace the three points by

$$a_1 = b_0 \quad b_1 = c_0 \quad c_1 = c_0 + 1.618\,(c_0 - b_0) \tag{6.78}$$

and repeat this step until

$$f(b_n) < f(c_n) \tag{6.79}$$

or n exceeds a given maximum number. In this case no minimum can be found and we should check if the initial step width was too large.

Brent's method can be improved by making use of derivatives and by combining the golden section search with parabolic interpolation [87].

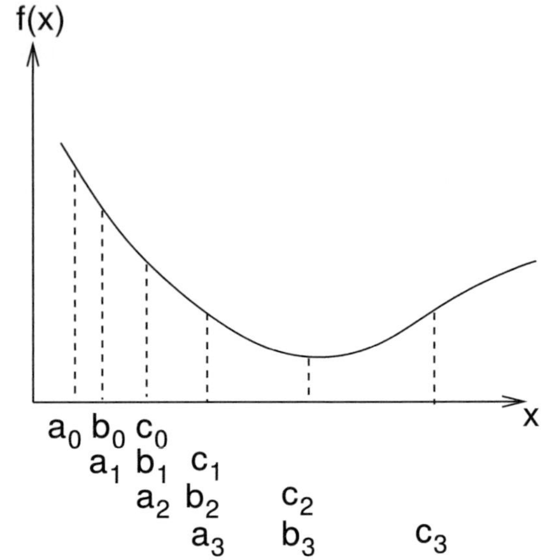

Fig. 6.20 Brent's exponential search

6.3 Minimization in Multidimensions

We search for local minima (or maxima) of a function

$$h(\mathbf{x})$$

which is at least two times differentiable. In the following we denote the gradient vector by

$$\mathbf{g}^T(\mathbf{x}) = (\frac{\partial h}{\partial x_1}, \cdots \frac{\partial h}{\partial x_n}) \tag{6.80}$$

and the matrix of second derivatives (Hessian) by

$$H = \left(\frac{\partial^2}{\partial x_i \partial x_j} h \right). \tag{6.81}$$

The very popular class of direction set methods proceeds as follows (Fig. 6.21). Starting from an initial guess $\mathbf{x}_0$ a set of direction vectors $\mathbf{s}_r$ and step lengths λ_r is determined such that the series of vectors

$$\mathbf{x}_{r+1} = \mathbf{x}_r + \lambda_r \mathbf{s}_r \tag{6.82}$$

approaches the minimum of $h(\mathbf{x})$. The method stops if the norm of the gradient becomes sufficiently small or if no lower function value can be found.

Fig. 6.21 Direction Set Minimization. Starting from an initial guess $\mathbf{x}_0$ a local minimum is approached by making steps along a set of direction vectors $\mathbf{s}_r$

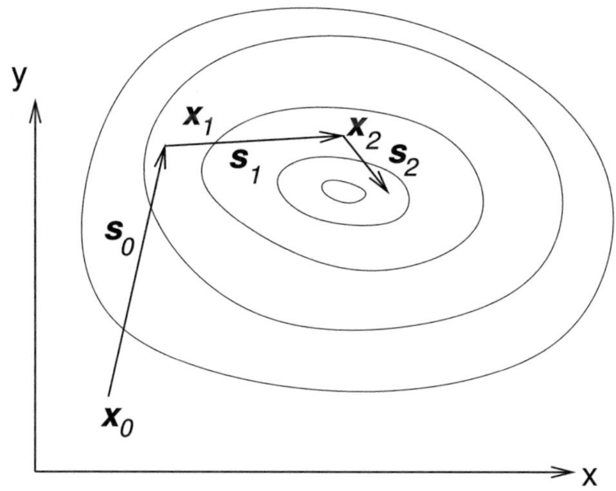

The optimal step length can be found from a line search

$$\lambda_r = \underset{\lambda}{\operatorname{argmin}}\, h(\mathbf{x}_r + \lambda \mathbf{s}_r). \tag{6.83}$$

A full line search can be very time consuming as it needs many function evaluations. Often only an inexact line search is applied to lower the function value sufficiently until the Armijo condition [89] (Fig. 6.22)

$$h(\mathbf{x}_{r+1}) \leq h(\mathbf{x}_r) + c_1 \lambda \mathbf{s}_r^T \mathbf{g}_r \tag{6.84}$$

together with the curvature condition

$$-\mathbf{s}_r^T \mathbf{g}_{r+1} \leq -c_2 \mathbf{s}_r^T \mathbf{g}_r \tag{6.85}$$

or its stronger version

$$|\mathbf{s}_r^T \mathbf{g}_{r+1}| \leq |c_2 \mathbf{s}_r^T \mathbf{g}_r| \tag{6.86}$$

are fulfilled (Fig. 6.23). Equations (6.84 and 6.85) together are also known as the Wolfe [90, 91] conditions, (6.84 and 6.86) together as the strong Wolfe conditions). The constants $c_{1,2}$ have to be chosen according to the method, usually c_1 is much smaller than c_2 [92].[2]

---

[2] For instance $c_1 \approx 10^{-3}$, $c_2 \approx 0.1$.

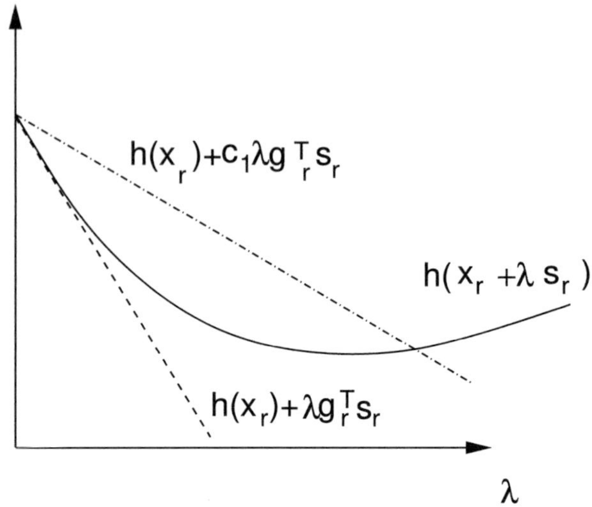

Fig. 6.22 Armijo condition. The Armijo condition demands that the new function value is below the reduced tangent (dash-dotted). This guarantees that the function value reduces at least a certain fraction of the linear prediction and keeps the step length from becoming too large

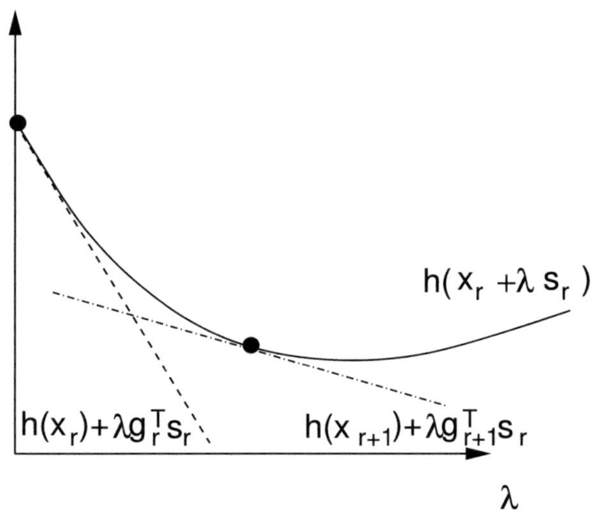

Fig. 6.23 Wolfe condition. The Wolfe curvature condition demands that the gradient becomes smaller and keeps the step length from becoming too small

6.3.1 Gradient Descent Methods

The simplest choice for the search direction, is the method of gradient descent or steepest descent[3] which follows the direction of the negative gradient

$$\mathbf{s}_r = -\frac{\mathbf{g}_r}{|\mathbf{g}_r|} \tag{6.87}$$

---

[3] Which should not be confused with the method of steepest descent for the approximate calculation of integrals.

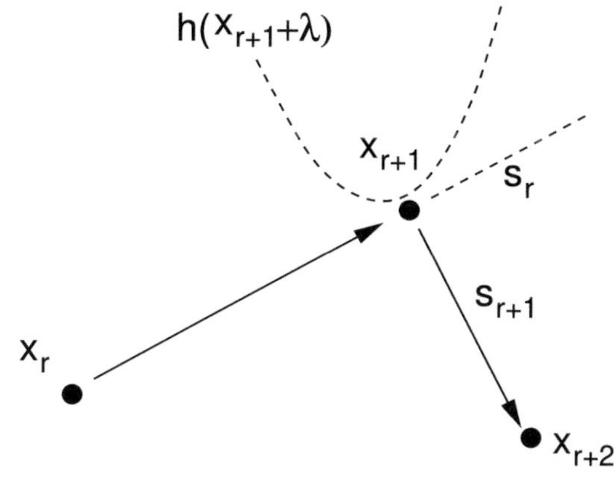

Fig. 6.24 Orthogonality of consecutive gradient descent steps

and determines the step length by minimizing $h(\mathbf{x}_r + \lambda \mathbf{s}_r)$ along this direction. Obviously two consecutive steps are orthogonal to each other since (Fig. 6.24)

$$0 = \frac{\partial}{\partial \lambda} h(\mathbf{x}_{r+1} + \lambda \mathbf{s}_r)_{|\lambda=0} = \mathbf{g}_{r+1}^T \frac{\partial}{\partial \lambda}(\mathbf{x}_{r+1} + \lambda \mathbf{s}_r) = \mathbf{g}_{r+1}^T \mathbf{s}_r. \tag{6.88}$$

This can lead to a zigzagging behavior and a very slow convergence of this method (Figs. 6.25 and 6.26).

For neural networks often a simplified gradient descent algorithm is used

$$\mathbf{x}_{r+1} = \mathbf{x}_r - \eta_r \mathbf{g}_r \tag{6.89}$$

where the learning rate η_r is chosen constant or according to a simple predetermined schedule [93].

6.3.2 Stochastic Gradient Descent

Artificial neural networks (ANN) involve the optimization of a huge set (10^6 or more) of parameters for a cost function which averages the losses of a large data set

$$L(\mathbf{x}) = \frac{1}{\mathrm{N_{data}}} \sum_{i=1}^{\mathrm{N_{data}}} L_i(\mathbf{x}). \tag{6.90}$$

Stochastic gradient descent (SGD) methods do not calculate the gradients for the whole batch of data but only for randomly chosen samples[4] or small subsets

[4] Usually the data set is scanned repeatedly and shuffled randomly between two scans.

(minibatches) [94–96]. This may reduce the computational effort but slow down convergence.

6.3.3 Gradient Descent with Momentum

The gradient descent method can get stuck in a local minimum. This can be avoided with a two-step method which is known as momentum method or heavy ball algorithm [97, 98][5]

$$\mathbf{x}_{r+1} = \mathbf{x}_r - \alpha \mathbf{g}_r + \beta(\mathbf{x}_r - \mathbf{x}_{r-1}). \tag{6.91}$$

If the change of $\mathbf{x}$ during a unit time step is interpreted as the velocity (or momentum)

$$\mathbf{x}_{r+1} - \mathbf{x}_r = \mathbf{v}_r \tag{6.92}$$

(6.91) can be formulated alternatively as

$$\mathbf{v}_r = -\alpha \mathbf{g}_r + \beta \mathbf{v}_{r-1} \tag{6.93}$$

$$\mathbf{x}_{r+1} - \mathbf{x}_r = \mathbf{v}_r. \tag{6.94}$$

For a better understanding we write

$$0 = \frac{1}{\alpha}(\mathbf{x}_{r+1} - \mathbf{x}_r) + \mathbf{g}_r - \frac{\beta}{\alpha}(\mathbf{x}_r - \mathbf{x}_{r-1}) \tag{6.95}$$

$$= \frac{1}{\alpha}\mathbf{x}_{r+1} - \mathbf{x}_r \frac{1+\beta}{\alpha} + \frac{\beta}{\alpha}\mathbf{x}_{r-1} + \mathbf{g}_r \tag{6.96}$$

$$= \frac{\beta}{\alpha}(\mathbf{x}_{r+1} - 2\mathbf{x}_r + \mathbf{x}_{r-1}) + (\mathbf{x}_{r+1} - \mathbf{x}_r)(\frac{1-\beta}{\alpha}) + \mathbf{g}_r. \tag{6.97}$$

This resembles the discretized equation of motion of a particle with mass β/α moving in a viscous medium with damping constant $(1-\beta)/\alpha$ under the influence of the force field $-\mathbf{g}_r$. Hence this two step algorithm introduces inertia. The single step method with $\beta = 0$ relaxes to a local minimum where the gradient vanishes. Inertia adds small oscillations which can help to escape from a local minimum.

Nesterov's accelerated gradient method (NAGD) [99, 100] makes a small but important modification, taking the gradient at an advanced configuration

$$\mathbf{g}_r = \nabla h(\mathbf{x}_r + \beta(\mathbf{x}_r - \mathbf{x}_{r-1})). \tag{6.98}$$

[5] β is usually chosen close to 0.9.

Nesterov introduces an additional sequence of intermediate vectors $\mathbf{y}_r$

$$\mathbf{y}_r = \mathbf{x}_r + \beta(\mathbf{x}_r - \mathbf{x}_{r-1}) \tag{6.99}$$

$$\mathbf{x}_{r+1} = \mathbf{x}_r - \alpha \nabla h(\mathbf{y}_r). \tag{6.100}$$

The choice of the learning rate is very important. If it is too small, convergence is very slow, if it is too large, the minimum cannot be located accurately enough. Several methods have been proposed to adjust the learning rate like Adagrad [101], Rmsprop [102], Adadelta [103]. The very famous Adam (adaptive momentum estimation) method [104] determines an adaptive learning rate for each parameter separately based on exponential averages of the gradients and their squares

$$g_{r+1,i} = \frac{\partial}{\partial x_i} h(\mathbf{x}_r) \tag{6.101}$$

$$m_{r+1,i} = \beta_1 m_{r,i} + (1-\beta_1) g_{r+1,i} \tag{6.102}$$

$$v_{r+1,i} = \beta_2 v_{r,i} + (1-\beta_2) g_{r+1,i}^2 \tag{6.103}$$

$$\hat{m}_{r+1,i} = \frac{m_{r+1,i}}{1-\beta_1^r} \tag{6.104}$$

$$\hat{v}_{r+1,i} = \frac{v_{r+1,i}}{1-\beta_2^r} \tag{6.105}$$

$$x_{r+1,i} = x_{r,i} - \alpha \frac{\hat{m}_{r+1,i}}{\sqrt{\hat{v}_{r+1,i}} + \epsilon}. \tag{6.106}$$

During the initial steps, the moments are rescaled to counteract the zero bias due to the initialization of the algorithm with zero values of m and v. The small parameter $\epsilon \approx 10^{-8}$ avoids division by zero if the second moment becomes too small.

Recently a modification AMAdam (absolute momentum Adam) has been proposed [105] to improve the convergence properties in the case of very small second moment $\hat{v}$ by using the absolute value of the moment instead of the squared gradient and re-initialization of $\hat{v}$. This method reduces the learning rate in the vicinity of a minimum.

$$g_{r+1,i} = \frac{\partial}{\partial x_i} h(\mathbf{x}_r) \tag{6.107}$$

$$m_{r+1,i} = \beta_1 m_{r,i} + (1-\beta_1) g_{r+1,i} \tag{6.108}$$

$$v_{r+1,i} = \text{abs}(\hat{m}_{r+1,i}) \tag{6.109}$$

$$\hat{v}_{r+1,i} = \max(v_{r+1,i}, \hat{v}_{r,i}) \tag{6.110}$$

$$x_{r+1,i} = x_{r,i} - \alpha \frac{\hat{m}_{r+1,i}}{\sqrt{\hat{v}_{r+1,i}} + \epsilon}. \tag{6.111}$$

6.3.4 Conjugate Gradient Method

This is also a two-step method where the search direction is iterated according to

$$\mathbf{s}_0 = -\mathbf{g}_0 \tag{6.112}$$

$$\mathbf{s}_{r+1} = -\mathbf{g}_{r+1} + \beta_{r+1} \mathbf{s}_r \tag{6.113}$$

and the step length is chosen from a full or inexact linesearch

$$\lambda_r = \text{argmin}_\lambda h(\mathbf{x}_r + \lambda \mathbf{s}_r) \tag{6.114}$$

to obtain

$$\mathbf{x}_{r+1} = \mathbf{x}_r + \lambda_r \mathbf{s}_r. \tag{6.115}$$

The simplest choice for β is made by Fletcher and Rieves [106]

$$\beta_{r+1} = \frac{g_{r+1}^2}{g_r^2}. \tag{6.116}$$

This method was devised to minimize a quadratic function and to solve the related system of linear equations, but it is also very efficient for more complicated functions (Sect. 5.6.4).

6.3.5 Newton-Raphson Method

The Newton-Raphson method uses the iteration.[6]

[6] This solves $\mathbf{g}(\mathbf{x}) = 0$ with the first order NR method 6.1.3.

$$\mathbf{x}_{r+1} = \mathbf{x}_r - H(\mathbf{x}_r)^{-1}\mathbf{g}(\mathbf{x}_r). \tag{6.117}$$

This corresponds to the search direction

$$\mathbf{s_r} = H_r^{-1}\mathbf{g_r} \tag{6.118}$$

and the step length $\lambda = 1$.[7] This method converges fast if the starting point is close to the minimum. However, calculation of the Hessian can be very time consuming (Figs. 6.25 and 6.26).

6.3.6 Quasi-Newton Methods

Calculation of the full Hessian matrix as needed for the Newton-Raphson method can be very time consuming. Quasi-Newton methods use instead an approximation to the Hessian which is updated during each iteration. From the Taylor series

$$h(\mathbf{x}) = h_0 + \mathbf{b}^T\mathbf{x} + \frac{1}{2}\mathbf{x}^T H\mathbf{x} + \cdots \tag{6.119}$$

we obtain the gradient

$$\mathbf{g}(\mathbf{x}_r) = \mathbf{b} + H\mathbf{x}_r + \cdots = \mathbf{g}(\mathbf{x}_{r-1}) + H(\mathbf{x}_r - \mathbf{x}_{r-1}) + \cdots . \tag{6.120}$$

Defining the differences

$$\mathbf{d}_r = \mathbf{x}_{r+1} - \mathbf{x}_r \tag{6.121}$$

$$\mathbf{y}_r = \mathbf{g}_{r+1} - \mathbf{g}_r \tag{6.122}$$

and neglecting higher order terms we obtain the quasi-Newton or secant condition

$$H\mathbf{d}_r = \mathbf{y}_r. \tag{6.123}$$

We want to approximate the true Hessian by a series of matrices H_r which are updated during each iteration to sum up all the information gathered so far. Since the Hessian is symmetric and positive definite, this also has to be demanded for the H_r.[8] This cannot be achieved by a rank one update matrix. Popular methods use a symmetric rank two update of the form

---

[7] This is not obligatory, the NR method can be combined with a line search to improve performance.

[8] This is a major difference to the Quasi Newton methods for root finding (6.1.9).

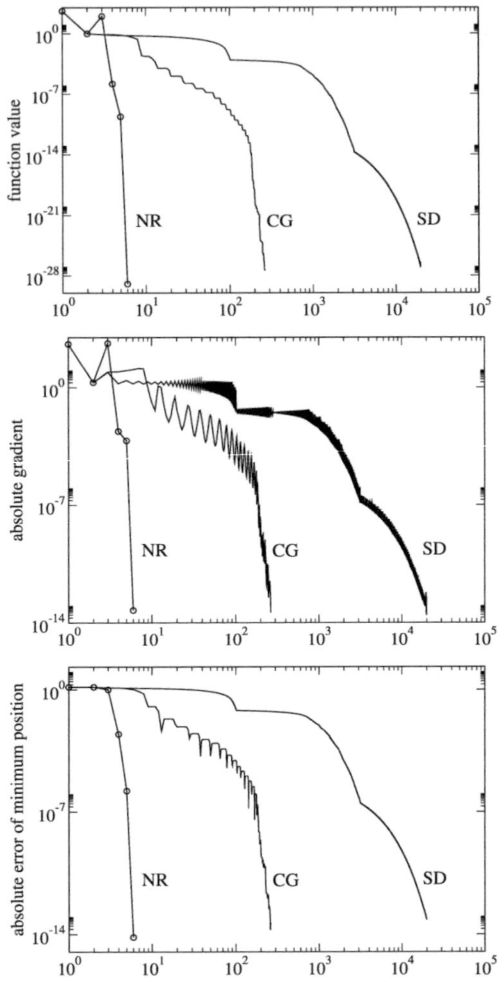

Fig. 6.25 Function minimization. The minimum of the Rosenbrock function $h(x, y) = 100(y - x^2)^2 + (1 - x)^2$ is determined with different methods. Conjugate gradients converge much faster than steepest descent. Starting at $(x, y) = (0, 2)$, Newton-Raphson reaches the minimum at $x = y = 1$ within only 5 iterations to machine precision

$$H_{r+1} = H_r + \alpha \mathbf{u}\mathbf{u}^T + \beta \mathbf{v}\mathbf{v}^T. \tag{6.124}$$

The Quasi-Newton condition then gives

$$H_{r+1}\mathbf{d}_r = H_r\mathbf{d}_r + \alpha(\mathbf{u}^T\mathbf{d}_r)\mathbf{u} + \beta(\mathbf{v}^T\mathbf{d}_r)\mathbf{v} = \mathbf{y}_r \tag{6.125}$$

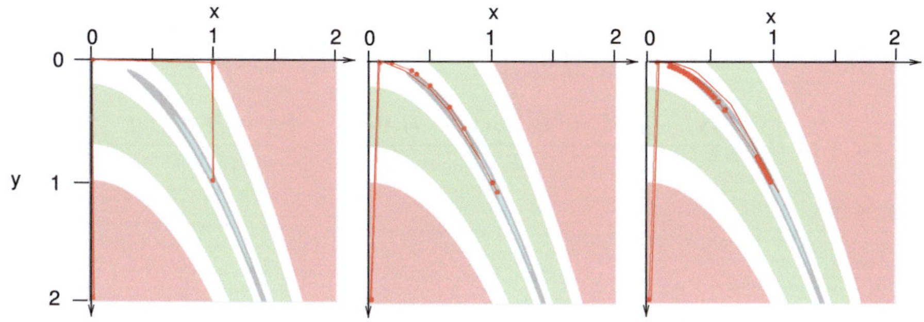

Fig. 6.26 Minimization of the Rosenbrock function. **Left:** Newton-Raphson finds the minimum after 5 steps within machine precision. **Middle:** Conjugate gradients reduce the gradient to 4×10^{-14} after 265 steps. **Right:** The steepest descent method needs 20000 steps to reduce the gradient to 5×10^{-14}. Red lines show the minimization pathway. Colored areas indicate the function value (light blue: < 0.1, grey: 0.1..0.5, green: 5..50, pink: > 100). Screen shots taken from Exercise 6.2

hence $H_r\mathbf{d}_r - \mathbf{y}_r$ must be a linear combination of $\mathbf{u}$ and $\mathbf{v}$. Making the simple choice

$$\mathbf{u} = \mathbf{y}_r \quad \mathbf{v} = H_r\mathbf{d}_r \tag{6.126}$$

and assuming that these two vectors are linearly independent, we find

$$\beta = -\frac{1}{(\mathbf{v}^T\mathbf{d}_r)} \quad = -\frac{1}{(\mathbf{d}_r^T H_r\mathbf{d}_r)} \tag{6.127}$$

$$\alpha = \frac{1}{(\mathbf{u}^T\mathbf{d}_r)} = \frac{1}{(\mathbf{y}_r^T\mathbf{d}_r)} \tag{6.128}$$

which together defines the very popular BFGS (Broyden, Fletcher, Goldfarb, Shanno) method [107–110]

$$H_{r+1} = H_r + \frac{\mathbf{y}_r\mathbf{y}_r^T}{\mathbf{y}_r^T\mathbf{d}_r} - \frac{(H_r\mathbf{d}_r)(H_r\mathbf{d}_r)^T}{\mathbf{d}_r^T H_r\mathbf{d}_r}. \tag{6.129}$$

Alternatively the DFP method by Davidson,Fletcher and Powell, directly updates the inverse Hessian matrix $B = H^{-1}$ according to

$$B_{r+1} = B_r + \frac{\mathbf{d}_r\mathbf{d}_r^T}{\mathbf{y}_r^T\mathbf{d}_r} - \frac{(B_r\mathbf{y}_r)(B_r\mathbf{y}_r)^T}{\mathbf{y}_r^T B_r\mathbf{y}_r.} \tag{6.130}$$

Both of these methods can be inverted with the help of the Sherman-Morrison formula to give

$$B_{r+1} = B_r + \frac{(\mathbf{d}_r - B_r\mathbf{y}_r)\mathbf{d}_r^T + \mathbf{d}_r(\mathbf{d}_r - B\mathbf{y}_r)^T}{\mathbf{y}_r^T\mathbf{d}_r} - \frac{(\mathbf{d}_r - B_r\mathbf{y}_r)^T\mathbf{y}}{(\mathbf{y}_r^T\mathbf{d})^2}\mathbf{d}\mathbf{d}^T \quad (6.131)$$

$$H_{r+1} = H_r + \frac{(\mathbf{y}_r - H_r\mathbf{d}_r)\mathbf{y}_r^T + \mathbf{y}_r(\mathbf{y}_r - H_r\mathbf{d}_r)^T}{\mathbf{y}_r^T\mathbf{d}_r} - \frac{(\mathbf{y}_r - H_r\mathbf{d}_r)^T\mathbf{d}_r}{(\mathbf{y}_r^T\mathbf{d}_r)^2}\mathbf{y}_r\mathbf{y}_r^T. \quad (6.132)$$

Exercises

Exercise 6.1 Root Finding Methods

This computer experiment searches roots of several test functions:

$f(x) = x^n - 2 \quad n = 1, 2, 3, 4$ (Fig. 6.10)
$f(x) = 5\sin(5x)$
$f(x) = (\cos(2x))^2 - x^2$
$f(x) = 5\left(\sqrt{|x+2|} - 1\right)$
$f(x) = e^{-x}\ln x$
$f(x) = (x-1)^3$ (Fig. 6.11)
$f(x) = x^{25}$ (Fig. 6.12)

You can vary the initial interval or starting value and compare the behavior of different methods:

- bisection
- regula falsi
- Dekker's method
- Brent's method
- Chandrupatla's method
- Newton-Raphson method

Exercise 6.2 Stationary Points

This computer experiment searches a local minimum of the Rosenbrock function[9]

$$h(x, y) = 100(y - x^2)^2 + (1 - x)^2. \quad (6.133)$$

- The method of steepest descent minimizes $h(x, y)$ along the search direction

$$s_x^{(n)} = -g_x^{(n)} = -400x(x_n^2 - y_n) - 2(x_n - 1) \quad (6.134)$$

[9] A well known test function for minimization algorithms.

$$s_y^{(n)} = -g_y^{(n)} = -200(y_n - x_n^2). \tag{6.135}$$

- Conjugate gradients make use of the search direction

$$s_x^{(n)} = -g_x^{(n)} + \beta_n s_x^{(n-1)} \tag{6.136}$$

$$s_y^{(n)} = -g_y^{(n)} + \beta_n s_y^{(n-1)}. \tag{6.137}$$

- The Newton-Raphson method needs the inverse Hessian

$$H^{-1} = \frac{1}{\det(H)} \begin{pmatrix} h_{yy} & -h_{xy} \\ xy & h_{xx} \end{pmatrix} \tag{6.138}$$

$$\det(H) = h_{xx}h_{yy} - h_{xy}^2 \tag{6.139}$$

$$h_{xx} = 1200x^2 - 400y + 2 \quad h_{yy} = 200 \quad h_{xy} = -400x \tag{6.140}$$

and iterates according to

$$\begin{pmatrix} x_{n+1} \\ y_{n+1} \end{pmatrix} = \begin{pmatrix} x_n \\ y_n \end{pmatrix} - H^{-1} \begin{pmatrix} g_x^n \\ q_y^n \end{pmatrix}. \tag{6.141}$$

You can choose an initial point (x_0, y_0). The iteration stops if the gradient norm falls below 10^{-14} or if the line search fails to find a lower function value.

Chapter 7
Fourier Transformation

Fourier transformation is a very important tool for signal analysis but also helpful to simplify the solution of differential equations or the calculation of convolution integrals. In this chapter we discuss the discrete Fourier transformation as a numerical approximation to the continuous Fourier integral. It can be realized efficiently by Goertzel's algorithm or the family of fast Fourier transformation methods. Computer experiments demonstrate trigonometric interpolation and nonlinear filtering as applications.

7.1 Fourier Integral and Fourier Series

We use the symmetric definition of the Fourier transformation:

$$\tilde{f}(\omega) = \mathcal{F}[f](\omega) = \frac{1}{\sqrt{2\pi}} \int_{-\infty}^{\infty} f(t) e^{-i\omega t} dt. \tag{7.1}$$

The inverse Fourier transformation

$$f(t) = \mathcal{F}^{-1}[\tilde{f}](t) = \frac{1}{\sqrt{2\pi}} \int_{-\infty}^{\infty} \tilde{f}(\omega) e^{i\omega t} d\omega \tag{7.2}$$

Supplementary Information The online version contains supplementary material available at https://doi.org/10.1007/978-3-032-07856-8_7.

P. O. J. Scherer, *Computational Physics I*, Graduate Texts in Physics,
https://doi.org/10.1007/978-3-032-07856-8_7

decomposes $f(t)$ into a superposition of oscillations. The Fourier transform of a convolution integral

$$g(t) = f(t) \otimes h(t) = \int_{-\infty}^{\infty} f(t')h(t-t')dt' \tag{7.3}$$

becomes a product of Fourier transforms:

$$\begin{aligned}\widetilde{g}(\omega) &= \frac{1}{\sqrt{2\pi}} \int_{-\infty}^{\infty} dt' f(t')\mathrm{e}^{-\mathrm{i}\omega t'} \int_{-\infty}^{\infty} h(t-t')\mathrm{e}^{-\mathrm{i}\omega(t-t')} d(t-t') \\ &= \sqrt{2\pi}\, \widetilde{f}(\omega)\widetilde{h}(\omega).\end{aligned} \tag{7.4}$$

A periodic function with $f(t+T) = f(t)$[1] is transformed into a Fourier series

$$f(t) = \sum_{n=-\infty}^{\infty} \mathrm{e}^{\mathrm{i}\omega_n t} \hat{f}(\omega_n) \text{ with } \omega_n = n\frac{2\pi}{T}, \quad \hat{f}(\omega_n) = \frac{1}{T}\int_0^T f(t)\mathrm{e}^{-\mathrm{i}\omega_n t} dt. \tag{7.5}$$

For a periodic function which in addition is real valued $f(t) = f(t)^*$ and even $f(t) = f(-t)$, the Fourier series becomes a cosine series

$$f(t) = \hat{f}(\omega_0) + 2\sum_{n=1}^{\infty} \hat{f}(\omega_n) \cos \omega_n t \tag{7.6}$$

with real valued coefficients

$$\hat{f}(\omega_n) = \frac{1}{T}\int_0^T f(t) \cos \omega_n t \, dt. \tag{7.7}$$

7.2 Discrete Fourier Transformation

We divide the time interval $0 \le t < T$ by introducing a grid of N equidistant points

$$t_n = n\Delta t = n\frac{T}{N} \quad \text{with } n = 0, 1, \ldots N-1. \tag{7.8}$$

The function values (samples)

$$f_n = f(t_n) \tag{7.9}$$

are arranged as components of a vector

$$\mathbf{f} = \begin{pmatrix} f_0 \\ \vdots \\ f_{N-1} \end{pmatrix}.$$

[1] This could also be the periodic continuation of a function which is only defined for $0 < t < T$.

With respect to the orthonormal basis

$$\mathbf{e}_n = \begin{pmatrix} \delta_{0,n} \\ \vdots \\ \delta_{N-1,n} \end{pmatrix}, \quad n = 0, 1, \cdots N-1 \tag{7.10}$$

f is expressed as a linear combination

$$\mathbf{f} = \sum_{n=0}^{N-1} f_n \mathbf{e}_n. \tag{7.11}$$

The discrete Fourier transformation is the transformation to an orthogonal base in frequency space

$$\mathbf{e}_{\omega_j} = \sum_{n=0}^{N-1} \mathrm{e}^{\mathrm{i}\omega_j t_n} \mathbf{e}_n = \begin{pmatrix} 1 \\ \mathrm{e}^{\mathrm{i}\frac{2\pi}{N}j} \\ \vdots \\ \mathrm{e}^{\mathrm{i}\frac{2\pi}{N}j(N-1)} \end{pmatrix} \tag{7.12}$$

with

$$\omega_j = \frac{2\pi}{T} j. \tag{7.13}$$

These vectors are orthogonal

$$\mathbf{e}_{\omega_j}\mathbf{e}^*_{\omega_{j'}} = \sum_{n=0}^{N-1} \mathrm{e}^{\mathrm{i}(j-j')\frac{2\pi}{N}n} = \begin{cases} \frac{1-\mathrm{e}^{\mathrm{i}(j-j')2\pi}}{1-\mathrm{e}^{\mathrm{i}(j-j')2\pi/N}} = 0 & for \quad j-j' \neq 0 \\ N & for \quad j-j' = 0 \end{cases} \tag{7.14}$$

$$\mathbf{e}_{\omega_j}\mathbf{e}^*_{\omega_{j'}} = N\delta_{j,j'}. \tag{7.15}$$

Alternatively a real valued basis can be defined:

$$\begin{aligned} &\cos(\frac{2\pi}{N} jn) \qquad j = 0, 1, \cdots j_{\max} \\ &\sin(\frac{2\pi}{N} jn) \qquad j = 1, 2 \cdots j_{\max} \\ &j_{\max} = \frac{N}{2} \text{ (even N)} \quad j_{\max} = \frac{N-1}{2} \text{ (odd N)}. \end{aligned} \tag{7.16}$$

The components of **f** in frequency space are given by the scalar product

$$\widetilde{f}_{\omega_j} = \mathbf{f}\mathbf{e}_{\omega_j} = \sum_{n=0}^{N-1} f_n \mathrm{e}^{-\mathrm{i}\omega_j t_n} = \sum_{n=0}^{N-1} f_n \mathrm{e}^{-\mathrm{i}j\frac{2\pi}{T}n\frac{T}{N}} = \sum_{n=0}^{N-1} f_n \mathrm{e}^{-\mathrm{i}\frac{2\pi}{N}jn}. \tag{7.17}$$

This can be written in matrix notation as[2]

$$\tilde{\mathbf{f}} = \begin{pmatrix} 1 & 1 & 1 & \cdots & 1 \\ 1 & \omega & \omega^2 & \cdots & \omega^{N-1} \\ 1 & \omega^2 & \omega^4 & \cdots & \omega^{2(N-1)} \\ \vdots & & & \ddots & \vdots \\ 1 & \omega^{N-1} & \omega^{2(N-1)} & \cdots & \omega^{(N-1)(N-1)} \end{pmatrix} \mathbf{f} \tag{7.18}$$

where

$$\omega = \mathrm{e}^{-2\pi \mathrm{i}/N} \tag{7.19}$$

is a primitive Nth root of unity.

From

$$\sum_{j=0}^{N-1} \widetilde{f}_{\omega_j} \mathrm{e}^{\mathrm{i}\omega_j t_n} = \sum_{n'} \sum_{\omega_j} f_{n'} \mathrm{e}^{-\mathrm{i}\omega_j t_{n'}} \mathrm{e}^{\mathrm{i}\omega_j t_n} = N f_n \tag{7.20}$$

we find the inverse transformation

$$f_n = \frac{1}{N} \sum_{j=0}^{N-1} \widetilde{f}_{\omega_j} \mathrm{e}^{\mathrm{i}\omega_j t_n} = \frac{1}{N} \sum_{j=0}^{N-1} \widetilde{f}_{\omega_j} \mathrm{e}^{\mathrm{i}\frac{2\pi}{N} nj}. \tag{7.21}$$

7.2.1 Trigonometric Interpolation

The last equation can be interpreted as an interpolation of the function $f(t)$ at the sampling points t_n by a linear combination of trigonometric functions

$$f(t) = \frac{1}{N} \sum_{j=0}^{N-1} \widetilde{f}_{\omega_j} \left(\mathrm{e}^{\mathrm{i}\frac{2\pi}{T} t}\right)^j \tag{7.22}$$

which is a polynomial of the root of unity

$$q = \mathrm{e}^{\mathrm{i}\frac{2\pi}{T} t}. \tag{7.23}$$

Since

$$\mathrm{e}^{-\mathrm{i}\omega_j t_n} = \mathrm{e}^{-\mathrm{i}\frac{2\pi}{N} jn} = \mathrm{e}^{\mathrm{i}\frac{2\pi}{N}(N-j)n} = \mathrm{e}^{\mathrm{i}\omega_{N-j} t_n} \tag{7.24}$$

[2] This matrix is the Vandermonde matrix for the roots of unity. It is known as the DFT matrix or Fourier matrix. Often a prefactor of $N^{-1/2}$is introduced which makes DFT and its inverse more symmetric and provides a unitary DFT matrix.

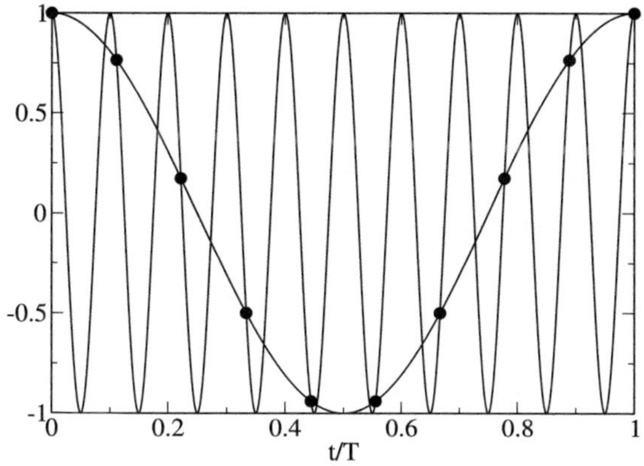

Fig. 7.1 Equivalence of ω_1 and ω_{N-1}. The two functions $\cos \omega t$ and $\cos(N-1)\omega t$ have the same values at the sample points t_n but are very different in between

the frequencies ω_j and ω_{N-j} are equivalent (Fig. 7.1)

$$\widetilde{f}_{\omega_{N-j}} = \sum_{n=0}^{N-1} f_n \mathrm{e}^{-\mathrm{i}\frac{2\pi}{N}(N-j)n} = \sum_{n=0}^{N-1} f_n \mathrm{e}^{\mathrm{i}\frac{2\pi}{N}jn} = \widetilde{f}_{\omega_{-j}}. \tag{7.25}$$

If we use trigonometric interpolation to approximate $f(t)$ between the grid points, the two frequencies are no longer equivalent and we have to restrict the frequency range to avoid unphysical high frequency components (Fig. 7.2):

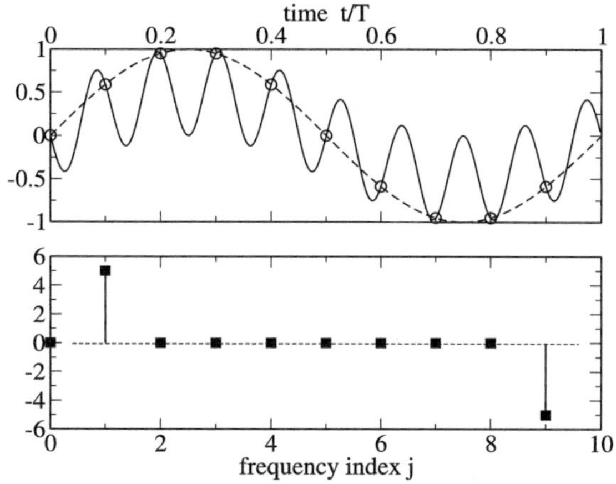

Fig. 7.2 Trigonometric interpolation. For trigonometric interpolation the high frequencies have to be replaced by the corresponding negative frequencies to provide meaningful results between the sampling points. The circles show sampling points which are fitted using only positive frequencies (full curve) or replacing the unphysical high frequency by its negative counterpart (broken curve). The squares show the calculated Fourier spectrum. See also Exercise 7.1

$$-\frac{2\pi}{T}\frac{N-1}{2} \leq \omega_j \leq \frac{2\pi}{T}\frac{N-1}{2} \qquad N \text{ odd}$$

$$-\frac{2\pi}{T}\frac{N}{2} \leq \omega_j \leq \frac{2\pi}{T}(\frac{N}{2}-1) \qquad N \text{ even.} \tag{7.26}$$

The interpolating function (N even) is

$$f(t) = \frac{1}{N}\sum_{j=-\frac{N}{2}}^{\frac{N}{2}-1} \widetilde{f}_{\omega_j} e^{i\omega_j t} \quad N \text{ even} \tag{7.27}$$

$$f(t) = \frac{1}{N}\sum_{j=-\frac{N-1}{2}}^{\frac{N-1}{2}} \widetilde{f}_{\omega_j} e^{i\omega_j t} \quad N \text{ odd.} \tag{7.28}$$

The maximum frequency is

$$\omega_{\max} = \frac{2\pi}{T}\frac{N}{2} \tag{7.29}$$

and hence

$$f_{\max} = \frac{1}{2\pi}\omega_{\max} = \frac{N}{2T} = \frac{f_s}{2} \tag{7.30}$$

This is known as the sampling theorem which states that the sampling frequency f_s must be larger than twice the maximum frequency present in the signal.

7.2.2 Real Valued Functions

For a real valued function

$$f_n = f_n^* \tag{7.31}$$

and hence

$$\widetilde{f}^*_{\omega_j} = \left(\sum_{n=0}^{N-1} f_n e^{-i\omega_j t_n}\right)^* = \sum_{n=0}^{N-1} f_n e^{i\omega_j t_n} = \widetilde{f}_{\omega_{-j}}. \tag{7.32}$$

Here it is sufficient to calculate the sums for $j = 0, \cdots N/2$. If the function is real valued and also even

$$f_{-n} = f_n \tag{7.33}$$

$$\widetilde{f}_{\omega_j} = \sum_{n=0}^{N-1} f_{-n} e^{-i\omega_j t_n} = \sum_{n=0}^{N-1} f_n e^{-i(-\omega_j)t_n} = \widetilde{f}_{\omega_{-j}} \tag{7.34}$$

and the Fourier sum (7.21) turns into a cosine sum

$$f_n = \frac{1}{2M-1}\tilde{f}_{\omega_0} + \frac{2}{2M-1}\sum_{j=1}^{M-1}\tilde{f}_{\omega_j}\cos\left(\frac{2\pi}{2M-1}jn\right) \quad \text{odd } N = 2M-1 \tag{7.35}$$

$$f_n = \frac{1}{2M}\tilde{f}_{\omega_0} + \frac{1}{M}\sum_{j=1}^{M-1}\tilde{f}_{\omega_j}\cos\left(\frac{\pi}{M}jn\right) + \frac{1}{2M}\tilde{f}_{\omega_M}\cos(n\pi) \quad \text{even } N = 2M \tag{7.36}$$

which correspond to two out of eight different versions [111] of the discrete cosine transformation [112, 113].

Equation (7.36) can be used to define the interpolating function

$$f(t) = \frac{1}{2M}\tilde{f}_{\omega_0} + \frac{1}{M}\sum_{j=1}^{M-1}\tilde{f}_{\omega_j}\cos\left(\omega_j t\right) + \frac{1}{2M}\tilde{f}_{\omega_M}\cos(\frac{2\pi M}{T}t). \tag{7.37}$$

The real valued Fourier coefficients are given by

$$\tilde{f}_{\omega_j} = f_0 + 2\sum_{n=1}^{M-1} f_n\cos(\omega_j t_n) \quad \text{odd } N = 2M-1 \tag{7.38}$$

$$\tilde{f}_{\omega_j} = f_0 + 2\sum_{n=1}^{M-1} f_n\cos(\omega_j t_n) + f_M\cos(j\pi) \quad \text{even } N = 2M. \tag{7.39}$$

7.2.3 Approximate Continuous Fourier Transformation

We continue the function $f(t)$ periodically by setting

$$f_N = f_0 \tag{7.40}$$

and write

$$\tilde{f}_{\omega_j} = \sum_{n=0}^{N-1} f_n e^{-i\omega_j n} = \frac{1}{2}f_0 + e^{-i\omega_j}f_1 + \cdots e^{-i\omega_j(N-1)}f_{N-1} + \frac{1}{2}f_N. \tag{7.41}$$

Comparing with the trapezoidal rule (4.13) for the integral

$$\int_0^T e^{-i\omega_j t}f(t)dt \approx \frac{T}{N}(\frac{1}{2}e^{-i\omega_j 0}f(0) + e^{-i\omega_j\frac{T}{N}}f(\frac{T}{N})$$

$$+\cdots + e^{-i\omega_j\frac{T}{N}(N-1)}f(\frac{T}{N}(N-1)) + \frac{1}{2}f(T)) \tag{7.42}$$

we find

$$\hat{f}(\omega_j) = \frac{1}{T}\int_0^T e^{-i\omega_j t} f(t)dt \approx \frac{1}{N}\widetilde{f}_{\omega_j} \tag{7.43}$$

which shows that the discrete Fourier transformation is an approximation to the Fourier series of a periodic function with period T which coincides with $f(t)$ in the interval $0 < t < T$. The range of the integral can be formally extended to $\pm\infty$ by introducing a windowing function

$$W(t) = \begin{cases} 1 & for \quad 0 < t < T \\ 0 & else \end{cases} . \tag{7.44}$$

The discrete Fourier transformation approximates the continuous Fourier transformation but windowing leads to a broadening of the spectrum (see Sect. 8.1). For practical purposes smoother windowing functions are used like a triangular window or one of the following [114]:

$$W(t_n) = e^{-\frac{1}{2}\left(\frac{n-(N-1)/2}{\sigma(N-1)/2}\right)^2} \quad \sigma \leq 0.5 \quad \text{Gaussian window}$$

$$W(t_n) = 0.53836 - 0.46164 \cos\left(\frac{2\pi n}{N-1}\right) \quad \text{Hamming window}$$

$$W(t_n) = 0.5\left(1 - \cos\left(\frac{2\pi n}{N-1}\right)\right) \quad \text{Hann window.}$$

7.3 Fourier Transform Algorithms

Straight Forward evaluation of the sum

$$\widetilde{f}_{\omega_j} = \sum_{n=0}^{N-1} \cos\left(\frac{2\pi}{N}jn\right) f_n + i\sin\left(\frac{2\pi}{N}jn\right) f_n \tag{7.45}$$

needs $O(N^2)$ additions, multiplications and trigonometric functions.

7.3.1 Goertzel's Algorithm

Goertzel's method [115] is very useful if not the whole Fourier spectrum is needed but only some of the Fourier components, for instance to demodulate a frequency shift key signal or the dial tones which are used in telephony.

The Fourier transform can be written as

$$\sum_{n=0}^{N-1} f_n \mathrm{e}^{-\mathrm{i}\frac{2\pi}{N}jn} = f_0 + \mathrm{e}^{-\frac{2\pi\mathrm{i}}{N}j}(f_1 + \mathrm{e}^{-\frac{2\pi\mathrm{i}}{N}j} f_2 \cdots (f_{N-2} + \mathrm{e}^{-\frac{2\pi\mathrm{i}}{N}j} f_{N-1})\cdots) \quad (7.46)$$

which can be evaluated recursively

$$y_{N-1} = f_{N-1}$$

$$y_n = f_n + \mathrm{e}^{-\frac{2\pi\mathrm{i}}{N}j} y_{n+1} \quad n = N-2, \cdots 0 \quad (7.47)$$

to give the result

$$\hat{f}_{\omega_j} = y_0. \quad (7.48)$$

Equation (7.47) is a simple discrete filter function. Its transmission function is obtained by application of the z-transform [116]

$$u(z) = \sum_{n=0}^{\infty} u_n z^{-n} \quad (7.49)$$

(the discrete version of the Laplace transform) which yields

$$y(z) = \frac{f(z)}{1 - z\mathrm{e}^{-\frac{2\pi\mathrm{i}}{N}j}}. \quad (7.50)$$

One disadvantage of this method is that it uses complex numbers. This can be avoided by the following more complicated recursion

$$u_{N+1} = u_N = 0$$

$$u_n = f_n + 2u_{n+1}\cos\frac{2\pi}{N}k - u_{n+2} \text{ for } n = N-1, \cdots 0 \quad (7.51)$$

with the transmission function

$$\frac{u(z)}{f(z)} = \frac{1}{1 - z\left(\mathrm{e}^{\frac{2\pi\mathrm{i}}{N}j} + \mathrm{e}^{-\frac{2\pi\mathrm{i}}{N}j}\right) + z^2}$$

$$= \frac{1}{\left(1 - z\mathrm{e}^{-\frac{2\pi\mathrm{i}}{N}j}\right)\left(1 - z\mathrm{e}^{\frac{2\pi\mathrm{i}}{N}j}\right)}. \quad (7.52)$$

A second filter removes one factor in the denominator

$$\frac{y(z)}{u(z)} = \left(1 - z e^{\frac{2\pi i}{N} j}\right) \tag{7.53}$$

which in the time domain corresponds to the simple expression

$$y_n = u_n - e^{\frac{2\pi i}{N} j} u_{n+1}.$$

The overall filter function finally again is (7.50).

$$\frac{y(z)}{f(z)} = \frac{1}{1 - z e^{-\frac{2\pi i}{N} j}}. \tag{7.54}$$

Hence the Fourier component of $\mathbf{f}$ is given by

$$\hat{f}_{\omega_j} = y_0 = u_0 - e^{\frac{2\pi i}{N} j} u_1. \tag{7.55}$$

The order of the iteration (7.46) can be reversed by writing

$$\hat{f}_{\omega_j} = f_0 \cdots e^{\frac{2\pi i}{N}(N-1)} f_{N-1} = e^{-\frac{2\pi i}{N} j(N-1)} (f_0 e^{\frac{2\pi i}{N} j(N-1)} \cdots f_{N-1}) \tag{7.56}$$

which is very useful for real time filter applications.

7.3.2 *Fast Fourier Transformation*

If the number of samples is $N = 2^p$, the Fourier transformation can be performed very efficiently by this method (Algorithms 11 and 12).[3] The phase factor

$$e^{-i\frac{2\pi}{N} jm} = W_N^{jm} \tag{7.57}$$

can take only N different values. The number of trigonometric functions can be reduced by reordering the sum. Starting from a sum with N samples

$$F_N(f_0 \cdots f_{N-1}) = \sum_{n=0}^{N-1} f_n W_N^{jn} \tag{7.58}$$

[3] There exist several Fast Fourier Transformation algorithms [117, 118]. We consider only the simplest one here [119].

we separate even and odd powers of the unit root

$$F_N(f_0 \cdots f_{N-1}) = \sum_{m=0}^{\frac{N}{2}-1} f_{2m} W_N^{j2m} + \sum_{m=0}^{\frac{N}{2}-1} f_{2m+1} W_N^{j(2m+1)}$$

$$= \sum_{m=0}^{\frac{N}{2}-1} f_{2m} e^{-i\frac{2\pi}{N/2}jm} + W_N^j \sum_{m=0}^{\frac{N}{2}-1} f_{2m+1} e^{-i\frac{2\pi}{N/2}jm}$$

$$= F_{N/2}(f_0, f_2 \cdots f_{N-2}) + W_N^j F_{N/2}(f_1, f_3 \cdots f_{N-1}). \tag{7.59}$$

This division is repeated until only sums with one summand remain

$$F_1(f_n) = f_n. \tag{7.60}$$

For example, consider the case $N = 8$:

$$F_8(f_0 \cdots f_7) = F_4(f_0 f_2 f_4 f_6) + W_8^j F_4(f_1 f_3 f_5 f_7)$$

$$---$$

$$F_4(f_0 f_2 f_4 f_6) = F_2(f_0 f_4) + W_4^j F_2(f_2 f_6)$$
$$F_4(f_1 f_3 f_5 f_7) = F_2(f_1 f_5) + W_4^j F_2(f_3 f_7)$$

$$---$$

$$F_2(f_0 f_4) = f_0 + W_2^j f_4$$
$$F_2(f_2 f_6) = f_2 + W_2^j f_6$$
$$F_2(f_1 f_5) = f_1 + W_2^j f_5$$
$$F_2(f_3 f_7) = f_3 + W_2^j f_7.$$

Expansion gives

$$F_8 = f_0 + W_2^j f_4 + W_4^j f_2 + W_4^j W_2^j f_6$$

$$+ W_8^j f_1 + W_8^j W_2^j f_5 + W_8^j W_4^j f_3 + W_8^j W_4^j W_2^j f_7. \tag{7.61}$$

Generally a summand of the Fourier sum can be written using the binary representation of n

$$n = \sum l_i \quad l_i = 1, 2, 4, 8 \cdots \tag{7.62}$$

in the following way:

$$f_n e^{-i\frac{2\pi}{N}jn} = f_n e^{-i\frac{2\pi}{N}(l_1+l_2+\cdots)j} = f_n W_{N/l_1}^j W_{N/l_2}^j \cdots. \tag{7.63}$$

The function values have to be reordered (Algorithm 11) to provide the input for the FFT (Algorithm 12)

Algorithm 11: Reordering of the function values

count from 0 to N-1 using binary numbers $m = 000, 001, 010, \cdots$;
bit reversal gives the binary numbers $n = 000, 100, 010, \cdots$;
$s_m \leftarrow f_n$;

Algorithm 12: Fast Fourier Transformation

```
Input: Data[k] are the input data in bit reversed order
Size ← 2 ;
First ← 0 ;
while First < NumberOfSamples do
    for N = 0 to Size/2 − 1 do
        J ← First + N;
        K ← J + Size/2 − 1;
        T ← exp{−2π i N/NumberOfSamples} ∗ Data[k];
        Data[J] ← Data[J] + T;
        Data[K] ← Data[K] − T;
    end
    First ← First ∗ 2;
    Size ← Size ∗ 2;
end
```

As an example for N=8 the function values are in the order

$$\begin{pmatrix} s_0 \\ s_1 \\ s_2 \\ s_3 \\ s_4 \\ s_5 \\ s_6 \\ s_7 \end{pmatrix} = \begin{pmatrix} f_0 \\ f_4 \\ f_2 \\ f_6 \\ f_1 \\ f_5 \\ f_3 \\ f_7 \end{pmatrix}. \tag{7.64}$$

Now calculate sums with two summands. Since W_2^j can take only two different values

$$W_2^j = \begin{cases} 1 \; for \; j = 0, 2, 4, 6 \\ -1 \; for \; j = 1, 3, 5, 7 \end{cases} \tag{7.65}$$

a total of 8 sums have to be calculated which can be stored again in the same workspace:

$$\begin{pmatrix} f_0 + f_4 \\ f_0 - f_4 \\ f_2 + f_6 \\ f_2 - f_6 \\ f_1 + f_5 \\ f_1 - f_5 \\ f_3 + f_7 \\ f_3 - f_7 \end{pmatrix} = \begin{pmatrix} s_0 + W_2^0 s_1 \\ s_0 + W_2^1 s_1 \\ s_2 + W_2^2 s_3 \\ s_2 + W_2^3 s_3 \\ s_4 + W_2^4 s_5 \\ s_4 + W_2^5 s_5 \\ s_6 + W_2^6 s_7 \\ s_6 + W_2^7 s_7 \end{pmatrix}. \tag{7.66}$$

Next calculate sums with four summands. W_4^j can take one of four values

$$W_4^j = \begin{cases} 1 & for \quad j = 0, 4 \\ -1 & for \quad j = 2, 6 \\ W_4 & for \quad j = 1, 5 \\ -W_4 & for \quad j = 3, 7 \end{cases}. \tag{7.67}$$

The following combinations are needed:

$$\begin{pmatrix} f_0 + f_4 + (f_2 + f_6) \\ f_0 + f_4 - (f_2 + f_6) \\ (f_0 - f_4) + W_4(f_2 - f_6) \\ (f_0 - f_4) - W_4(f_2 - f_6) \\ f_1 + f_5 + (f_3 + f_7) \\ f_1 + f_5 - (f_3 + f_7) \\ (f_1 - f_5) \pm W_4(f_3 - f_7) \\ (f_1 - f_5) \pm W_4(f_3 - f_7) \end{pmatrix} = \begin{pmatrix} s_0 + W_4^0 s_2 \\ s_1 + W_4^1 s_3 \\ s_0 + W_4^2 s_2 \\ s_1 + W_4^3 s_3 \\ s_4 + W_4^4 s_6 \\ s_5 + W_4^5 s_7 \\ s_4 + W_4^6 s_6 \\ s_5 + W_4^7 s_7 \end{pmatrix}. \tag{7.68}$$

The next step gives the sums with eight summands. With

$$W_8^j = \begin{cases} 1 & j = 0 \\ W_8 & j = 1 \\ W_8^2 & j = 2 \\ W_8^3 & j = 3 \\ -1 & j = 4 \\ -W_8 & j = 5 \\ -W_8^2 & j = 6 \\ -W_8^3 & j = 7 \end{cases} \tag{7.69}$$

we calculate

$$\begin{pmatrix} f_0 + f_4 + (f_2 + f_6) + (f_1 + f_5 + (f_3 + f_7)) \\ f_0 + f_4 - (f_2 + f_6) + W_8(f_1 + f_5 - (f_3 + f_7)) \\ (f_0 - f_4) + W_4(f_2 - f_6) + W_8^2(f_1 - f_5) \pm W_4(f_3 - f_7) \\ (f_0 - f_4) - W_4(f_2 - f_6) + W_8^3((f_1 - f_5) \pm W_4(f_3 - f_7)) \\ f_0 + f_4 + (f_2 + f_6) - (f_1 + f_5 + (f_3 + f_7)) \\ f_0 + f_4 - (f_2 + f_6) - W_8(f_1 + f_5 - (f_3 + f_7)) \\ (f_0 - f_4) + W_4(f_2 - f_6) - W_8^2((f_1 - f_5) \pm W_4(f_3 - f_7)) \\ (f_0 - f_4) - W_4(f_2 - f_6) - W_8^3((f_1 - f_5) \pm W_4(f_3 - f_7)) \end{pmatrix} = \begin{pmatrix} s_0 + W_8^0 s_4 \\ s_1 + W_8^1 s_5 \\ s_2 + W_8^2 s_6 \\ s_3 + W_8^3 s_7 \\ s_0 + W_8^4 s_4 \\ s_1 + W_8^5 s_5 \\ s_2 + W_8^6 s_6 \\ s_3 + W_8^7 s_7 \end{pmatrix} \tag{7.70}$$

which is the final result.

Algorithm 12 shows a simple Fast Fourier Transformation algorithm. The number of trigonometric function evaluations can be reduced but this reduces the readability.

Exercises

Exercise 7.1 Discrete Fourier Transformation

In this computer experiment for a given set of input samples

$$f_n = f(n\frac{T}{N}) \quad n = 0 \cdots N-1 \tag{7.71}$$

- the Fourier coefficients

$$\widetilde{f}_{\omega_j} = \sum_{n=0}^{N-1} f_n e^{-i\omega_j t_n} \quad \omega_j = \frac{2\pi}{T} j, \quad j = 0 \cdots N-1 \tag{7.72}$$

  are calculated with Görtzel's method (7.3.1).
- The results from the inverse transformation

$$f_n = \frac{1}{N} \sum_{j=0}^{N-1} \widetilde{f}_{\omega_j} e^{i\frac{2\pi}{N} nj} \tag{7.73}$$

  are compared with the original function values $f(t_n)$.
- The Fourier sum is used for trigonometric interpolation with only positive frequencies

$$f(t) = \frac{1}{N} \sum_{j=0}^{N-1} \widetilde{f}_{\omega_j} \left(e^{i\frac{2\pi}{T} t}\right)^j. \tag{7.74}$$

- Finally the unphysical high frequencies are replaced by negative frequencies (7.26). The results can be studied for several kinds of input data.

Exercise 7.2 Noise Filter

This computer experiment demonstrates a nonlinear filter.
First a noisy input signal is generated.
The signal can be chosen as

- monochromatic $\sin(\omega t)$
- the sum of two monochromatic signals $a_1 \sin \omega_1 t + a_2 \sin \omega_2 t$
- a rectangular signal with many harmonic frequencies $\mathrm{sign}(\sin \omega t)$

Different kinds of white noise can be added

- dichotomous ± 1
- constant probability density in the range $[-1, 1]$
- Gaussian probability density

The amplitudes of signal and noise can be varied. All Fourier components are removed which are below a threshold value and the filtered signal is calculated by inverse Fourier transformation. Figure 7.3 shows a screenshot from the program.

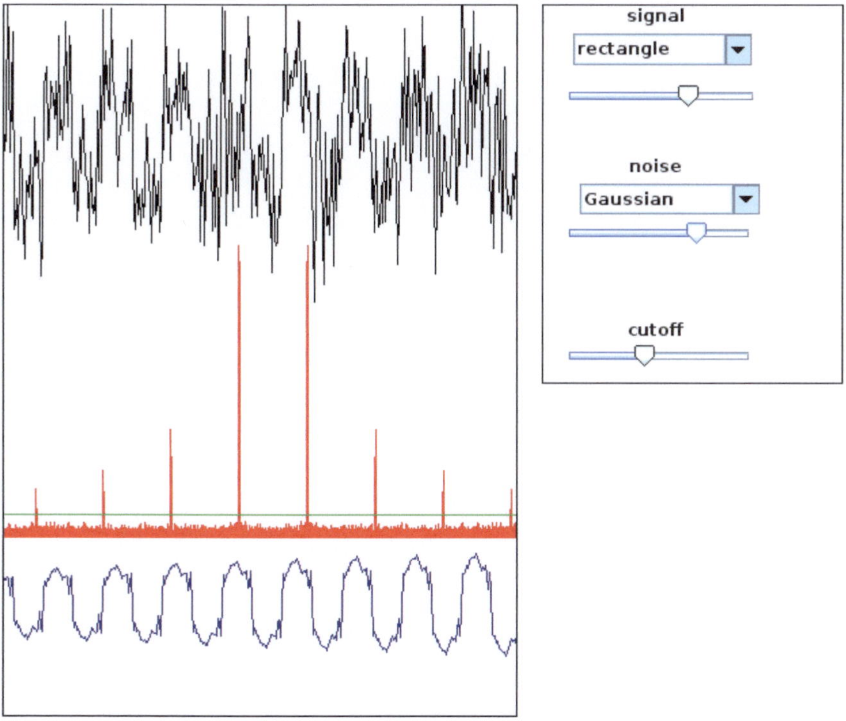

Fig. 7.3 screenshot from computer Exercise 7.2. **Top:** The input signal is rectangular with Gaussian noise. **Middle:** The Components of the Fourier spectrum (red) below the threshold (green line) are dropped. **Bottom:** The filtered signal is reconstructed (blue)

Chapter 8
Time-Frequency Analysis

Fourier-analysis provides a description of a given data set in terms of monochromatic oscillations without any time information. It is thus mostly useful for stationary signals. If the spectrum changes in time it is desirable to obtain information about the time at which certain frequencies appear. This can be achieved by applying Fourier analysis to a short slice of the data (short time Fourier analysis) which is shifted along the time axis. The frequency resolution is the same for all frequencies and therefore it can be difficult to find a compromise between time and frequency resolution. Wavelet analysis uses a frequency dependent window and keeps the relative frequency resolution constant. This is achieved by scaling and shifting a prototype wavelet—the so called mother wavelet. Depending on the application, wavelets can be more general and need not be sinusoidal or even continuous functions. Multiresolution analysis provides orthonormal wavelet bases which simplify the wavelet analysis. The fast wavelet transform connects a set of sampled data with its wavelet coefficients and is very useful for processing audio and image data.

8.1 Short Time Fourier Transform (STFT)

Fourier analysis transforms a function in the time domain $f(t)$ into its spectrum

$$\widetilde{f}(\omega) = \mathcal{F}[f](\omega) = \frac{1}{\sqrt{2\pi}} \int_{-\infty}^{\infty} f(t)\mathrm{e}^{-\mathrm{i}\omega t}dt \tag{8.1}$$

Supplementary Information The online version contains supplementary material available at https://doi.org/10.1007/978-3-032-07856-8_8.

P. O. J. Scherer, *Computational Physics I*, Graduate Texts in Physics,
https://doi.org/10.1007/978-3-032-07856-8_8

thereby losing all time information. Short time Fourier analysis applies a windowing function[1] (Sect. 7.2.3) e.g. a simple rectangle (Fig. 8.1)[2]

$$W_R(t) = \begin{cases} 1 \text{ for } |t| \le d \\ 0 \text{ else} \end{cases} \tag{8.2}$$

$$\tilde{W}_R(\omega) = \frac{2d}{\sqrt{2\pi}} \frac{\sin \omega d}{\omega d} = \frac{2d}{\sqrt{2\pi}} \text{sinc}\,(\omega d) \tag{8.3}$$

or triangle (Fig. 8.2)

$$W_{Tr}(t) = \begin{cases} \left(1 - \frac{|t|}{d}\right) \text{ for } t \le d \\ 0 \text{ else} \end{cases} \tag{8.4}$$

$$\tilde{W}_{Tr}(\omega) = \frac{d}{\sqrt{2\pi}} \frac{2(1 - \cos \omega d)}{\omega^2 d^2} = \frac{d}{\sqrt{2\pi}} \left(\text{sinc}(\frac{\omega d}{2}) \right)^2. \tag{8.5}$$

A smoother window is the Gaussian (Fig. 8.3)

$$W_G(t) = \frac{1}{d\sqrt{2\pi}} \exp\left\{-\frac{t^2}{2d^2}\right\} \tag{8.6}$$

with

$$\tilde{W}_G(\omega) = \frac{1}{\sqrt{2\pi}} \exp\left\{-\frac{\omega^2 d^2}{2}\right\}. \tag{8.7}$$

For the Gaussian window the standard deviations[3]

$$\sigma_t = d \quad \sigma_\omega = \frac{1}{d} \tag{8.8}$$

---

[1] Also known as apodization function or tapering function.

[2] There are two different definitions of the sinc function in the literature.

[3] Here we use the definition $\sigma^2 = \int_{-\infty}^{\infty} dt\, W(t) t^2$. If instead $\sigma^2 = \int_{-\infty}^{\infty} dt\, |W(t)|^2 t^2$ is used then $\sigma_t = \frac{d}{\sqrt{2}}$ and $\sigma_\omega = \frac{1}{\sqrt{2}d}$.

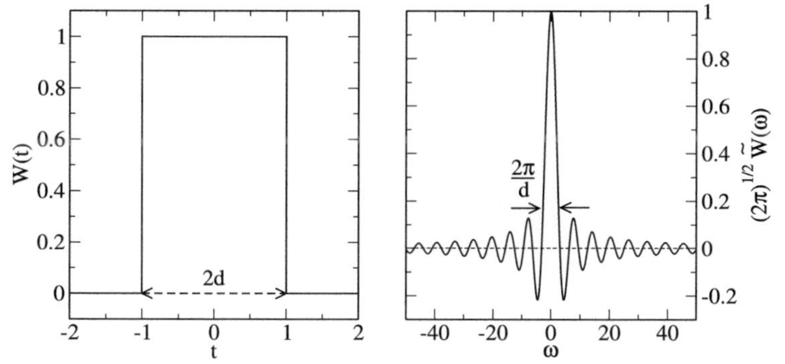

Fig. 8.1 Rectangular window. The rectangular (uniform) window and its Fourier transform are shown for $d = 1$

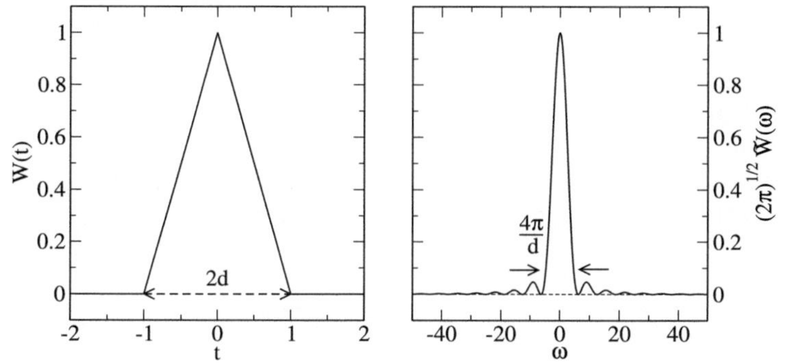

Fig. 8.2 Triangular window. The triangular (Bartlett) window and its Fourier transform are shown for $d = 1$

obey the uncertainty relation[4]

$$\sigma_t \sigma_\omega = 1. \tag{8.9}$$

Since the Gaussian extends to infinity, it has to be cut off for practical calculations.

Quite popular are the Hann(ing) and Hamming windows (Fig. 8.4)

$$W_{Hann}(t) = \cos^2\left(\frac{\pi t}{2d}\right) = \left(\frac{1}{2} + \frac{1}{2}\cos\frac{\pi t}{d}\right) \tag{8.10}$$

$$\tilde{W}_{Hann}(\omega) = \frac{d}{\sqrt{2\pi}} \frac{\text{sinc}\,(\omega d)}{1 - \frac{\omega^2 d^2}{\pi^2}} \tag{8.11}$$

$$W_{Hamm}(t) = \left(\frac{27}{54} + \frac{23}{54}\cos\frac{\pi t}{d}\right) \tag{8.12}$$

[4] For a Gaussian the time-bandwidth product is minimal.

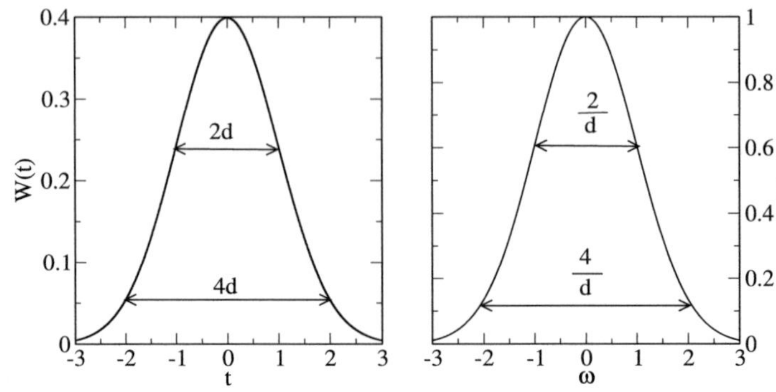

Fig. 8.3 Gaussian window. The Gaussian window and its Fourier transform are shown for $d = 1$

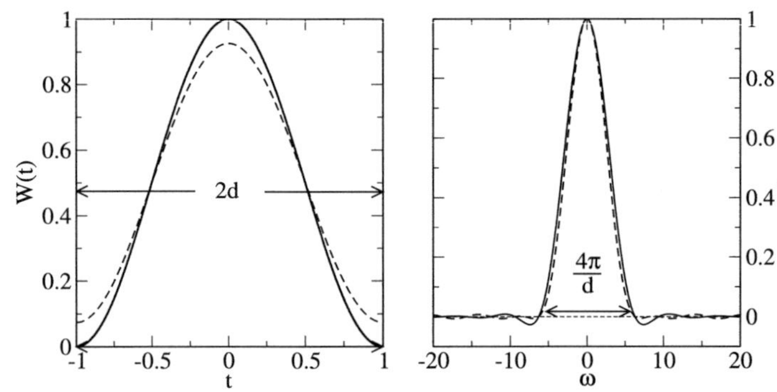

Fig. 8.4 The Hann (full curves) and Hamming (dashed curves) windows together with their Fourier transforms are shown for $\Delta t = 1$. The Hamming window is optimized to reduce the side lobes in the spectrum

$$\tilde{W}_{Hamm}(\omega) = \frac{d}{\sqrt{2\pi}} \frac{1 - \frac{4}{27}\frac{\omega^2 d^2}{\pi^2}}{1 - \frac{\omega^2 d^2}{\pi^2}} \operatorname{sinc}(\omega d). \tag{8.13}$$

For a general real valued function

$$\tilde{W}(\omega)^* = \left(\frac{1}{\sqrt{2\pi}} \int_{-\infty}^{\infty} W(t) \mathrm{e}^{-\mathrm{i}\omega t} dt \right)^* = \frac{1}{\sqrt{2\pi}} \int_{-\infty}^{\infty} W(t) \mathrm{e}^{\mathrm{i}\omega t} dt = \tilde{W}(-\omega) \tag{8.14}$$

and for an even function

$$\tilde{W}(\omega) = \frac{1}{\sqrt{2\pi}} \int_{-\infty}^{\infty} W(t) \mathrm{e}^{-\mathrm{i}\omega t} dt = \frac{1}{\sqrt{2\pi}} \int_{\infty}^{-\infty} W(t) \mathrm{e}^{\mathrm{i}\omega t} d(-t) = \tilde{W}(-\omega). \tag{8.15}$$

If $W(t)$ is real and even than this holds also for its Fourier transform

$$\tilde{W}(-\omega) = \tilde{W}(\omega) = \tilde{W}(\omega)^* = \frac{1}{\sqrt{2\pi}} \int_{-\infty}^{\infty} W(t) \cos(\omega t) dt. \tag{8.16}$$

The short time Fourier Transform

$$X(t_0, \omega) = \mathcal{F}[W^*(t - t_0) f(t)](\omega) = \frac{1}{\sqrt{2\pi}} \int_{-\infty}^{\infty} W^*(t - t_0) f(t) e^{-i\omega t} dt \tag{8.17}$$

depends on two variables $t_0$and ω. Since it has the form of a convolution integral it becomes a product in Fourier space, where

$$\begin{aligned}
&\frac{1}{\sqrt{2\pi}} \int dt_0 e^{-i\omega_0 t_0} X(t_0, \omega) = \frac{1}{\sqrt{2\pi}} \int_{-\infty}^{\infty} dt_0 e^{-i\omega_0 t_0} \frac{1}{\sqrt{2\pi}} \int_{-\infty}^{\infty} dt\, W^*(t - t_0) f(t) e^{-i\omega t} \\
&= \frac{1}{2\pi} \int_{-\infty}^{\infty} \int_{-\infty}^{\infty} dt\, d(t - t_0) e^{i\omega_0 (t - t_0)}\, W^*(t - t_0) f(t) e^{-i\omega_0 t} e^{-i\omega t} \\
&= \frac{1}{2\pi} \left(\int_{-\infty}^{\infty} d(t - t_0) e^{-i\omega_0 (t - t_0)} W(t - t_0) \int_{-\infty}^{\infty} \right)^* dt\ \ f(t) e^{-i\omega_0 t} e^{-i\omega t} \\
&= \tilde{W}^*(\omega_0) \tilde{f}(\omega + \omega_0).
\end{aligned} \tag{8.18}$$

For a real valued an even windowing function like the Gaussian (8.6) the STFT can therefore be calculated from

$$X(t_0, \omega) = \frac{1}{\sqrt{2\pi}} \int d\omega_0 e^{i\omega_0 t_0} \tilde{W}(\omega_0) \tilde{f}(\omega_0 + \omega). \tag{8.19}$$

Alternatively, the STFT can be formulated as

$$\begin{aligned}
X(t_0, \omega) &= \frac{1}{\sqrt{2\pi}} e^{i\omega t_0} \int_{-\infty}^{\infty} f(t)\, W^*(t - t_0) e^{-i\omega (t - t_0)} dt \\
&= \frac{1}{\sqrt{2\pi}} e^{i\omega t_0} \int_{-\infty}^{\infty} f(t)\, \Omega^*(t - t_0) dt
\end{aligned} \tag{8.20}$$

which involves a convolution of $f(t)$ with the wave packet (Fig. 8.5)

$$\Omega(t - t_0) = W(t - t_0) e^{i\omega (t - t_0)} \tag{8.21}$$

which is localized around t_0. In frequency space the wave packet becomes a band pass filter

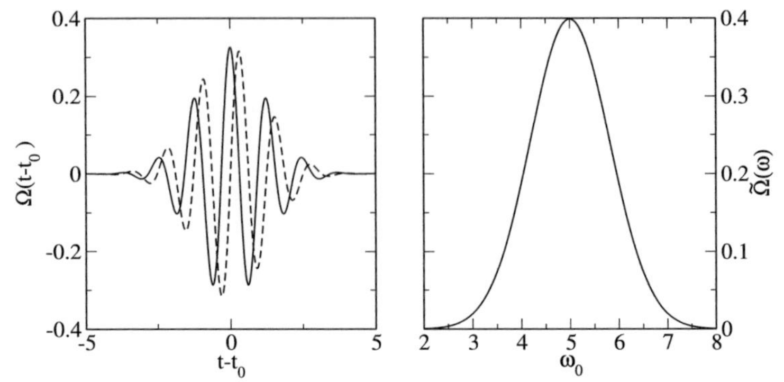

Fig. 8.5 Gabor wave packet. **Left:** Real (full curve) and imaginary (dashed curve) parts of the wave packet (8.21) are shown for a Gaussian windowing function with $\omega = 5$ and $2d^2 = 3$. **Right:** In the frequency domain the wave packet acts as a bandpass filter at $\omega_0 = \omega$

$$\begin{aligned}\tilde{\Omega}(\omega_0) &= \frac{1}{\sqrt{2\pi}} \int_{-\infty}^{\infty} dt e^{-i\omega_0 t} W(t) e^{i\omega t} \\ &= \frac{1}{\sqrt{2\pi}} \int_{-\infty}^{\infty} dt e^{-i(\omega_0-\omega)t} W(t) \\ &= \tilde{W}(\omega_0 - \omega)\end{aligned}$$

$$\begin{aligned}&\frac{1}{2\pi} \int_{-\infty}^{\infty} dt_0 e^{-i\omega_0 t_0} \int_{-\infty}^{\infty} f(t)\, \Omega^*(t-t_0) dt \\ &= \frac{1}{2\pi} \int_{-\infty}^{\infty} dt \int_{-\infty}^{\infty} d(t-t_0)\, f(t) e^{-i\omega_0 t}\, \Omega^*(t-t_0) e^{i\omega_0(t-t_0)} \\ &= \tilde{f}(\omega_0)\tilde{\Omega}^*(\omega_0)\end{aligned} \tag{8.22}$$

$$\begin{aligned}X(t_0,\omega) &= \frac{1}{\sqrt{2\pi}} e^{i\omega t_0} \int_{-\infty}^{\infty} d\omega_0 e^{i\omega_0 t_0} \tilde{f}(\omega_0)\tilde{\Omega}^*(\omega_0) \\ &= \frac{1}{\sqrt{2\pi}} \int_{-\infty}^{\infty} d\omega_0 e^{i(\omega+\omega_0)t_0} \tilde{f}(\omega_0)\tilde{\Omega}^*(\omega_0) \\ &= \frac{1}{\sqrt{2\pi}} \int_{-\infty}^{\infty} d\omega_0 e^{i\omega t_0} \tilde{f}(\omega_0-\omega)\tilde{\Omega}^*(\omega_0-\omega).\end{aligned}$$

The STFT can be inverted by

$$\int d\omega X(t_0,\omega) e^{i\omega t_0} = \frac{1}{\sqrt{2\pi}} \int dt \int d\omega e^{i\omega t_0} W^*(t-t_0) f(t) e^{-i\omega t}$$

$$= \frac{1}{\sqrt{2\pi}} \int dt W^*(t-t_0) f(t) 2\pi\delta(t-t_0) = \sqrt{2\pi} W^*(0) f(t_0) \qquad (8.23)$$

or alternatively

$$\int dt_0 \int d\omega X(t_0,\omega)\mathrm{e}^{\mathrm{i}\omega t} = \int dt_0 \int d\omega \frac{1}{\sqrt{2\pi}} \mathrm{e}^{\mathrm{i}\omega t} \int_{-\infty}^{\infty} W^*(t'-t_0) f(t') \mathrm{e}^{-\mathrm{i}\omega t'} dt'$$
$$= \int dt_0 \frac{1}{\sqrt{2\pi}} \int_{-\infty}^{\infty} 2\pi\delta(t-t') W^*(t'-t_0) f(t') dt'$$
$$= \int dt_0 \frac{1}{\sqrt{2\pi}} 2\pi W^*(t-t_0) f(t) = \sqrt{2\pi} f(t) \int W^*(t') dt'. \qquad (8.24)$$

STFT with a Gaussian window is also known as Gabor transform [120] which is conventionally defined as

$$\mathcal{G}[f](t_0,\omega) = \int_{-\infty}^{\infty} dt\, \mathrm{e}^{-\alpha\pi(t-t_0)^2} \mathrm{e}^{-\mathrm{i}\omega t} f(t). \qquad (8.25)$$

Example: Spectrogram

The STFT is often used to analyze audio signals. Let us consider as a simple example a monochromatic signal, which is switched on at time $t = 0$

$$f(t) = \begin{cases} 0 & t < 0 \\ \sin(\omega_s t) & t \geq 0. \end{cases} \qquad (8.26)$$

Using a Gaussian window, the Fourier transform can be calculated explicitly (an algebra program is very helpful)

$$X(t_0,\omega) = \frac{1}{2\pi\Delta t} \int_0^{\infty} dt\, \mathrm{e}^{-\mathrm{i}\omega t} \mathrm{e}^{-(t-t_0)^2/2d^2} \sin(\omega_s t)$$

$$= -\frac{\mathrm{i}}{4\sqrt{2\pi}} \mathrm{e}^{-\mathrm{i}t_0(\omega-\omega_s)} \mathrm{e}^{-d^2(\omega-\omega_s)^2/2} \left(\mathrm{erf}\left(\frac{\mathrm{i}\Delta t^2(\omega-\omega_s)-t_0}{\sqrt{2}d}\right) - 1 \right)$$

$$+ \frac{\mathrm{i}}{4\sqrt{2\pi}} \mathrm{e}^{-\mathrm{i}t_0(\omega+\omega_s)} \mathrm{e}^{-d^2(\omega+\omega_s)^2/2} \left(\mathrm{erf}\left(\frac{\mathrm{i}d^2(\omega+\omega_s)-t_0}{\sqrt{2}d}\right) - 1 \right). \qquad (8.27)$$

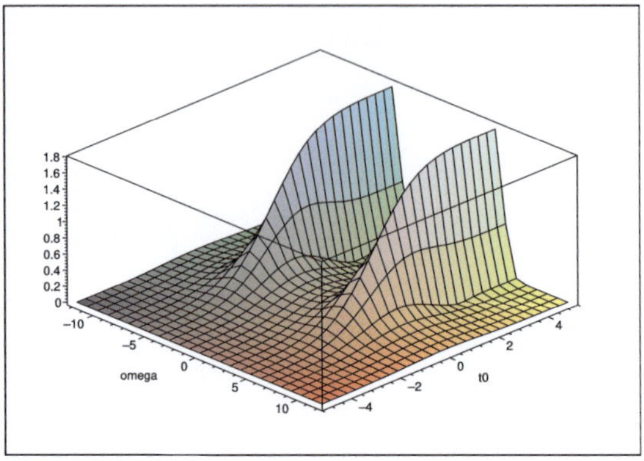

Fig. 8.6 3D spectrogram. The squared magnitude of the STFT (8.27) is shown for $\omega_s = 5$ and $2d = 1$

There are two contributions since the real function $f(t)$ contains oscillations with $\pm\omega_s$. The squared magnitude[5]

$|X(t_0, \omega)|^2$is shown as a 3D spectrogram in Fig. 8.6. The width of the window determines the resolution both in time and frequency. Neglecting interference terms, at resonance $\omega = \omega_s$ the squared magnitude rises according to

$$|X(t_0, \omega_s)|^2 = \frac{1}{32\pi}\left(\text{erf}\,\frac{t_0}{\sqrt{2}d} - 1\right)^2 \tag{8.28}$$

and reaches its stationary value within a time of $\approx 2d$, whereas in the stationary state at $t_0 \gg d$, the dependency on the frequency mismatch $\Delta\omega = \omega - \omega_s$ is given by a Gaussian with a width of $2\sigma_\omega = \sqrt{2}/d$. The dependence of time and frequency resolution on the window width is shown qualitatively as two-dimensional spectrograms in Fig. 8.7.

8.2 Discrete Short Time Fourier Transform

The continuous STFT is very redundant and not useful for the analysis of data which are sampled at discrete times. Therefore we introduce a series of overlapping windows centered at equidistant times $t_n = n\Delta t$ (Fig. 8.8)

$$W_n(t) = W(n\Delta t - t). \tag{8.29}$$

Assuming that the windowing function $W_n(t) = 0$ outside the interval $[t_n - d, t_n + d]$ we apply (7.5) and expand $W_n(t)f(t)$ inside the interval as a

[5] This is a measure of the spectral power distribution.

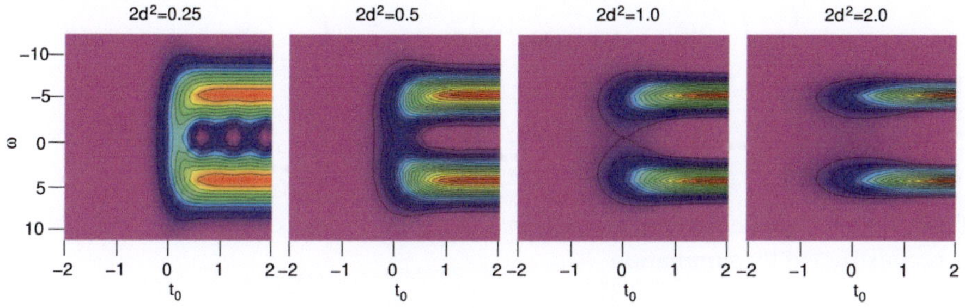

Fig. 8.7 Spectrograms with different window width Δt. The squared magnitude of the STFT (8.27) is shown for $\omega_s = 5$ and $2d^2 = 0.25, 0.5, 1.0, 2.0$. For larger values of the time window d the frequency resolution becomes higher but the time resolution lower

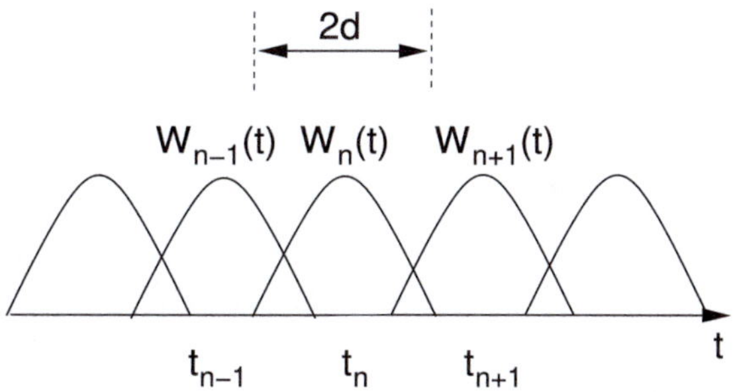

Fig. 8.8 Discrete STFT

Fourier series

$$g_n(t) = W_n(t)f(t) = \sum_{m=-\infty}^{\infty} e^{i\omega_m t}\hat{g}_{nm} \text{ with } \omega_m = m\frac{\pi}{d} \quad |t - t_n| \le d. \tag{8.30}$$

We extend this expression to all times t by introducing the characteristic function of the interval

$$\chi_n(t) = \begin{cases} 1 & \text{for } |t - t_n| \le d \\ & 0 \text{ else} \end{cases} \tag{8.31}$$

$$g_n(t) = W_n(t)f(t) = \chi_n(t) \sum_{m=-\infty}^{\infty} e^{i\omega_m t}\hat{g}_{nm}. \tag{8.32}$$

The Fourier coefficients, given by the integral

$$\hat{g}_{nm} = \frac{1}{2d}\int_{t_n-d}^{t_n+d} W(t - t_n)f(t)e^{-i\omega_m t}dt \tag{8.33}$$

obviously correspond to the STFT at times t_n and frequencies ω_m

$$X(t_n,\omega_m) = \frac{1}{\sqrt{2\pi}}\int_{-\infty}^{\infty} W(t-t_n)f(t)e^{-i\omega_m t}dt = \frac{1}{\sqrt{2\pi}}\int_{t_n-d}^{t_n+d} W(t-t_n)f(t)e^{-i\omega_m t}dt$$

$$= \frac{2d}{\sqrt{2\pi}}\hat{g}_{nm}. \tag{8.34}$$

If the windows are dense enough such that there union spans all times, the signal can be reconstructed by summation

$$\sum_n g_n(t) = f(t)\sum_n W_n(t) = \sum_{nm} \chi_n(t)e^{i\omega_m t}\hat{g}_{nm}. \tag{8.35}$$

This expression simplifies, if

$$\sum_n W_n(t) = \text{const} \tag{8.36}$$

which is e.g. the case for triangular windows as well as the Hann and Hamming windows with $\Delta t = d$ (Fig. 8.9).

For practical applications, we assume that the function $g(t)$ has been sampled at N equidistant times within the interval $[t_n - d, t_n + d]$

$$\tau_{n,s} = t_n - d + s\frac{2d}{N} \quad s = 0, 1, \ldots N-1 \tag{8.37}$$

$$\omega_j = \frac{\pi}{d}j, \quad \omega_j(\tau_{n,s} - t_n + d) = js\frac{2\pi}{N} \tag{8.38}$$

and apply the discrete Fourier transformation method (Sect. 7.2)

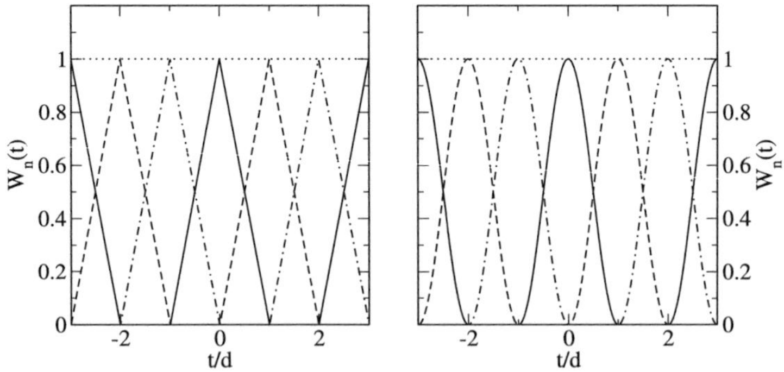

Fig. 8.9 Window functions with constant sum. The sum $\sum_n W_n(t)$ becomes constant (dotted lines) if the windows are shifted by half their width $\Delta t = d$. **Left:** Triangular window; **Right:** Hann window

$$\widetilde{g}_{n,\omega_j} = \sum_{s=0}^{N-1} g_{n,s} e^{-i\omega_j(\tau_{ns}-t_n+d)} = \sum_{s=0}^{N-1} g_{n,s} e^{-ijs\frac{2\pi}{N}} \tag{8.39}$$

$$\frac{1}{N}\sum_{j=0}^{N-1} \tilde{g}_{n,\omega_j} e^{ijs\frac{2\pi}{N}} = \frac{1}{N}\sum_{j=0}^{N-1}\sum_{s=0}^{N-1} g_{n,s'} e^{-ijs'\frac{2\pi}{N}} e^{ijs\frac{2\pi}{N}} = g_{n,s}. \tag{8.40}$$

Example: FM Signal

Figures 8.10 and 8.11 show the STFT analysis of a frequency modulated signal

$$f(t) = \sin\Phi(t) = \sin\left(\omega_0 t + \frac{a\omega_0}{\omega_1}(1-\cos\omega_1 t)\right) \tag{8.41}$$

with a momentaneous frequency of

$$\omega(t) = \frac{\partial\Phi}{\partial t} = \omega_0(1 + a\sin\omega_1 t) \tag{8.42}$$

for carrier frequency $\frac{\omega_0}{2\pi} = 10\,\text{kHz}$, modulation frequency $\frac{\omega_1}{2\pi} = 25\ \text{Hz}(100\ \text{Hz})$ and modulation depth $a = 0.3$.

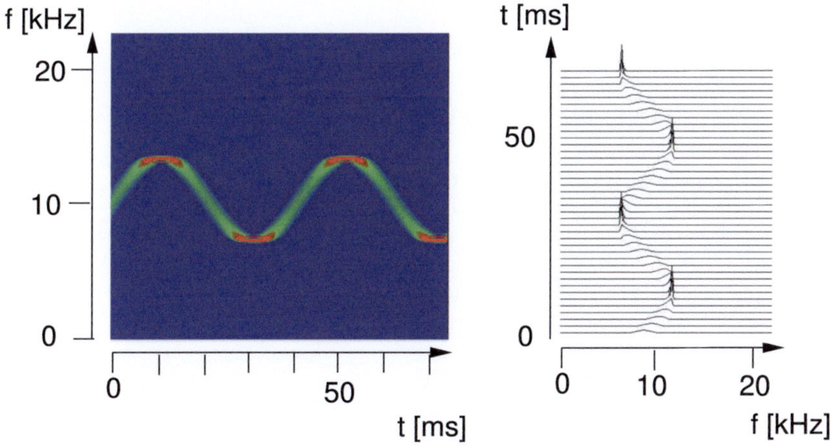

Fig. 8.10 STFT analysis of a FM signal. Screenshots from Exercise 8.1. **Left:** Spectrogram; **Right:** STFT spectra. Sampling frequency 44100 Hz, number of samples 512, Hann windows are used with a shift of 8 samples (0.18 ms) between neighbor windows. 6 ms time resolution and 1.1 kHz frequency resolution are sufficient to resolve 25 Hz modulation. The time dependent spectra have their smallest width at the stationary points of the momentaneous frequency

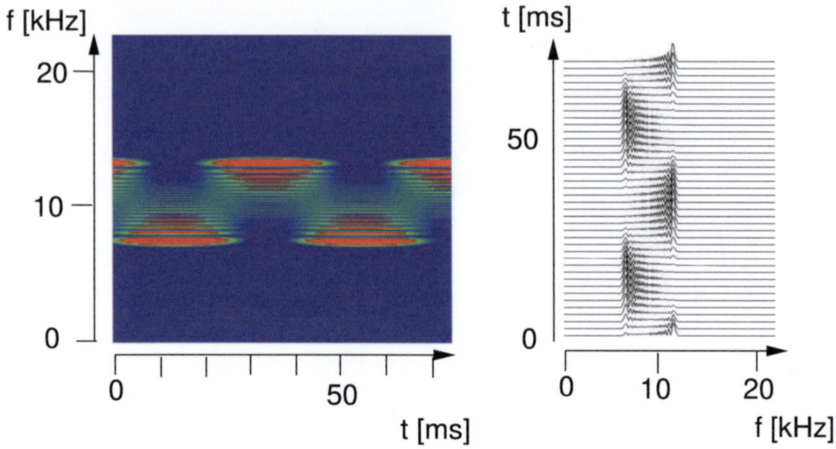

Fig. 8.11 STFT analysis of a FM signal. This figure shows screenshots from Exercise 8.1. **Left:** Spectrogram; **Right:** STFT spectra. Sampling frequency 44100 Hz, number of samples 512, Hann windows are used with a shift of 2 samples (0.045 ms) between neighbor windows. 6 ms time resolution and 1.1 kHz frequency resolution are not sufficient to resolve 100 Hz modulation. Only minimum and maximum frequencies are observed

8.3 Gabor Expansion

For the special case of rectangular windows with distance $\Delta t = 2d$[6]

$$W_n(t) = W_R(t - 2d\,n) = \chi_n(t) \tag{8.43}$$

$$\sum_n W_n(t) = 1 = \text{const} \tag{8.44}$$

we have

$$f(t) = \sum_n g_n(t) \tag{8.45}$$

$$= \sum_n \sum_{m=-\infty}^{\infty} \chi_n(t) e^{i\omega_m t} \hat{g}_{nm}. \tag{8.46}$$

This equation expands $f(t)$ and its Fourier transform as linear combinations of elementary "signals" which are located at t_n in time and ω_m in frequency

$$h_{n,m} = \chi_n e^{i\omega_m t}. \tag{8.47}$$

[6] For simplicity, we do not normalize the window here.

$$\tilde{h}_{n,m} = \frac{2d}{\sqrt{2\pi}} \mathrm{e}^{-\mathrm{i}t_n(\omega-\omega_m)} \operatorname{sinc}\ (d(\omega-\omega_m)) \tag{8.48}$$

$$f(t) = \sum_{nm} h_{n,m}\hat{g}_{n,m} \quad \tilde{f}(\omega) = \sum_{nm} \tilde{h}_{nm}\hat{g}_{nm}. \tag{8.49}$$

A similar expansion is obtained if we use rectangular windows in Fourier space with width and distance $\Delta\omega$ [121]

$$\tilde{h}_{nm}(\omega) = \chi_{\mathrm{m}}\mathrm{e}^{-\mathrm{i}t_n(\omega-\omega_m)}$$

$$h_{nm}(t) = \frac{2\Delta\omega}{\sqrt{2\pi}} \operatorname{sinc}\ ((t-t_n)\Delta\omega)\mathrm{e}^{\mathrm{i}\omega_m t}$$

and sample the spectrum at times

$$t_n = n\frac{\pi}{\Delta\omega} \tag{8.50}$$

to obtain

$$\tilde{f}(\omega) = \sum \chi_m(\omega)\tilde{f}_m(\omega) = \sum_{n=-\infty}^{\infty} \chi_m(\omega)\mathrm{e}^{\mathrm{i}\omega t_n}\hat{f}_{nm} \tag{8.51}$$

$$\hat{f}_{nm} = \frac{1}{2\Delta\omega}\int_{\omega_m-\Delta\omega}^{\omega_m+\Delta\omega} \tilde{f}(\omega)\mathrm{e}^{\mathrm{i}\omega t_n}d\omega. \tag{8.52}$$

Gabor [120] discussed an expansion with Gaussian signals (Fig. 8.5). In general, however, the elementary signals are not orthogonal which makes the determination of the coefficients $a_{n,m}$ complicated. Bastiaans [121, 122] introduced another auxiliary set of elementary signals

$$\gamma_{n,m} = \gamma(t-n\Delta t)\mathrm{e}^{\mathrm{i}\omega_m t} \tag{8.53}$$

which are biorthogonal, i.e.

$$\int \gamma^*_{n',m'}(t)h_{n,m}(t)dt = \delta_{n,n'}\delta_{m,m'} \tag{8.54}$$

and allow the calculation of the Gabor expansion coefficients from a scalar product

$$\int \gamma^*_{n,m}(t)f(t)dt = \sum_{nm}\int a_{nm}\gamma^*_{n,m}h_{nm}(t)dt = a_{nm}. \tag{8.55}$$

Determination of γ for a given windowing function can be simplified by application of the Zak transform [123]. Discrete versions of the Gabor transform [124] are popular in signal, speech and image processing.

8.4 Wavelet Analysis

The STFT method uses constant frequency and time resolution. Therefore the lowest frequency of interest determines the minimum width of the window whereas at higher frequencies shorter time windows could be more appropriate to increase time resolution while keeping the relative uncertainty in frequency constant (Fig. 8.12). This is the basic idea of the wavelet transform. Whereas STFT uses wave packets of the form (8.21)

$$\Omega_{t_0,\omega}(t) = W(t - t_0)\mathrm{e}^{\mathrm{i}\omega(t-t_0)} \tag{8.56}$$

where only the oscillating part is scaled with frequency ω, wavelets scale the whole function like in

$$\Omega_{t_0,s}(t) = W(\frac{t - t_0}{s})\mathrm{e}^{\mathrm{i}\omega_0(t-t_0)/s} \tag{8.57}$$

or, more generally

$$\Omega_{t_0,s}(t) = \frac{1}{\sqrt{|s|}}\Psi(\frac{t - t_0}{s}) \tag{8.58}$$

$$\tilde{\Omega}(\omega) = \frac{1}{\sqrt{2\pi}}\int_{-\infty}^{\infty} \mathrm{e}^{-\mathrm{i}\omega t}\frac{1}{\sqrt{|s|}}\Psi(\frac{t - t_0}{s})dt$$

$$= \frac{1}{\sqrt{2\pi}}\int_{-\infty\mathrm{sign}\, s}^{\infty\mathrm{sign}\, s} \mathrm{e}^{-\mathrm{i}\omega(st'+t_0)}\frac{1}{\sqrt{|s|}}\Psi(t')d(st' + t_0)$$

$$= \sqrt{|s|}\mathrm{e}^{-\mathrm{i}\omega t_0}\frac{1}{\sqrt{2\pi}}\int_{-\infty}^{\infty} \mathrm{e}^{-\mathrm{i}\omega st'}\Psi(t')dt' = \sqrt{|s|}\mathrm{e}^{-\mathrm{i}\omega t_0}\tilde{\Psi}(s\omega) \tag{8.59}$$

where a whole family of wavelets is derived from the "mother wavelet" $\Psi(t)$ by shifting and rescaling. The prefactor has been introduced to keep the norm invariant

$$\int |\Omega_{t_0,s}(t)|^2 dt = \frac{1}{|s|}\int |\Psi(\frac{t - t_0}{s})|^2 dt$$

$$= \frac{1}{|s|}\int_{-\infty\mathrm{sign}\, s}^{\infty\,\mathrm{sign}\, s} |\Psi(t')|^2 d(st' + t_0) = \int |\Psi(t')|^2 dt'. \tag{8.60}$$

Closely related to the short time Fourier analysis is the Morlet (or Gabor) wavelet, which is also very useful in quantum physics [125]. It is defined as[7]

$$\Psi(t) = W_G(t)\mathrm{e}^{\mathrm{i}\omega_0 t} = \frac{1}{\pi^{1/4}\sqrt{d}}\exp\left\{-\frac{t^2}{2d^2}\right\}\mathrm{e}^{\mathrm{i}\omega_0 t} \tag{8.61}$$

[7] The conventional normalization is $\int dt|\Psi(t)|^2 = 1$.

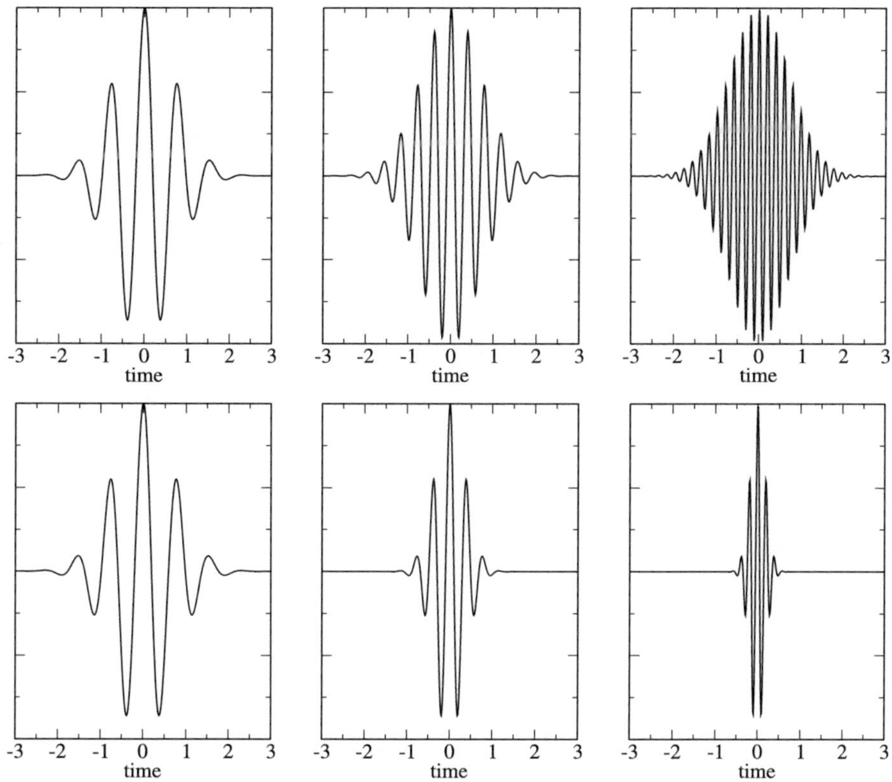

Fig. 8.12 Morlet wavelets and STFT wave packets. **Top:** STFT uses the same window for all frequencies. **Bottom:** Wavelets use a variable window width to keep the form of the wave packet and the relative frequency resolution constant (only the real part is shown)

$$\tilde{\Psi}(\omega) = \tilde{W}_G(\omega - \omega_0) = \frac{\sqrt{d}}{\pi^{1/4}} \exp\left\{-\frac{d^2}{2}(\omega - \omega_0)^2\right\}. \tag{8.62}$$

The similarity of a signal $f(t)$ to a wavelet with scale s centered at t_0 is measured by the correlation integral

$$C(t_0, s) = \int_{-\infty}^{\infty} f(t)\Omega^*_{t_0,s}(t)dt = \frac{1}{\sqrt{|s|}} \int_{-\infty}^{\infty} \Psi^*(\frac{t - t_0}{s}) f(t)dt \tag{8.63}$$

which becomes a product after Fourier transformation with respect to t_0

$$\tilde{C}(\omega, s) = \frac{1}{\sqrt{2\pi}} \int C(t_0, s)\mathrm{e}^{-\mathrm{i}\omega t_0} dt_0$$

$$= \frac{1}{\sqrt{2\pi}} \int dt_0 e^{-i\omega t_0} \int dt \frac{1}{2\pi} \int d\omega' \sqrt{|s|} \tilde{\Psi}^*(s\omega') e^{-i\omega'(t-t_0)} \int d\omega'' \tilde{f}(\omega'') e^{i\omega'' t}$$

$$= \sqrt{2\pi |s|} \int dt \int d\omega' \tilde{\Psi}^*(s\omega') e^{-i\omega' t} \int d\omega'' \tilde{f}(\omega'') e^{i\omega'' t} \delta(\omega - \omega') \delta(\omega' - \omega'')$$

$$= \sqrt{2\pi |s|} \int d\omega' \tilde{\Psi}^*(s\omega') e^{-i\omega' t} \tilde{f}(\omega') e^{i\omega' t}$$

$$= \sqrt{2\pi |s|} \tilde{\Psi}^*(s\omega) \tilde{f}(\omega). \tag{8.64}$$

For the Morlet wavelet this becomes

$$\tilde{C}(\omega, s) = \pi^{1/4} \sqrt{2 |s| d} \exp\left\{ -\frac{d^2}{2} (s\omega - \omega_0)^2 \right\} \tilde{f}(\omega)$$

$$= \pi^{1/4} \sqrt{2 |s| d} \exp\left\{ -\frac{(sd)^2}{2} (\omega - \frac{\omega_0}{s})^2 \right\} \tilde{f}(\omega) \tag{8.65}$$

i.e. $\tilde{C}(\omega, s)$ averages the spectrum $\tilde{f}$ over a range with a width of $\sigma_\omega = 1/sd$ around $\omega = \omega_0/s$ and a constant ratio

$$\frac{\sigma_\omega}{\omega} = \frac{1}{\omega_0 d}. \tag{8.66}$$

8.5 Wavelet Synthesis

For data processing it is necessary to reconstruct the data from the wavelet coefficients $C(t_0 s)$. This can be achieved with the help of the integral[8]

$$\int_{-\infty}^{\infty} \int_{-\infty}^{\infty} \frac{1}{s^2} C(t_0, s) \Omega_{t_0,s}(t) dt_0 ds = \int \frac{1}{s^2} dt_0 ds \frac{1}{\sqrt{|s|}} \Psi(\frac{t - t_0}{s}) C(t_0, s)$$

$$= \frac{1}{\sqrt{2\pi}} \int \frac{1}{s^2} ds \int dt_0 \int d\omega e^{i\omega(t-t_0)/s} \tilde{\Psi}(\omega) \int d\omega' e^{i\omega' t_0} \tilde{\Psi}^*(s\omega) \tilde{f}(\omega)$$

$$= \frac{1}{\sqrt{2\pi}} \int \frac{1}{s^2} ds \int dt_0 \int s d\omega'' e^{i\omega''(t-t_0)} \tilde{\Psi}(\omega'' s) \int d\omega' e^{i\omega' t_0} \tilde{\Psi}^*(s\omega') \tilde{f}(\omega')$$

$$= \frac{1}{\sqrt{2\pi}} \int \frac{1}{s} ds \int d\omega'' e^{i\omega'' t} \tilde{\Psi}(\omega'' s) \int d\omega' \tilde{\Psi}^*(s\omega') \tilde{f}(\omega') 2\pi \delta(\omega' - \omega'')$$

[8] A more rigorous treatment introducing the concept of frames in Hilbert space can be found in [126].

$$= \sqrt{2\pi} \int \frac{1}{s} ds \int s d\omega'' e^{i\omega'' t} \tilde{\Psi}(\omega'' s) \Psi(s\omega'') \tilde{f}(\omega'')$$

$$= \sqrt{2\pi} \int \frac{1}{s} ds \int d\omega'' e^{i\omega'' t} \tilde{\Psi}(\omega'' s) \tilde{\Psi}^*(s\omega'') \tilde{f}(\omega''). \tag{8.67}$$

If the admissibility condition is fulfilled, which states that the integral

$$C_\Psi = \int_{-\infty}^{\infty} \frac{|\tilde{\Psi}(\omega)|^2}{\omega} d\omega < \infty \tag{8.68}$$

exists and is final, then

$$\int_{-\infty}^{\infty} \frac{1}{s} ds \tilde{\Psi}(\omega s) \tilde{\Psi}^*(\omega s) = \int_{-\infty}^{\infty} \frac{1}{\omega'} d\omega' \tilde{\Psi}(\omega') \tilde{\Psi}^*(\omega') = C_\Psi \tag{8.69}$$

and we obtain

$$f(t) = \frac{1}{2\pi C_\Psi} \int_{-\infty}^{\infty} \int_{-\infty}^{\infty} \frac{1}{s^2} C(t_0, s) \Omega_{t_0,s}(t) dt_0 ds. \tag{8.70}$$

The admissibility condition implies that $\tilde{\Psi}(0) = 0$ and thus $\int dt \Psi(t) = 0$. Hence, the Morlet wavelet (8.61) has to be modified[9]

$$\Psi(t) = \frac{1}{\pi^{1/4}\sqrt{d}} N_d \exp\left\{-\frac{t^2}{2d^2}\right\} \left[e^{i\omega_0 t} - \exp\left\{-\frac{\omega_0^2 d^2}{2}\right\} \right] \tag{8.71}$$

$$\tilde{\Psi}(\omega) = \frac{\sqrt{d}}{\pi^{1/4}} N_d \left[\exp\left\{-\frac{d^2}{2}(\omega - \omega_0)^2\right\} - \exp\left\{-\frac{d^2}{2}(\omega^2 + \omega_0^2)\right\} \right] \tag{8.72}$$

$$N_d = \left[\left(1 + \exp\left\{-\omega^2 d^2\right\} - 2\exp\left\{-\frac{3}{4}\omega^2 d^2\right\} \right) \right]^{-1/2}. \tag{8.73}$$

Another popular (continuous) wavelet is the "Mexican hat" (also known as Ricker wavelet or Marr wavelet) which is given by the normalized negative second derivative of a Gaussian (Fig. 8.13)

$$\Psi(t) = \frac{2}{\pi^{1/4}\sqrt{3d}} \left(1 - \frac{t^2}{d^2}\right) \exp\left\{-\frac{t^2}{2d^2}\right\} \tag{8.74}$$

$$\tilde{\Psi}(\omega) = \frac{2\sqrt{d}}{\pi^{1/4}\sqrt{3}} \omega^2 \exp\left\{-\frac{\omega^2 d^2}{2}\right\}. \tag{8.75}$$

---

[9] This correction is often neglected, if the width is large.

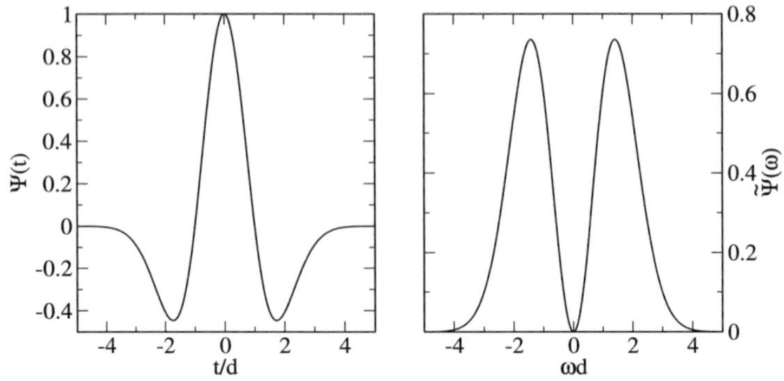

Fig. 8.13 Mexican hat wavelet. **Left:** The mexican hat wavelet is essentially the second derivative of a Gaussian. **Right:** Its Fourier transform is a band pass filter around $\omega_{\max} = \pm 2/d$

Example: Wavelet Analysis of a Nonstationary Signal

The following example shows screen shots from Exercise 8.2. The signal consists of two sweeps with linearly increasing frequency of the form

$$f_{1,2}(t) = \sin\left[\omega_{1,2}t + \frac{\alpha_{1,2}}{2}t^2\right] \tag{8.76}$$

and another component which switches between a 5kHz oscillation and the sum of 300 Hz and a 20 kHz oscillation at a rate 20 Hz

$$f_3(t) = \begin{cases} \sin(\omega_{2kHz}t) + \sin(\omega_{300Hz}t) & \text{if } \sin(\omega_{20Hz}t) < 0 \\ \sin(\omega_{5kHz}t) & \text{else.} \end{cases} \tag{8.77}$$

The signal is sampled with a rate of 44 kHz and analyzed with Morlet wavelets over 6 octaves (Fig. 8.14). The parameter d of the mother wavelet (8.61) determines frequency and time resolution. The frequency ω_0 of the mother wavelet is taken as the Nyquist frequency which is half the sampling rate. The convolution with the daughter wavelets

$$\Psi_{m,n}(t) = \frac{1}{\sqrt{s_m}}\Psi\left(\frac{t-t_n}{s_m}\right) \tag{8.78}$$

is calculated at 400 times with a step size of 0.726 msec (corresponding to 32 samples)

$$t_n = t_0 + n\Delta t \tag{8.79}$$

and for 300 different values of the scaling parameter

$$s_m = 1.015^m. \tag{8.80}$$

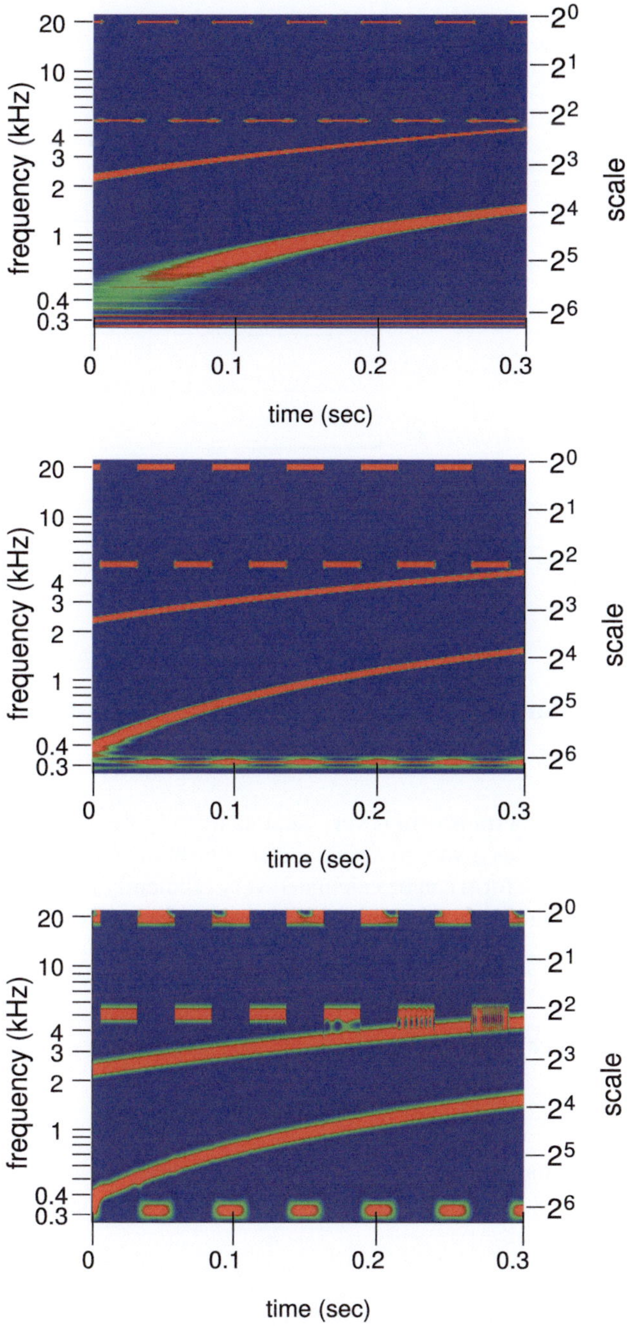

Fig. 8.14 Wavelet analysis. **Top:** For $d = 1$ms the frequency resolution is high for the stationary parts of the signal and time resolution is low. **Middle:** For $d = 0.25$ms the pulsating component 300 Hz can be resolved but time resolution is still poor. **Bottom:** For $d = 0.0625$ ms time resolution is sufficient to show all the modulations while frequency resolution is rather poor

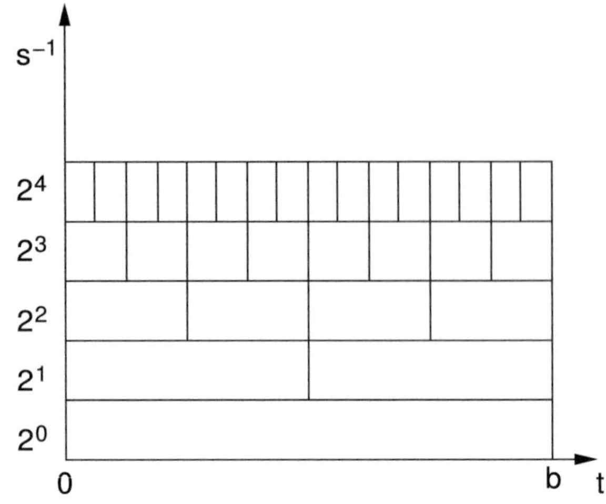

Fig. 8.15 Multiresolution analysis. Data are analyzed with decreasing time window $\Delta t = b/2^m$

In a logarithmic plot, the relative frequency uncertainty has the same size for all stationary signals.

8.6 Discrete Wavelet Transform and Multiresolution Analysis

The continuous wavelet transform is very redundant and time consuming. Multiresolution analysis provides a way to define a discrete set of orthogonal wavelets, for which the wavelet transform can be calculated very efficiently from a scalar product. A discrete wavelet transform uses discrete values of shift and scaling parameters

$$s = a^{-m} \quad t_0 = na^{-m}b \tag{8.81}$$

to define the daughter wavelets[10]

$$\Psi_{m,n}(t) = a^{m/2}\Psi\left(a^m t - nb\right). \tag{8.82}$$

For integer a, in most cases $a = 2$, this equation defines wavelets of a multiresolution analysis (Fig. 8.15) where m corresponds to the resolution 2^m.

[10] Equation (8.78), in contrast, describes the continuous wavelet transform, which has to be discretized for numerical calculations.

8.6.1 Scaling Function and Multiresolution Approximation

At the basic resolution 2^0 the function $f(t)$ is approximated as a linear combination

$$f(t) \approx f(t)^{(0)} = \sum_n f_{0,n} \Phi_{0,n}(t) \tag{8.83}$$

of a scaling function and its translations,

$$\Phi_{0,n} = \Phi(t - nb) \quad , n = 0, \pm 1 \ldots \tag{8.84}$$

which is chosen [127, 128] such, that the $\Phi_{0,n}$form an orthonormal basis of the space of linear combinations

$$V_0 = \text{span}\{\Phi_{0,n}, n = 0, \pm 1, \ldots\} \tag{8.85}$$

$$\int \Phi_{0,n}^*(t)\Phi_{0,n'}(t)dt = \delta_{n,n'}. \tag{8.86}$$

The best approximation is found by minimizing the norm

$$||f(t) - \sum_n f_{0,n}\Phi_{0n}(t)||^2 = \int (f^*(t) - \sum_n f_{0,n}^*\Phi_{0n}^*(t))(f(t) - \sum_{n'} f_{0,n'}\Phi_{0n'}(t))dt$$

$$= \int |f(t)|^2 dt - \sum_{n'} f_{0n'} \int f^*(t)\Phi_{0n'}(t)dt - \sum_n f_{0n}^*(t) \int \Phi_{0n}^*(t) f(t)dt + \sum_n |f_{0n}|^2 \tag{8.87}$$

hence by choosing

$$f_{0n} = \int \Phi_{0n}^*(t) f(t)dt \tag{8.88}$$

i.e., the orthogonal projection of $f(t)$ onto V_0. Approximation at the higher resolution 2^m similarly is given by linear combination

$$f(t) \approx f(t)^{(m)} = \sum_n f_{m,n}\Phi_{m,n}(t) \tag{8.89}$$

of the scaled functions

$$\Phi_{m,n} = 2^{m/2}\Phi(2^m t - nb) \tag{8.90}$$

which form an orthonormal basis for the space

$$V_m = \text{span}\{\Phi_{m,n}, n = 0, \pm 1, \ldots\} \tag{8.91}$$

since

$$\int \Phi^*_{m,n}(t)\Phi_{m,n'}(t)dt = 2^m \int \Phi^*(2^m t - nb)\Phi(2^m t - n'b)dt$$
$$= 2^m \int \Phi^*(t' - nb)\Phi(t' - n'b)\frac{dt'}{2^m} = \delta_{n,n'}. \tag{8.92}$$

The sequence of spaces V_m is called a multiresolution approximation to the space of square integrable functions $L^2(\mathbb{R})$, if [127]

$$\text{(i)} \quad \cdots \subset V_{-1} \subset V_0 \subset V_1 \subset V_2 \ldots \tag{8.93}$$

$$\text{(ii)} \quad \bigcup_{m=-\infty}^{\infty} V_m \quad \text{is dense in } L^2(\mathbb{R}) \tag{8.94}$$

$$\text{(iii)} \quad \bigcap_{m=-\infty}^{\infty} V_m = \{0\}. \tag{8.95}$$

Property (ii) has as a consequence, that the approximations $f^{(m)}(t)$ converge to $f(t)$ for large m. Hence, due to orthonormality

$$f^{(m)}(t) = \sum_n \Phi_{m,n}(t) \int_{-\infty}^{\infty} \Phi^*_{m,n}(t')f(t')dt' \to f(t) \tag{8.96}$$

and the projection operator onto V_m

$$P_m = \sum_n \Phi_{m,n}(t)\Phi^*_{m,n}(t') = 2^m \sum_n \Phi(2^m t - nb)\Phi^*(2^m t' - nb) \to 1 \tag{8.97}$$

converges to the unit operator. Now, with $a > 0$ choose the function

$$f_a(t) = \begin{cases} 1 \text{ if } -a \le x \le a \\ 0 \text{ else.} \end{cases} \tag{8.98}$$

Then,

$$(P_m f_{ab})(t) = \sum_n \Phi(2^m t - nb) \int_{-a}^{a} 2^m dt' \Phi^*(2^m t' - nb)$$

$$= \sum_n \Phi(2^m t - nb) \int_{-2^m a}^{2^m a} \Phi^*(t' - nb)dt'. \tag{8.99}$$

For large a, the integrals become more and more independent on n, and

$$\left(\int_{-\infty}^{\infty} \Phi^*(t')dt'\right) \sum_n \Phi(2^m t - nb) \to 1. \tag{8.100}$$

Now we integrate the sum over one period $0 \leq t \leq 2^{-m}b$ and find

$$\int_0^{2^{-m}b} \sum_n \Phi(2^m t - nb)dt = \sum_n \int_{-nb}^{(1-n)b} \Phi(t)\, 2^{-m}dt = 2^{-m} \int_{-\infty}^{\infty} \Phi(t)dt \tag{8.101}$$

and therefore

$$\left(\int_{-\infty}^{\infty} \Phi^*(t')dt'\right)\left(\int_{-\infty}^{\infty} \Phi(t')dt'\right) = b \tag{8.102}$$

or

$$\left|\int_{-\infty}^{\infty} \Phi(t')dt'\right| = \sqrt{b} \tag{8.103}$$

as well as

$$|\tilde{\Phi}(0)| = \frac{1}{\sqrt{2\pi}}\left|\int_{-\infty}^{\infty} \Phi(t')dt'\right| = \sqrt{\frac{b}{2\pi}}. \tag{8.104}$$

Fourier transformation of (8.86) gives

$$\delta_{nn'} = \int \Phi^*(t - nb)\Phi(t - n'b)dt$$

$$= \frac{1}{2\pi} \int dt \int \tilde{\Phi}^*(\omega)\mathrm{e}^{-\mathrm{i}\omega(t-nb)}d\omega \int \tilde{\Phi}(\omega')\mathrm{e}^{\mathrm{i}\omega'(t-n'b)}d\omega'$$

$$= \int d\omega d\omega' \tilde{\Phi}^*(\omega)\tilde{\Phi}(\omega')\mathrm{e}^{\mathrm{i}(\omega n - \omega' n')b}\delta(\omega - \omega') = \int d\omega |\tilde{\Phi}(\omega)|^2 \mathrm{e}^{\mathrm{i}\omega(n-n')b}$$

$$= \sum_{j=-\infty}^{\infty} \int_{2\pi j/b}^{2\pi(j+1)/b} d\omega |\tilde{\Phi}(\omega)|^2 \mathrm{e}^{\mathrm{i}\omega(n-n')b} = \int_0^{2\pi/b} d\omega \sum_{j=-\infty}^{\infty} |\tilde{\Phi}(\omega + 2\pi j/b)|^2 \mathrm{e}^{-\mathrm{i}\omega\Delta nb}$$

$$= \int_0^{2\pi/b} d\omega F(\omega)\mathrm{e}^{-\mathrm{i}\omega\Delta nb}. \tag{8.105}$$

$F(\omega)$ is periodic with period $\Omega_0 = 2\pi/b$ and can be represented as a Fourier sum

$$F(\omega) = \sum_{n=-\infty}^{\infty} F_n \mathrm{e}^{\mathrm{i}2\pi n\omega/\Omega_0} = \sum_{n=-\infty}^{\infty} F_n \mathrm{e}^{\mathrm{i}nb\omega} \tag{8.106}$$

where the Fourier coefficients

$$F_n = \frac{1}{\Omega_0} \int_0^{\Omega_0} F(\omega)\mathrm{e}^{-\mathrm{i}2\pi n\omega/\Omega_0} = \frac{b}{2\pi} \int_0^{2\pi/b} F(\omega)\mathrm{e}^{-\mathrm{i}nb\omega} \tag{8.107}$$

are found from comparison with (8.105)

$$F_n = \frac{b}{2\pi}\delta_{n,0} \tag{8.108}$$

Finally, evaluation of the Fourier sum (8.106) gives

$$F(\omega) = \sum_j |\tilde{\Phi}(\omega + j\Omega_0)|^2 = \frac{1}{\Omega_0} \tag{8.109}$$

which is the equivalent of the orthonormality of Φ_{0n} in Fourier space.

Equation (8.93) implies that $\Phi_{m,n}$can be represented as linear combination of the $\Phi_{m+1,n}$. Starting from

$$\Phi(t) = \Phi_{0,0}(t) = \sum_n h_n \Phi_{1,n}(t) = \sqrt{2}\sum_n h_n \Phi(2t - nb) \tag{8.110}$$

scaling and translation gives

$$\Phi_{m,n}(t) = \sum_{n'} h_{n'-2n}\Phi_{m+1,n'}(t). \tag{8.111}$$

Fourier transformation of (8.90) gives

$$\tilde{\Phi}_{m,n}(\omega) = \mathrm{e}^{-2n\pi\mathrm{i}\omega/\Omega_m}\tilde{\Phi}_{m,0}(\omega) = \frac{1}{\sqrt{2^m}}\mathrm{e}^{-2n\pi\mathrm{i}\omega/\Omega_m}\tilde{\Phi}(\omega/2^m) \tag{8.112}$$

$$\tilde{\Phi}_{m+1,n}(\omega) = \frac{1}{\sqrt{2}}\tilde{\Phi}_{mn}(\omega/2) \tag{8.113}$$

with

$$\Omega_m = 2^m\frac{2\pi}{b} = 2^m\Omega_0 \tag{8.114}$$

and (8.110) becomes

$$\tilde{\Phi}(\omega) = \sum_n \frac{h_n}{\sqrt{2}}\mathrm{e}^{-2n\pi\mathrm{i}\omega/\Omega_1}\tilde{\Phi}(\omega/2^m) = M_0(\omega/2)\tilde{\Phi}(\omega/2) \tag{8.115}$$

where

$$M_0(\omega/2) = \sum_n \frac{h_n}{\sqrt{2}}\mathrm{e}^{-2n\pi\mathrm{i}(\omega/2)/\Omega_0} \tag{8.116}$$

is Ω_0-periodic. Similarly, we find

$$\tilde{\Phi}_{m0}(\omega) = \sum_n h_n \tilde{\Phi}_{m+1n}(\omega) = \sum_n \frac{h_n}{\sqrt{2^{m+1}}} \mathrm{e}^{-2n\pi \mathrm{i}(\omega/2^{m+1})/\Omega_0} \tilde{\Phi}(\omega/2^{m+1})$$

$$= \frac{1}{\sqrt{2^m}} M_0(\omega/2^{m+1}) \tilde{\Phi}(\omega/2^{m+1}) \tag{8.117}$$

Eq. (8.115) can be iterated to obtain

$$\tilde{\Phi}(\omega) = M_0(\omega/2)\tilde{\Phi}(\omega/2) = M_0(\omega/2)M_0(\omega/4)\tilde{\Phi}(\omega/4) = \ldots$$

$$= \prod_{j=1}^{\infty} M_0(\omega/2^j)\tilde{\Phi}(0) = \prod_{j=1}^{\infty} M_0(\omega/2^j)\sqrt{\frac{b}{2\pi}}. \tag{8.118}$$

This equation shows that knowledge of M_0 is sufficient to determine the scaling function (see also Sect. 8.7.4).

From the orthogonality condition (8.109) we obtain

$$\frac{1}{\Omega_0} = \sum_j |\tilde{\Phi}(\omega + j\Omega_0)|^2 = \sum_j |M_0(\omega/2 + j\frac{\Omega}{2})|^2 |\tilde{\Phi}(\omega/2 + j\frac{\Omega}{2})|^2$$

$$= \sum_j |M_0(\omega/2 + j\Omega_0)|^2 |\tilde{\Phi}(\omega/2 + j\Omega_0)|^2 + \sum_j |M_0(\omega/2 + (j + \frac{1}{2})\Omega_0)|^2 |\tilde{\Phi}(\omega/2$$
$$+ (j + \frac{1}{2})\Omega_0)|^2$$

$$= |M_0(\omega/2)|^2 \sum_j |\tilde{\Phi}(\omega/2 + j\Omega_0)|^2 + |M_0(\omega/2 + \frac{\Omega_0}{2})|^2 \sum_j |\tilde{\Phi}((\omega + \Omega_0)/2 + j\Omega_0)|^2$$

$$= \frac{1}{\Omega_0}\left[|M_0(\omega/2)|^2 + |M_0(\omega/2 + \frac{\Omega_0}{2})|^2 \right]. \tag{8.119}$$

Example: Rectangular Scaling Function

The simplest example of a scaling function is the rectangular function

$$\Phi(t) = \begin{cases} \frac{1}{\sqrt{b}} & \text{for } |t - \frac{b}{2}| \le \frac{b}{2} \\ 0 & \text{else} \end{cases} \tag{8.120}$$

with the scaled and translated functions

$$\Phi_{0,n}(t) = \Phi(t - nb) = \begin{cases} \frac{1}{\sqrt{b}} & \text{for } |t - (n + \frac{1}{2})b| \le \frac{b}{2} \\ 0 & \text{else} \end{cases} \tag{8.121}$$

$$\Phi_{1,n}(t) = \sqrt{2}\Phi(2t - nb) = \begin{cases} \frac{1}{\sqrt{b/2}} & \text{for } |t - (n + \frac{1}{2})\frac{b}{2}| \le \frac{b}{4} \\ 0 & \text{else} \end{cases} \tag{8.122}$$

$$\vdots$$

$$\Phi_{m,n}(t) = \sqrt{2^m}\Phi(2^m t - nb) = \begin{cases} \frac{1}{\sqrt{b/2^m}} & \text{for } |t - (n + \frac{1}{2})\frac{b}{2^m}| < \frac{b}{2^{m+1}} \\ 0 & \text{else.} \end{cases} \tag{8.123}$$

Obviously, the $\Phi_{mn}(t)$ for fixed m are orthonormal and can be represented as linear combination

$$\Phi_{m,n}(t) = \frac{1}{\sqrt{2}}\Phi_{m+1,2n}(t) + \frac{1}{\sqrt{2}}\Phi_{m+1,2n+1}(t). \tag{8.124}$$

V_m is the space of functions which are piecewise constant on intervals $|t - (n + 1/2)b/2^m| < b/2^{m+1}$. Figure 8.16 shows the approximation of the parabola $f(t) = t^2$ by functions in $V_0 \dots V_3$.

The Fourier transform of the scaling function is

$$\tilde{\Phi}(\omega) = \frac{1}{\sqrt{2\pi}} \frac{2\sin\left(\frac{\omega b}{2}\right)}{\omega\sqrt{b}} \mathrm{e}^{-\mathrm{i}\omega b/2} = \frac{\sqrt{b}}{\sqrt{2\pi}} \operatorname{sinc}\left(\frac{\omega b}{2}\right) \mathrm{e}^{-\mathrm{i}\omega b/2} \tag{8.125}$$

and from

$$\tilde{\Phi}(\omega) = \frac{1}{\sqrt{2\pi}} \frac{2\left[2\sin\left(\frac{\omega b}{4}\right)\cos\left(\frac{\omega b}{4}\right)\right]}{\omega\sqrt{b}} \mathrm{e}^{-\mathrm{i}\omega b/2} = \tilde{\Phi}\left(\frac{\omega b}{2}\right)\cos\left(\frac{\omega b}{4}\right)\mathrm{e}^{-\mathrm{i}\omega b/4} \tag{8.126}$$

we find

$$M_0\left(\frac{\omega}{2}\right) = \cos\left(\frac{\omega b}{4}\right)\mathrm{e}^{-\mathrm{i}\omega b/4}. \tag{8.127}$$

8.6.2 Construction of an Orthonormal Wavelet Basis

The approximation $f^{(m+1)}(t)$ contains more details than $f^{(m)}(t)$. We would like to extract these details by dividing the space

$$V_{m+1} = V_m + W_m \tag{8.128}$$

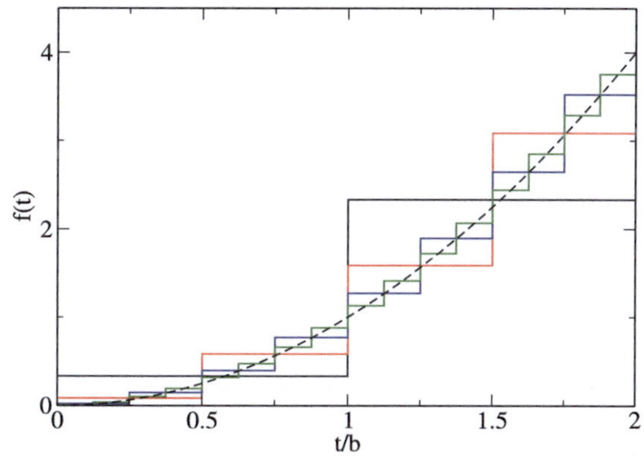

Fig. 8.16 Approximation by piecewise constant functions. The parabola $f(t) = t^2$(dashed curve) is approximated by linear combination of orthonormal rectangular functions (8.123) $f_m(t) = \sum_n \Phi_{mn}(t) \int_{-\infty}^{\infty} \Phi^*_{mn}(t) f(t) dt$ for $m = 0$ (black) $m = 1$ (red) $m = 2$ (blue) $m = 3$ (green)

into the sum of V_m and an orthogonal complement $W_m \perp V_m$. The approximation $f^{(m+1)}(t)$ then can be divided into the approximation $f^{(m)}$ plus the projection onto W_m, which provides the details. In the following we will construct an orthonormal basis of W_m in terms of wavelet functions $\Psi(t)$ which have the properties

$$\text{(i)} \quad \Psi \in V_{m+1} \tag{8.129}$$

or

$$\Psi = \sum_n C_n \Phi_{m+1,n} \tag{8.130}$$

and

$$\text{(ii)} \quad \Psi \perp V_m \tag{8.131}$$

or

$$\int_{-\infty}^{\infty} \Psi^*(t)\Phi_{mn}(t)dt = 0 \quad \forall n \tag{8.132}$$

which is equivalent to

$$\int_{-\infty}^{\infty} \tilde{\Psi}^*(\omega)\tilde{\Phi}_{mn}(\omega)d\omega = 0 \quad \forall n. \tag{8.133}$$

Example: Haar Wavelet

With the rectangular scaling function

$$\Phi(t) = \begin{cases} 1 \text{ if } 0 \le x \le 1 \\ 0 \text{ else} \end{cases} \tag{8.134}$$

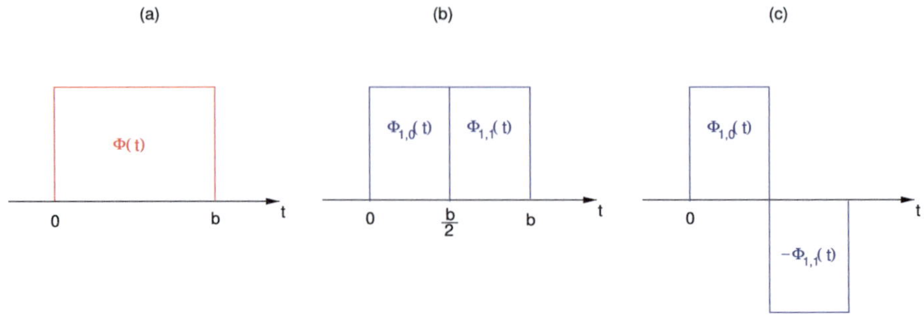

Fig. 8.17 Haar wavelet. The rectangular scaling function (**a**) can be written as a linear combination of translated scaling functions at the next higher resolution (**b**). This is also the case for the wavelet function (**c**) which is orthogonal to the scaling function

the Haar wavelet [129] (Fig. 8.17)

$$\Psi(t) = \frac{1}{\sqrt{2}}\Phi_{1,0}(t) - \frac{1}{\sqrt{2}}\Phi_{1,1}(t) \tag{8.135}$$

is a linear combination of the translated functions $\Phi_{1,n}$ and orthogonal to all $\Phi_{0,n}$. The family of scaled and translated daughter wavelets

$$\Psi_{m,n}(t) = \frac{1}{\sqrt{2}}\Phi_{m,n}(t) - \frac{1}{\sqrt{2}}\Phi_{m,n+1}(t) \tag{8.136}$$

obeys

$$\Psi_{m,n} \in V_{m+1} \quad \Psi_{m,n} \perp V_m. \tag{8.137}$$

Orthogonality Condition

After Fourier transformation (8.133) becomes

$$0 = \int \tilde{\Psi}^*(\omega)\mathrm{e}^{-ni\omega 2\pi/\Omega_m}\tilde{\Phi}_{m0}(\omega)d\omega$$

$$= \sum_j \int_{j\Omega_m}^{(j+1)\Omega_m} \tilde{\Psi}^*(\omega)\mathrm{e}^{-ni\omega 2\pi/\Omega_m}\tilde{\Phi}_{m0}(\omega)d\omega$$

$$= \sum_j \int_0^{\Omega_m} \tilde{\Psi}^*(\omega + j\Omega_m)\mathrm{e}^{-ni(\omega+j2\pi/t_m)t_m}\tilde{\Phi}_{m0}(\omega + j\Omega_m)d\omega$$

$$= \int_0^{\Omega_m} e^{-ni\omega 2\pi/\Omega_m} \sum_j \tilde{\Psi}^*(\omega + j\Omega_m)\tilde{\Phi}_{m0}(\omega + j\Omega_m)d\omega$$

$$= \int_0^{\Omega_m} e^{-ni\omega 2\pi/\Omega_m} G(\omega)d\omega = \Omega_m \hat{G}(t_n). \tag{8.138}$$

This expression looks like the Fourier coefficient of an Ω_m-periodic function with the Fourier sum (7.5) with ω and t exchanged

$$G(\omega) = \sum_{n=-\infty}^{\infty} e^{it_n\omega}\hat{G}(t_n) \text{ with } t_n = n\frac{2\pi}{\Omega_{m.}} \tag{8.139}$$

But, since $\hat{G}(t) = 0$, we obtain the orthogonality condition

$$\sum_j \tilde{\Psi}^*(\omega + j\Omega_m)\tilde{\Phi}_{m0}(\omega + j\Omega_m) = 0. \tag{8.140}$$

Construction of the Wavelet

Now, Ψ and Φ_{m0} both are in V_{m+1}, therefore (8.115)

$$\tilde{\Phi}_{m0} = M_{m0}(\omega/2^{m+1})\tilde{\Phi}(\omega/2^{m+1}) \tag{8.141}$$

$$\tilde{\Psi} = M_\Psi(\omega/2^{m+1})\tilde{\Phi}(\omega/2^{m+1}) \tag{8.142}$$

where $M_{m,0}$ and M_Ψ are Ω_0-periodic.

Hence, from (8.140)

$$0 = \sum_j M_\Psi^*((\omega + j\Omega_m)/2^{m+1})M_{m0}((\omega + j\Omega_m)/2^{m+1})|\tilde{\Phi}((\omega + j\Omega_m)/2^{m+1})|^2$$

$$= \sum_j M_\Psi^*(\omega/2^{m+1} + j\Omega_0/2)M_{m0}(\omega/2^{m+1} + j\Omega_0/2)|\tilde{\Phi}(\omega/2^{m+1} + j\Omega_0/2)|^2$$

$$= \sum_{j\,even} M_\Psi^*(\omega/2^{m+1})M_{m0}(\omega/2^{m+1}2)|\tilde{\Phi}(\omega/2^{m+1} + \Omega_0 j/2)|^2$$

$$+ \sum_{j\,odd} M_\Psi^*(\omega/2^{m+1} + \frac{\Omega_0}{2})M_{m0}(\omega/2^{m+1} + \frac{\Omega_0}{2})|\tilde{\Phi}(\omega/2^{m+1} + \Omega_0 j/2)|^2$$

$$= M_\Psi^*(\omega/2^{m+1})M_{m0}(\omega/2^{m+1}2)\sum_{j\,even}|\tilde{\Phi}(\omega/2^{m+1}+\Omega_0 j/2)|^2$$

$$+ M_\Psi^*(\omega/2^{m+1}+\frac{\Omega_0}{2})M_{m0}(\omega/2^{m+1}+\frac{\Omega_0}{2})\sum_{j\,odd}|\tilde{\Phi}(\omega/2^{m+1}+\Omega_0 j/2)|^2. \tag{8.143}$$

From orthogonality of $\Phi_{m+1,n}$

$$\sum_j |\tilde{\Phi}(\omega/2^{m+1}+j\Omega_0)|^2 = \frac{1}{\Omega_0} \tag{8.144}$$

we see that both sums have the same value

$$\sum_{j\,even}|\tilde{\Phi}(\omega/2^{m+1}+\Omega_0 j/2)|^2 = \sum_k |\tilde{\Phi}(\omega/2^{m+1}+k\Omega_0)|^2 = \frac{1}{\Omega_0} \tag{8.145}$$

$$\sum_{j\,odd}|\tilde{\Phi}(\omega/2^{m+1}+\Omega_0 j/2)|^2 = \sum_k |\tilde{\Phi}(\omega/2^{m+1}+k\Omega_0+\Omega_0/2)|^2 = \frac{1}{\Omega_0} \tag{8.146}$$

and therefore

$$M_\Psi^*(\omega/2^{m+1})M_{m0}(\omega/2^{m+1}2) + M_\Psi^*(\omega/2^{m+1}+\frac{\Omega_0}{2})M_{m0}(\omega/2^{m+1}+\frac{\Omega_0}{2}) = 0 \tag{8.147}$$

which can be satisfied by choosing [127]

$$M_\Psi(\omega/2^{m+1}) = M_{m0}^*(\omega/2^{m+1}+\frac{\Omega_0}{2})\mathrm{e}^{\mathrm{i}\omega 2\pi/\Omega_{m+1}} \tag{8.148}$$

which implies

$$M_\Psi(\omega/2^{m+1}+\frac{\Omega_0}{2}) = M_{m0}^*(\omega/2^{m+1}+\Omega_0)\mathrm{e}^{\mathrm{i}(\omega+\Omega_{m+1}/2)2\pi/\Omega_{m+1}}$$

$$= -M_{m0}^*(\omega/2^{m+1})\mathrm{e}^{\mathrm{i}\omega 2\pi/\Omega_{m+1}}. \tag{8.149}$$

Hence we obtain the solution

$$\tilde{\Psi}_m(\omega) = \mathrm{e}^{\mathrm{i}\omega 2\pi/\Omega_{m+1}} M_{m0}^*(\omega/2^{m+1}+\frac{\Omega_0}{2})\tilde{\Phi}(\omega/2^{m+1})$$

$$= \sum_{n'} \frac{h_{n'}^*}{\sqrt{2^{m+1}}}\mathrm{e}^{\mathrm{i}n'\pi}\mathrm{e}^{\mathrm{i}(n'+1)\omega 2\pi/\Omega_{m+1}}\tilde{\Phi}(\omega/2^{m+1}) \tag{8.150}$$

which becomes in the time domain

$$\Psi_m(t) = \sum_{n'} \frac{h^*_{n'}}{\sqrt{2^{m+1}}}(-1)^{n'}\frac{1}{\sqrt{2\pi}}\int d\omega e^{i\omega t} e^{i(n'+1)\omega 2\pi/\Omega_{m+1}}\tilde{\Phi}(\omega/2^{m+1})$$

$$= \sum_{n'}(-1)^{n'} h^*_{n'}\sqrt{2^{m+1}}\Phi(2^{m+1}t + (n'+1)b)$$

$$= \sum_{n'}(-1)^{n'} h^*_{n'}\Phi_{m+1,-n'-1}(t) \tag{8.151}$$

$$= \sum_{n}(-1)^{-n-1} h^*_{-n-1}\Phi_{m+1,n}(t). \tag{8.152}$$

From the orthogonality condition (8.109) we obtain

$$\sum_j |\tilde{\Psi}_0(\omega + j\Omega_0)|^2 = \sum |M_{00}(\omega/2 + (j+1)\frac{\Omega_0}{2})|^2 |\tilde{\Phi}(\omega/2 + j\frac{\Omega_0}{2})|^2$$

$$= \sum |M_{00}(\omega/2 + (2j+1)\frac{\Omega_0}{2})|^2 |\tilde{\Phi}(\omega/2 + 2j\frac{\Omega_0}{2})|^2$$

$$+ \sum |M_{00}(\omega/2 + (2j+2)\frac{\Omega_0}{2})|^2 |\tilde{\Phi}(\omega/2 + (2j+1)\frac{\Omega_0}{2})|^2$$

$$= |M_{00}(\omega/2 + \frac{\Omega_0}{2})|^2 \sum |\tilde{\Phi}(\omega/2 + j\Omega_0)|^2 + |M_{00}(\omega/2)|^2 \sum |\tilde{\Phi}((\omega + \Omega_0)/2 + j\Omega_0)|^2$$

$$= \frac{1}{\Omega_0}\left(|M_{00}(\omega/2 + \frac{\Omega_0}{2})|^2 + |M_{00}(\omega/2)|^2\right). \tag{8.153}$$

But, since the scaling function obeys the orthonormality condition (8.109),

$$\frac{1}{\Omega_0} = \sum |\tilde{\Phi}(\omega + j\Omega_0)|^2 = \sum |M_{00}(\omega/2 + j\frac{\Omega}{2})|^2 |\tilde{\Phi}(\omega/2 + j\frac{\Omega}{2})|^2$$

$$= \sum |M_{00}(\omega/2 + j\Omega_0)|^2 |\tilde{\Phi}(\omega/2 + j\Omega_0)|^2 + \sum |M_{00}(\omega/2 + (j+\frac{1}{2})\Omega_0)|^2 |\tilde{\Phi}(\omega/2 + (j+\frac{1}{2})\Omega_0)|^2$$

$$= |M_{00}(\omega/2)|^2 \sum |\tilde{\Phi}(\omega/2 + j\Omega_0)|^2 + |M_{00}(\omega/2 + \frac{\Omega_0}{2})|^2 \sum |\tilde{\Phi}((\omega + \Omega_0)/2 + j\Omega_0)|^2$$

$$= \frac{1}{\Omega_0}\left[|M_{00}(\omega/2)|^2 + |M_{00}(\omega/2 + \frac{\Omega_0}{2})|^2\right] \tag{8.154}$$

hence the wavelet also fulfills the orthonormality condition

$$\sum_j |\tilde{\Psi}_0(\omega + j\Omega_0)|^2 = \frac{1}{\Omega_0} \tag{8.155}$$

therefore the translated wavelet functions

$$\Psi_{mn}(t) = \Psi_m(t - n2^{-m}b) = \sum_{n'}(-1)^{-n'-1}h^*_{-n'-1}\sqrt{2^{m+1}}\Phi(2^{m+1}(t - n2^{-m}b) - n'b)$$

$$= \sum_{n'}(-1)^{-n'-1}h^*_{-n'-1}\Phi_{m+1,2n+n'}(t) \tag{8.156}$$

are orthonormal

$$\int \Psi^*_{mn}(t)\Psi_{mn'}(t)dt = \delta_{n.n'}. \tag{8.157}$$

Wavelets for different resolution m are orthogonal since they are by construction in orthogonal spaces. The $\Psi_{mn}(t)$ with $m, n = -\infty \ldots \infty$
provide an orthonormal basis of

$$L^2(\mathbb{R}) = \bigcup_{m=-\infty}^{\infty} W_m. \tag{8.158}$$

Alternatively, (8.158) is replaced by

$$L^2(\mathbb{R}) = V_0 + \bigcup_{m=0}^{\infty} W_m \tag{8.159}$$

which is more useful for practical applications with limited total observation time. According to (8.159), starting from a basic approximation in V_0, more and more details are added to obtain approximations with increasing accuracy.

Example: Meyer Wavelet

Meyer introduced the first non trivial wavelet (Fig. 8.18) which, in contrast to the Haar wavelet is differentiable [130, 131]. It was originally defined by its scaling function in Fourier space[11] (here, $b = 1$)

[11] There are different variants of the Meyer wavelet in the literature.

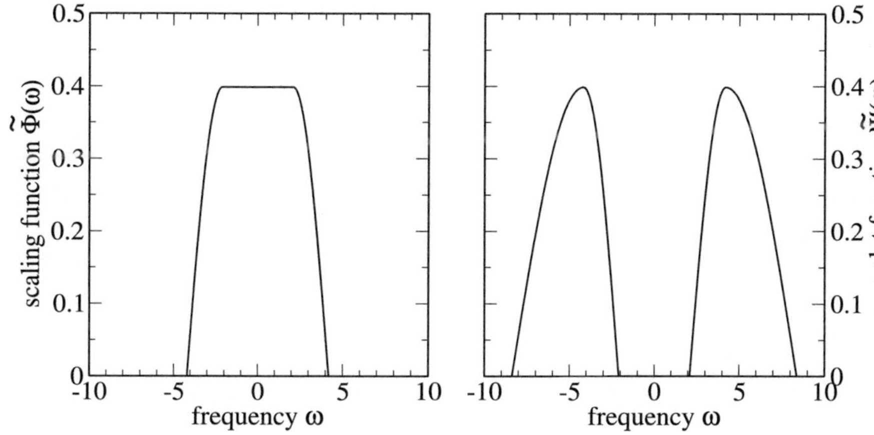

Fig. 8.18 Meyer wavelet in frequency space. **Left:** Scaling function; **Right:** Magnitude of the wavelet function

$$\tilde{\Phi}(\omega) = \begin{cases} \frac{1}{\sqrt{2\pi}} \quad \text{if } \omega \le \frac{2\pi}{3} \\ \frac{1}{\sqrt{2\pi}} \cos\left(\frac{\pi}{2}\left(\frac{3|\omega|}{2\pi} - 1\right)\right) \quad \text{if } \frac{2\pi}{3} < |\omega| < \frac{4\pi}{3} \\ 0 \quad \text{if } |\omega| > \frac{4\pi}{3} \end{cases} \tag{8.160}$$

from which the mother wavelet can be derived

$$\tilde{\Psi}(\omega) = \begin{cases} \frac{1}{\sqrt{2\pi}} \sin\left(\frac{\pi}{2}\left(\frac{3|\omega|}{2\pi} - 1\right)\right) \mathrm{e}^{\mathrm{i}\omega/2} \quad \text{if } \frac{2\pi}{3} \le |\omega| \le \frac{4\pi}{3} \\ \frac{1}{\sqrt{2\pi}} \cos\left(\frac{\pi}{2}\left(\frac{3|\omega|}{4\pi} - 1\right)\right) \mathrm{e}^{\mathrm{i}\omega/2} \quad \text{if } \frac{4\pi}{3} \le |\omega| \le \frac{8\pi}{3} \\ 0 \quad \text{else.} \end{cases} \tag{8.161}$$

Explicit expressions in the time domain (Fig. 8.19) were given in 2015 [132]:

$$\Phi(t) = \begin{cases} \frac{2}{3} + \frac{4}{3\pi} \quad \text{if } t = 0 \\ \frac{\sin \frac{2\pi}{3} t + \frac{4}{3} t \cos \frac{4\pi}{3} t}{\pi t - \frac{16\pi}{9} t^3} \quad \text{else} \end{cases} \tag{8.162}$$

$$\Psi(t) = \frac{\frac{4}{3\pi}\left(t - \frac{1}{2}\right) \cos\left[\frac{2\pi}{3}\left(t - \frac{1}{2}\right)\right] - \frac{1}{\pi} \sin\left[\frac{4\pi}{3}\left(t - \frac{1}{2}\right)\right]}{\left(t - \frac{1}{2}\right) - \frac{16}{9}\left(t - \frac{1}{2}\right)^3}$$

$$+ \frac{\frac{8}{3\pi}\left(t - \frac{1}{2}\right) \cos\left[\frac{8\pi}{3}\left(t - \frac{1}{2}\right)\right] + \frac{1}{\pi} \sin\left[\frac{4\pi}{3}\left(t - \frac{1}{2}\right)\right]}{\left(t - \frac{1}{2}\right) - \frac{64}{9}\left(t - \frac{1}{2}\right)^3} \tag{8.163}$$

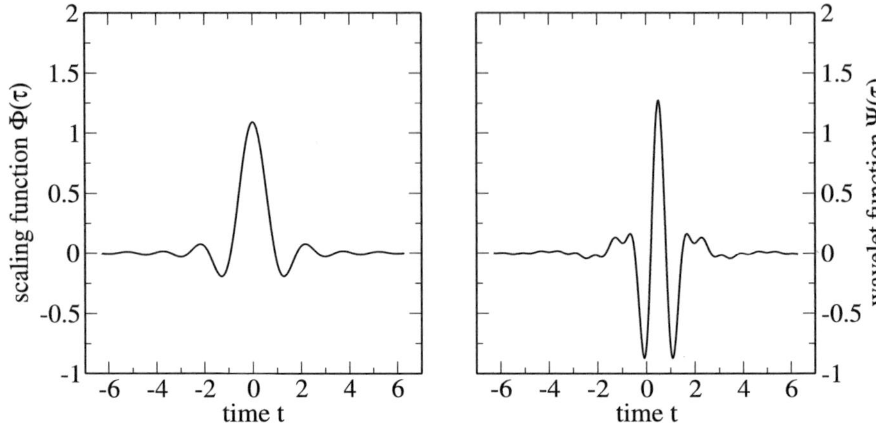

Fig. 8.19 Meyer wavelet in the time domain. **Left:** Scaling function; **Right:** Wavelet function

8.7 Discrete Data and Fast Wavelet Transform

Mallet's algorithm [128] starts with function values

$$f_n = f(n\Delta t_s) \tag{8.164}$$

sampled at multiples of

$$\Delta t_s = 1/f_s = b/2^{m_{\max}}. \tag{8.165}$$

We do not really approximate the function but from the series of sample values we construct the linear combination

$$\sum_n f_n \Phi_{m_{\max},n}(t) \tag{8.166}$$

which is an element of

$$V_{m_{\max}} = V_0 + \bigcup_{m=0}^{m_{\max}-1} W_m \tag{8.167}$$

and can therefore be represented as a coarse approximation in V_0 and a series of details with increasing resolution

$$\sum_n f_n \Phi_{m,n}(t) = \sum_n c_n \Phi_{0,n}(t) + \sum_{m=0}^{m_{\max}-1} \sum_n d_{mn} \Psi_{mn}(t). \tag{8.168}$$

8.7.1 Recursive Wavelet Transformation

The approximation coefficients c_n and detail coefficients d_{mn} are determined recursively which avoids the calculation of scalar products. Starting with

$$c_{m_{\max},n} = f_n \tag{8.169}$$

the details are extracted by expanding

$$\sum_n c\Phi_{m_{\max},n}(t) = \sum_n c_{m_{\max}-1,n}\Phi_{m_{\max}-1,n}(t) + \sum_n d_{m_{\max}-1,n}\Psi_{m_{\max}-1,n}(t). \tag{8.170}$$

Due to orthogonality, the coefficients at the next lower resolution can be determined from

$$c_{m_{\max}-1,n'} = \sum_n c_{m_{\max},n} < \Phi_{m_{\max}-1,n'}|\Phi_{m_{\max},n} >= \sum_n c_{m_{\max},n} h^*_{n-2n'}$$

$$= \sum_n c_{m_{\max},n+2n'} h^*_n \tag{8.171}$$

$$d_{m_{\max}-1,n'} = \sum_n c_{m_{\max},n} < \Psi_{m_{\max}-1,n'}|\Phi_{m_{\max},n} >= \sum_n c_{m_{\max},n}(-1)^{n-1} h_{2n'-n-1} \tag{8.172}$$

which can be written as

$$d_{m_{\max}-1,n'} = \sum_n c_{m_{\max},n} g^*_{n-2n'} = \sum_n c_{m_{\max},n+2n'} g^*_n \quad \text{with } g^*_n = (-1)^{n-1} h_{-n-1}. \tag{8.173}$$

Iterating this recursion allows the calculation of the wavelet coefficients even without explicit knowledge of the scaling and wavelet functions. Equations (8.171) and (8.173) have the form of discrete digital filter functions with subsequent down-sampling by a factor of two (dropping samples with odd n').[12] This can be seen by defining the down sampled coefficients

$$c^{\downarrow}_{n'/2} = \sum_n c_n h^*_{n-n'} \tag{8.174}$$

$$d^{\downarrow}_{n'/2} = \sum_n c_n (-1)^{n-1} h_{n'-n-1} \tag{8.175}$$

---

[12] For the more general class of bi-orthogonal wavelets, a different filter pair is used for reconstruction.

and applying the z-transform to (8.171, 8.173). For the approximation filter we obtain

$$f_c(z) = \sum_{n'=-\infty}^{\infty} \sum_{n=-\infty}^{\infty} h_n^* c_{n+n'} z^{-n'}$$

$$= \sum_n h_n^* z^n \sum_{n'} c_{n+n'} z^{-n'-n} = (\sum_n h_n z^{-n})^* \sum_{n'} c_{n'} z^{-n'} = h^*(z)\, c(z) \tag{8.176}$$

hence in frequency space the signal is multiplied with the filter function

$$h(\mathrm{e}^{\mathrm{i}\omega\Delta t}) = \sum_n h_n \mathrm{e}^{-n\mathrm{i}\omega\Delta t} = \sqrt{2} M_0(\omega). \tag{8.177}$$

Similarly we obtain for the detail filter

$$f_d(z) = \sum_{nn'} c_n (-1)^{n-1} h_{n'-n-1} z^{-n'} = \sum c_n z^{-n} (-1)^{n-1} h_{n'-n-1} z^{n-n'}. \tag{8.178}$$

Since only even values of n' are relevant, we may change the sign by $(-1)^{n'}$ to obtain

$$\sum c_n z^{-n} (-1)^{n'-n-1} h_{n'-n-1} z^{n-n'} = z^* h(-z)\, c(z) = g^*(z)\, c(z) \tag{8.179}$$

where

$$g(z) = \sum_n (-1)^{n-1} h_{-n-1}^* z^{-n} = \sum_n (-1)^{-n-2} h_n^* z^{n+1} = z \sum_n h_n^* (-z)^n$$

$$= z(\sum_n h_n (-z)^{-n})^* = z h^*(-z). \tag{8.180}$$

8.7.2 Example: Haar Wavelet

For the Haar wavelet with[13]

$$h_0 = h_1 = \frac{1}{\sqrt{2}} \quad h_n = 0 \text{ else} \tag{8.181}$$

$$g_{-1} = \frac{1}{\sqrt{2}} \quad g_{-2} = -\frac{1}{\sqrt{2}} \quad g_n = 0 \text{ else} \tag{8.182}$$

[13] The standard form of the Haar wavelet with $g_0 = 1/\sqrt{2}$, $g_1 = -1/\sqrt{2}$ differs from (8.182) by a shift and time reversal. The resulting wavelet basis, however, is the same.

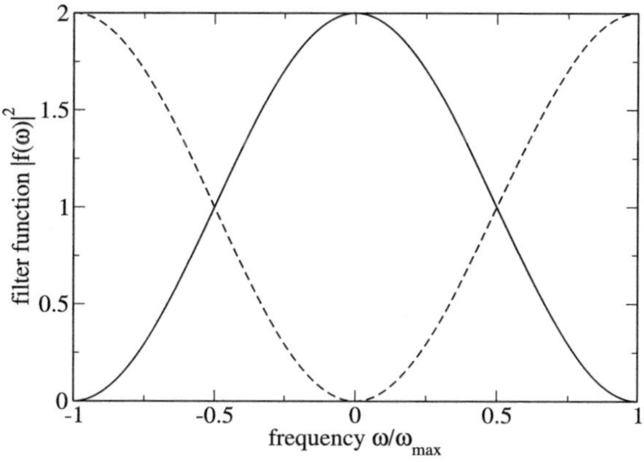

Fig. 8.20 Haar filter pair. Low pass filter (8.184, full curve); high pass filter (8.185, broken curve)

we obtain the filter functions

$$h(z) = \frac{1}{\sqrt{2}}(1 + \frac{1}{z}) \quad g(z) = \frac{1}{\sqrt{2}}(z - z^2). \tag{8.183}$$

On the unit circle,

$$|h(e^{i\omega\Delta t})|^2 = 1 + \cos\omega\Delta t \tag{8.184}$$

$$|g(e^{i\omega\Delta t})|^2 = 1 - \cos\omega\Delta t \tag{8.185}$$

which describes a low and a high pass forming a so called quadrature mirror filter pair (Fig. 8.20) [133].

8.7.3 Signal Reconstruction

The wavelet transformation can be inverted using the expansion

$$\sum_n c_{m,n}\Phi_{m,n}(t) = \sum_n c_{m-1,n}\Phi_{m-1,n}(t) + \sum_n d_{m-1,n}\Psi_{m-1,n}(t) \tag{8.186}$$

where the coefficients at the higher level of approximation are obtained from

$$c_{m,n'} = \sum_n c_{m-1,n} < \Phi_{m,n'}|\Phi_{m-1,n} > + \sum_n d_{m-1,n} < \Phi_{m,n'}|\Psi_{m-1,n} >$$

$$= \sum_n c_{m-1,n}h_{n'-2n} + \sum_n d_{m-1,n}(-1)^{n'-1}h^*_{2n-n'-1}$$

$$= \sum_n c_{m-1,n} h_{n'-2n} + \sum_n d_{m-1,n} (-1)^{n'-1} h^*_{2n-n'-1}. \tag{8.187}$$

This can be formulated as upsampling and subsequent filtering. Formally, we insert zeros and define the up sampled coefficients

$$c^{\uparrow}_{2n} = c_{m-1,n} \quad c^{\uparrow}_{2n+1} = 0 \tag{8.188}$$

$$d^{\uparrow}_{2n} = d_{m-1,n} \quad d^{\uparrow}_{2n+1} = 0. \tag{8.189}$$

Then,

$$\sum_n c_{m-1,n} h_{n'-2n} = \sum_n c^{\uparrow}_{2n} h_{n'-2n} = \sum_n c^{\uparrow}_n h_{n'-n} \tag{8.190}$$

$$\sum_n d_{m-1,n} (-1)^{n'-1} h^*_{2n-n'-1} = (-1)^{n'-1} \sum_n d^{\uparrow}_{2n} h^*_{2n-n'-1}$$

$$= (-1)^{n'-1} \sum_n d^{\uparrow}_n h^*_{n-n'-1} = \sum (-1)^n d_n g_{n'-n} \tag{8.191}$$

where due to (8.189) the alternating sign can be omitted. Z-transformation then gives

$$\sum_{n,n'} c_n h_{n'-n} z^{-n'} = \sum_{n,n'} c_n z^{-n} h_{n'-n} z^{n-n'} = h(z)\, c(z) \tag{8.192}$$

$$\sum_{nn'} d_n g_{n'-n} = g(z) d(z) = z h^*(-z)\, d(z). \tag{8.193}$$

8.7.4 Example: Analysis with Compactly Supported Wavelets

Wavelet analysis has become quite popular for processing of audio and image data. In Exercise 8.3 we use Daubechies wavelets [134] to analyze a complex audio signal consisting of a mixture of short tones, sweeps and noise (Figs. 8.21 and 8.22). Daubechies satisfies (8.119) by taking

$$M_0(\omega/2) = \left[\frac{1}{2}\left(1 + \mathrm{e}^{-\mathrm{i}\omega/2}\right)\right]^N Q(\mathrm{e}^{-\mathrm{i}\omega/2}) \tag{8.194}$$

with a trigonometric polynomial Q. This leads to a class of compactly supported orthonormal wavelet bases, which for N=1 include the Haar wavelet as the simplest member. For N=2,

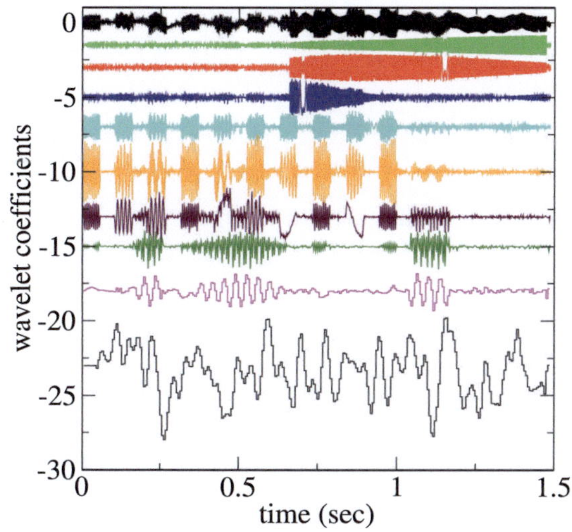

Fig. 8.21 Wavelet coefficients of a complex audio signal. From top to bottom: The black curve shows the input signal. The finest details in light green, red and blue correspond to a high frequency sweep from 5000 to 15000 Hz starting at 0.7 s plus some time dependent noise. Cyan, orange and maroon represent a sequence of dial tones 1000 Hz, dark green and magenta show the signature of several 100 Hz bursts with many harmonics. The black curve at the bottom shows the coefficients of the coarse approximation, which essentially describes random low frequency fluctuations. The curves are vertically shifted relative to each other

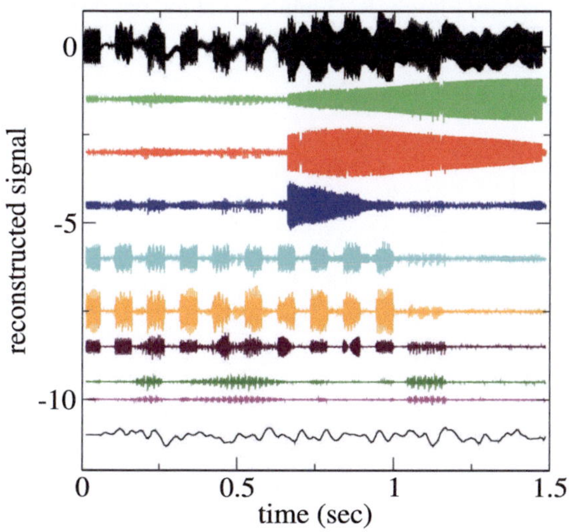

Fig. 8.22 Wavelet reconstruction. The different contributions to the signal are reconstructed from the wavelet coefficients. Color code as in Fig. 8.21. The original signal (top black curve) is exactly the sum of the coarse approximation (bottom black curve) and all details (colored curves). The curves are vertically shifted relative to each other

$$M_0(\omega/2) = \left[\frac{1}{2}\left(1 + e^{-i\omega/2}\right)\right]^2 \frac{1}{2}\left[\left(1+\sqrt{3}\right) + \left(1-\sqrt{3}\right)e^{-i\omega/2}\right] \tag{8.195}$$

$$= \frac{1}{8}\left[\left(1+\sqrt{3}\right) + \left(3+\sqrt{3}\right)e^{-i\omega/2} + \left(3-\sqrt{3}\right)e^{-2i\omega/2} + \left(1-\sqrt{3}\right)e^{-3i\omega/2}\right] \tag{8.196}$$

with the four nonzero scaling parameters

$$h_0 = \frac{\sqrt{2}}{8}(1+\sqrt{3}) \approx 0.48296 \tag{8.197}$$

$$h_1 = \frac{\sqrt{2}}{8}(3+\sqrt{3}) \approx 0.83652 \tag{8.198}$$

$$h_2 = \frac{\sqrt{2}}{8}(3-\sqrt{3}) \approx 0.22414 \tag{8.199}$$

$$h_3 = \frac{\sqrt{2}}{8}(1-\sqrt{3}) \approx -0.12941. \tag{8.200}$$

This defines the wavelet basis which is known as Daubechies 2. There are no analytic expressions for the scaling and wavelet functions available. They can be calculated numerically from the infinite product (8.118) or a corresponding (infinitely) nested convolution in real space. Figures 8.23 and 8.24 show the fast convergence.

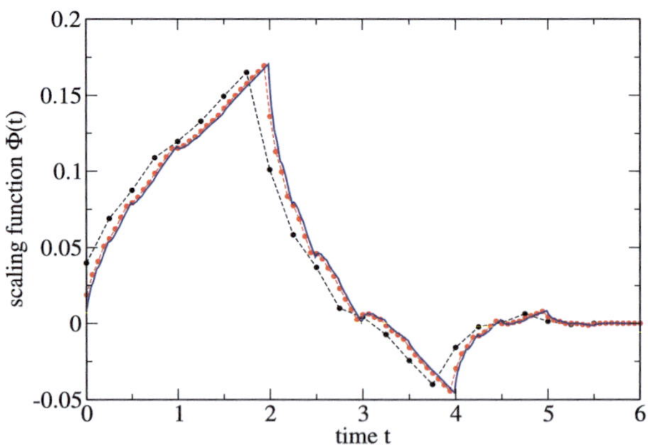

Fig. 8.23 Daubechies 2 scaling function. The scaling function is calculated numerically in the time domain from the Fourier transform of (8.118) with a finite number of factors. The blue curve shows the result for $j_{max} = 7$, red dots show results for $j_{max} = 5$, black dots for $j_{max} = 3$. Delta functions are replaced by rectangular functions of equal area

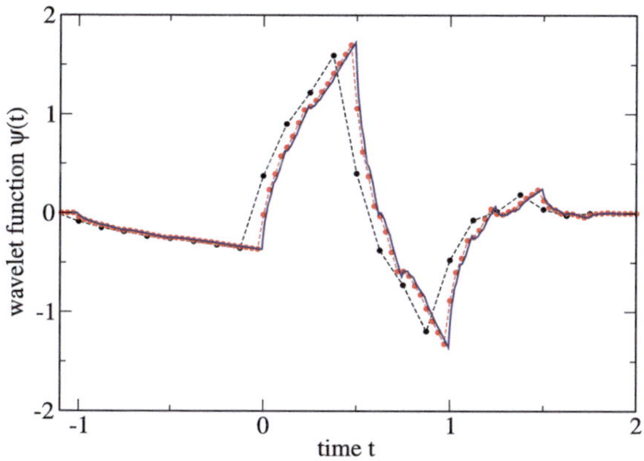

Fig. 8.24 Daubechies 2 wavelet function. The wavelet function is calculated numerically in the time domain from the Fourier transform of (8.118) and (8.150) with a finite number of factors. The blue curve shows the result for $j_{\max} = 7$, red dots show results for $j_{\max} = 5$, black dots for $j_{\max} = 3$. Delta functions are replaced by rectangular functions of equal area

Exercises

Exercise 8.1 Short Time Fourier Transformation

In this computer experiment STFT analysis of a frequency modulated signal

$$f(t) = \sin \Phi(t) = \sin\left(\omega_0 t + \frac{a\omega_0}{\omega_1}(1 - \cos \omega_1 t)\right) \tag{8.201}$$

with a momentaneous frequency of

$$\omega(t) = \frac{\partial \Phi}{\partial t} = \omega_0(1 + a \sin \omega_1 t) \tag{8.202}$$

is performed and shown as a spectrogram (Figs. 8.10 and 8.11). Sampling frequency 44100 Hz, number of samples 512.

You can vary the carrier frequency ω_0, modulation frequency ω_1 and depth a as well as the distance between the windows. Study time and frequency resolution.

Exercise 8.2 Wavelet Analysis of a Nonstationary Signal

In this computer experiment, a complex signal is analyzed with Morlet wavelets over 6 octaves (Fig. 8.14). The signal is sampled with a rate of 44 kHz. The parameter d of the mother wavelet (8.61) determines frequency and time resolution. The frequency ω_0 of the mother wavelet is taken as the Nyquist frequency which is half the sampling rate. The convolution with the daughter wavelets (8.78) is calculated at 400 times with a step size of 0.726 ms (corresponding to 32 samples)

$$t_n = t_0 + n\Delta t \tag{8.203}$$

and for 300 different values of the scaling parameter

$$s_m = 1.015^m. \tag{8.204}$$

The signal consists of two sweeps with linearly increasing frequency of the form

$$f_{1,2}(t) = \sin\left[\omega_{1,2}t + \frac{\alpha_{1,2}}{2}t^2\right] \tag{8.205}$$

and another component which switches between a 5 kHz oscillation and the sum of 300 Hz and a 20kHz oscillation at a rate 20 Hz

$$f_3(t) = \begin{cases} \sin(\omega_{2kHz}t) + \sin(\omega_{300Hz}t) & \text{if } \sin(\omega_{20Hz}t) < 0 \\ \sin(\omega_{5kHz}t) & else. \end{cases} \tag{8.206}$$

Study time and frequency resolution as a function of d

Exercise 8.3 Discrete Wavelet Transformation

In this computer experiment the discrete wavelet transformation is applied to a complex audio signal. You can switch on and off different components like sweeps, dial tones and noise. The wavelet coefficients and the reconstructed signals are shown (Figs. 8.21 and 8.22).

Chapter 9
Random Numbers and Monte Carlo Methods

Many-body problems often involve the calculation of integrals of very high dimension which can not be treated by standard methods. For the calculation of thermal averages Monte Carlo methods [135–138] are very useful which sample the integration volume at randomly chosen points. In this chapter we discuss algorithms for the generation of pseudo-random numbers with given probability distribution which are essential for all Monte Carlo methods. We show how the efficiency of Monte Carlo integration can be improved by sampling preferentially the important configurations. The famous Metropolis-Hastings algorithm is applied to classical many-particle systems and nonlinear optimization problems and the Gibbs sampler is discussed as a special variant which scans the coordinates sequentially.

9.1 Some Basic Statistics

In the following we discuss some important concepts which are used to analyze experimental data sets [139]. Repeated measurements of some observable usually give slightly different results due to fluctuations of the observable in time and errors of the measurement process. The distribution of the measured data is described by a probability distribution, which in many cases approximates a simple mathematical form like the Gaussian normal distribution. The moments of the probability density give important information about the statistical properties, especially the mean and the standard deviation of the distribution. If the errors of different measurements are uncorrelated, the average value of a larger number of measurements is a good approximation to the "exact" value.

Supplementary Information The online version contains supplementary material available at https://doi.org/10.1007/978-3-032-07856-8_9.

P. O. J. Scherer, *Computational Physics I*, Graduate Texts in Physics,
https://doi.org/10.1007/978-3-032-07856-8_9

9.1.1 Probability Density and Cumulative Probability Distribution

Consider an observable ξ, which is measured in a real or a computer experiment. Repeated measurements give a statistical distribution of values. The cumulative probability distribution (Fig. 9.1) is given by the function

$$F(x) = P\{\xi \leq x\} \tag{9.1}$$

and has the following properties:

- $F(x)$ is monotonously increasing
- $F(-\infty) = 0$, $F(\infty) = 1$
- $F(x)$ can be discontinuous (if there are discrete values of ξ).

The probability to measure a value in the interval $x_1 < \xi \leq x_2$ is

$$P(x_1 < \xi \leq x_2) = F(x_2) - F(x_1). \tag{9.2}$$

The height of a jump gives the probability of a discrete value

$$P(\xi = x_0) = F(x_0 + 0) - F(x_0 - 0). \tag{9.3}$$

In regions where $F(x)$ is continuous, the probability density can be defined as

$$f(x_0) = F'(x_0) = \lim_{\Delta x \to 0} \frac{1}{\Delta x} P(x_0 < \xi \leq x_0 + \Delta x). \tag{9.4}$$

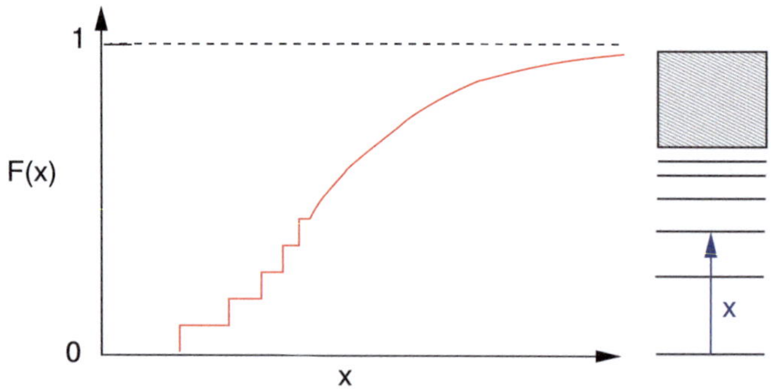

Fig. 9.1 Cumulative probability distribution of transition energies. This figure shows schematically the distribution of transition energies for an atom which has a discrete and a continuous part

9.1.2 Histogram

From an experiment $F(x)$ cannot be determined directly. Instead a finite number N of values x_i are measured. By

$$Z_N(x)$$

we denote the number of measurements with $x_i \leq x$. The cumulative probability distribution is the limit

$$F(x) = \lim_{N \to \infty} \frac{1}{N} Z_N(x). \tag{9.5}$$

A histogram (Fig. 9.2) counts the number of measured values which are in the interval $x_i < x \leq x_{i+1}$:

$$\frac{1}{N}(Z_N(x_{i+1}) - Z_N(x_i)) \approx F(x_{i+1}) - F(x_i) = P(x_i < \xi \leq x_{i+1}). \tag{9.6}$$

Contrary to $Z_N(x)$ itself, the histogram depends on the choice of the intervals.

9.1.3 Expectation Values and Moments

The expectation value of the random variable ξ is defined by

$$E[\xi] = \int_{-\infty}^{\infty} x dF(x) = \lim_{a \to -\infty, b \to \infty} \int_a^b x dF(x) \tag{9.7}$$

with the Riemann-Stieltjes-Integral [140]

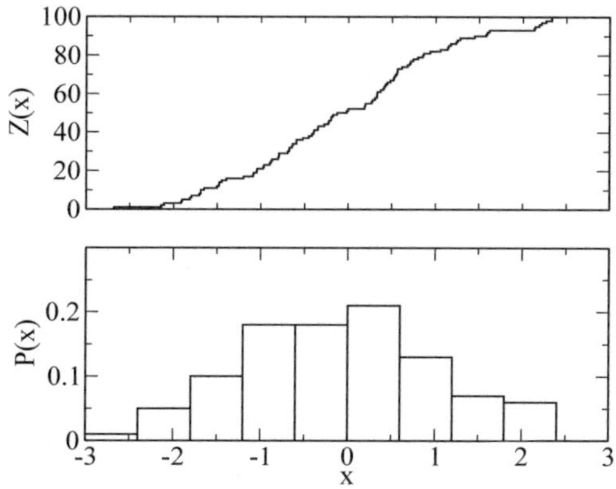

Fig. 9.2 Histogram. The cumulative distribution of 100 Gaussian random numbers is shown together with a histogram with bin width $\Delta x = 0.6$

$$\int_a^b x dF(x) = \lim_{N\to\infty} \sum_{i=1}^{N} x_i (F(x_i) - F(x_{i-1}))|_{x_i = a + \frac{b-a}{N} i}. \tag{9.8}$$

Higher moments are defined as

$$E[\xi^k] = \int_{-\infty}^{\infty} x^k dF(x) \tag{9.9}$$

if these integrals exist. Most important are the expectation value

$$\overline{x} = E[\xi] \tag{9.10}$$

and the variance, which results from the first two moments

$$\sigma^2 = \int_{-\infty}^{\infty} (x - \overline{x})^2 dF = \int x^2 dF + \int \overline{x}^2 dF - 2\overline{x} \int x dF$$

$$= E[\xi^2] - (E[\xi])^2. \tag{9.11}$$

The standard deviation σ is a measure of the width of the distribution. The expectation value of a function $\varphi(x)$ is defined by

$$E[\varphi(x)] = \int_{-\infty}^{\infty} \varphi(x) dF(x). \tag{9.12}$$

For continuous $F(x)$ we have with $\mathrm{d}F(x) = f(x)\mathrm{d}x$ the ordinary integral

$$E[\xi^k] = \int_{-\infty}^{\infty} x^k f(x) dx \tag{9.13}$$

$$E[\varphi(x)] = \int_{-\infty}^{\infty} \varphi(x) f(x) dx \tag{9.14}$$

whereas for a pure step function $F(x)$ (only discrete values x_i are observed with probabilities $p(x_i) = F(x_i + 0) - F(x_i - 0)$)

$$E[\xi^k] = \sum_i x_i^k p(x_i) \tag{9.15}$$

$$E[\varphi(x)] = \sum_i \varphi(x_i) p(x_i). \tag{9.16}$$

9.1.4 Example: Fair Die

When a six-sided fair die is rolled, each of its sides shows up with the same probability of 1/6. The cumulative probability distribution $F(x)$ is a pure step function (Fig. 9.3) and

$$\overline{x} = \int_{-\infty}^{\infty} x dF = \sum_{i=1}^{6} x_i (F(x_i + 0) - F(x_i - 0)) = \frac{1}{6}\sum_{i=1}^{6} x_i = \frac{21}{6} = 3.5 \quad (9.17)$$

$$\overline{x^2} = \sum_{i=1}^{6} x_i^2 (F(x_i + 0) - F(x_i - 0)) = \frac{1}{6}\sum_{i=1}^{6} x_i^2 = \frac{91}{6} = 15.1666\cdots \quad (9.18)$$

$$\sigma = \sqrt{\overline{x^2} - \overline{x}^2} = 2.9. \quad (9.19)$$

9.1.5 Normal Distribution

The Gaussian normal distribution is defined by the cumulative probability distribution

$$\Phi(x) = \frac{1}{\sqrt{2\pi}} \int_{-\infty}^{x} \mathrm{e}^{-t^2/2} dt \quad (9.20)$$

and the probability density

$$\varphi(x) = \frac{1}{\sqrt{2\pi}} \mathrm{e}^{-x^2/2} \quad (9.21)$$

with the properties

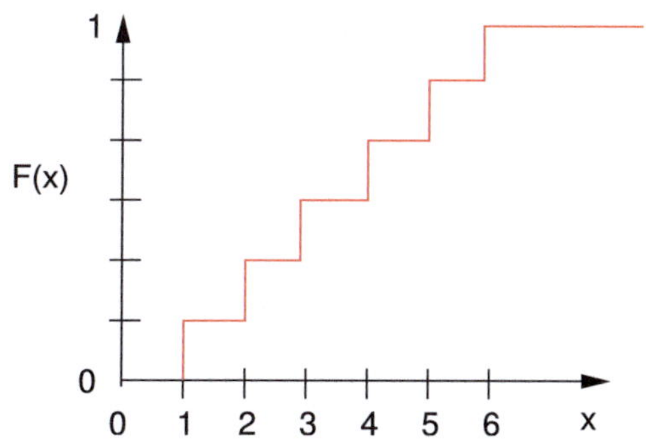

Fig. 9.3 Cumulative probability distribution of a fair die

$$\int_{-\infty}^{\infty} \varphi(x)dx = \Phi(\infty) = 1 \tag{9.22}$$

$$\overline{x} = \int_{-\infty}^{\infty} x\varphi(x)dx = 0 \tag{9.23}$$

$$\sigma^2 = \overline{x^2} = \int_{-\infty}^{\infty} x^2\varphi(x)dx = 1. \tag{9.24}$$

Since $\Phi(0) = \frac{1}{2}$ and with the definition

$$\Phi_0(x) = \frac{1}{\sqrt{2\pi}} \int_0^x \mathrm{e}^{-t^2/2} dt \tag{9.25}$$

we have

$$\Phi(x) = \frac{1}{2} + \Phi_0(x) \tag{9.26}$$

which can be expressed in terms of the error function[1]

$$\operatorname{erf}(x) = \frac{2}{\sqrt{\pi}} \int_0^x \mathrm{e}^{-t^2 dt} dt = 2\Phi_0(\sqrt{2}x) \tag{9.27}$$

as

$$\Phi_0(x) = \frac{1}{2}\operatorname{erf}(\frac{x}{\sqrt{2}}). \tag{9.28}$$

A general Gaussian distribution with mean value $\overline{x}$ and standard deviation σ has the probability distribution

$$\varphi_{\overline{x},\sigma} = \frac{1}{\sigma\sqrt{2\pi}} \exp\left(-\frac{(x' - \overline{x})^2}{2\sigma^2}\right) \tag{9.29}$$

and the cumulative distribution

$$\Phi_{\overline{x},\sigma}(x) = \Phi\left(\frac{x - \overline{x}}{\sigma}\right) = \int_{-\infty}^{x} dx' \frac{1}{\sigma\sqrt{2\pi}} \exp\left(-\frac{(x' - \overline{x})^2}{2\sigma^2}\right) \tag{9.30}$$

$$= \frac{1}{2}\left(1 + \operatorname{erf}\left(\frac{x - \overline{x}}{\sigma\sqrt{2}}\right)\right). \tag{9.31}$$

[1] erf(x) is an intrinsic function in FORTRAN or C.

9.1.6 Multivariate Distributions

Consider now two quantities which are measured simultaneously. ξ and η are the corresponding random variables. The cumulative distribution function is

$$F(x, y) = P(\xi \leq x \quad \text{and} \quad \eta \leq y). \tag{9.32}$$

Expectation values are defined as

$$E[\varphi(x, y)] = \int_{-\infty}^{\infty} \int_{-\infty}^{\infty} \varphi(x, y) d^2 F(x, y). \tag{9.33}$$

For continuous $F(x, y)$ the probability density is

$$f(x, y) = \frac{\partial^2 F}{\partial x \partial y} \tag{9.34}$$

and the expectation value is simply

$$E[\varphi(x, y)] \int_{-\infty}^{\infty} dx \int_{-\infty}^{\infty} dy \varphi(x, y) f(x, y). \tag{9.35}$$

The moments of the distribution are the expectation values

$$M_{k,l} = E[\xi^k \eta^l]. \tag{9.36}$$

Most important are the averages

$$\overline{x} = E[\xi] \quad \overline{y} = E[\eta] \tag{9.37}$$

and the covariance matrix

$$\begin{pmatrix} E[(\xi - \overline{x})^2] & E[(\xi - \overline{x})(\eta - \overline{y})] \\ E[(\xi - \overline{x})(\eta - \overline{y})] & E[(\eta - \overline{y})^2] \end{pmatrix} = \begin{pmatrix} \overline{x^2} - \overline{x}^2 & \overline{xy} - \overline{x}\,\overline{y} \\ \overline{xy} - \overline{x}\,\overline{y} & \overline{y^2} - \overline{y}^2 \end{pmatrix}. \tag{9.38}$$

The correlation coefficient is defined as

$$\rho = \frac{\overline{xy} - \overline{x}\,\overline{y}}{\sqrt{\left(\overline{x^2} - \overline{x}^2\right)\left(\overline{y^2} - \overline{y}^2\right)}}. \tag{9.39}$$

If there is no correlation then $\rho = 0$ and $F(x, y) = F_1(x) F_2(y)$.

9.1.7 Central Limit Theorem

Consider N independent random variables ξ_i with the same cumulative distribution function $F(x)$, for which $E[\xi] = 0$ and $E[\xi^2] = 1$. Define a new random variable

$$\eta_N = \frac{\xi_1 + \xi_2 + \cdots \xi_N}{\sqrt{N}} \tag{9.40}$$

with the cumulative distribution function $F_N(x)$. In the limit $N \to \infty$ this distribution approaches (Fig. 9.4) a cumulative normal distribution [141]

$$\lim_{N\to\infty} F_N(x) = \Phi(x) = \frac{1}{\sqrt{2\pi}} \int_{-\infty}^{x} \mathrm{e}^{-t^2/2} dt. \tag{9.41}$$

9.1.8 Example: Binomial Distribution

Toss a coin N times giving $\xi_i = 1$ (heads) or $\xi_i = -1$ (tails) with equal probability $P = \frac{1}{2}$. Then $E[\xi_i] = 0$ and $E[\xi_i^2] = 1$. The distribution of

$$\eta = \frac{1}{\sqrt{N}} \sum_{i=1}^{N} \xi_i \tag{9.42}$$

can be derived from the binomial distribution

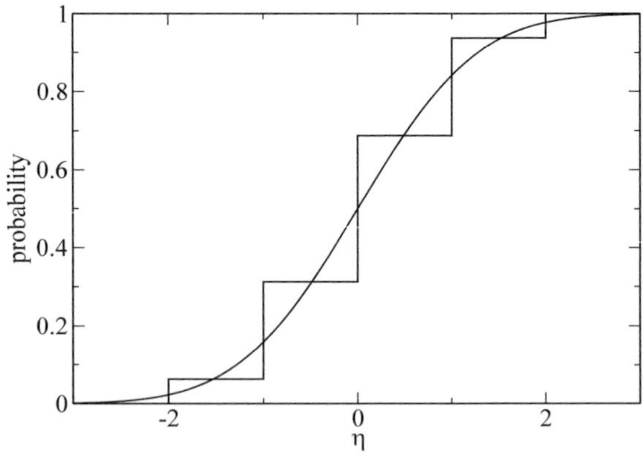

Fig. 9.4 Central limit theorem. The cumulative distribution function of η (9.42) is shown for $N = 4$ and compared to the normal distribution (9.20)

$$1 = \left[\frac{1}{2} + \left(-\frac{1}{2}\right)\right]^N = 2^{-N} \sum_{p=0}^{N} (-1)^{N-p} \binom{N}{N-p} \tag{9.43}$$

where p counts the number of tosses with $\xi = +1$. Since

$$n = p \cdot 1 + (N - p) \cdot (-1) = 2p - N \in [-N, N] \tag{9.44}$$

the probability of finding $\eta = \frac{n}{\sqrt{N}}$ is given by the binomial coefficient

$$P(\eta = \frac{2p - N}{\sqrt{N}}) = 2^{-N} \binom{N}{N-p} \tag{9.45}$$

or

$$P(\eta = \frac{n}{\sqrt{N}}) = 2^{-N} \binom{N}{\frac{N-n}{2}}. \tag{9.46}$$

9.1.9 *Average of Repeated Measurements*

A quantity X is measured N times. The results $X_1 \cdots X_N$ are independent random numbers with the same distribution function $f(X_i)$. Their expectation value is the exact value $E[X_i] = \int dX_i\, X_i\, f(X_i) = X$ and the standard deviation due to measurement uncertainties is $\sigma_X = \sqrt{E[X_i^2] - X^2}$. The new random variables

$$\xi_i = \frac{X_i - X}{\sigma_X} \tag{9.47}$$

have zero mean

$$E[\xi_i] = \frac{E[X_i] - X}{\sigma_X} = 0 \tag{9.48}$$

and unit standard deviation

$$\sigma_\xi^2 = E[\xi_i^2] - E[\xi_i]^2 = E\left[\frac{X_i^2 + X^2 - 2XX_i}{\sigma_X^2}\right] = \frac{E[X_i^2] - X^2}{\sigma_X^2} = 1. \tag{9.49}$$

Hence the quantity

$$\eta = \frac{\sum_1^N \xi_i}{\sqrt{N}} = \frac{\sum_1^N X_i - NX}{\sqrt{N}\sigma_X} = \frac{\sqrt{N}}{\sigma_X}(\overline{X} - X) \tag{9.50}$$

obeys a normal distribution

$$f(\eta) = \frac{1}{\sqrt{2\pi}} e^{-\eta^2/2}. \tag{9.51}$$

From

$$f(\overline{X})d\overline{X} = f(\eta)d\eta = f(\eta(\overline{X}))\frac{\sqrt{N}}{\sigma_X}d\overline{X} \tag{9.52}$$

we obtain

$$f(\overline{X}) = \frac{\sqrt{N}}{\sqrt{2\pi}\sigma_X} \exp\left\{-\frac{N}{2\sigma_X^2}(\overline{X} - X)^2\right\}. \tag{9.53}$$

The average of N measurements obeys a Gaussian distribution around the exact value X with a reduced standard deviation of

$$\sigma_{\overline{X}} = \frac{\sigma_X}{\sqrt{N}}. \tag{9.54}$$

9.2 Random Numbers

True random numbers of high quality can be generated using physical effects like thermal noise in a diode or atmospheric noise [142]. Computers very often make use of pseudo random numbers which have comparable statistical properties but are not totally unpredictable. For cryptographic purposes sophisticated algorithms are available which are slow but cryptographically secure, e.g. the Yarrow [143] and Fortuna [144] algorithms. In computational physics, usually simpler methods are sufficient which are not cryptographically secure, but pass important statistical tests like Marsaglia's DIEHARD collection [145, 146] and TestU01 [147, 148]. Most methods use an iterated function (Sect. 9.1 in Vol. 2). A set of numbers $\mathcal{Z}$ (e.g. 32-bit integers) is mapped onto itself by an invertible function $f(r)$ and, starting from a random seed number $r_0 \in \mathcal{Z}$, the sequence

$$r_{i+1} = f(r_i) \tag{9.55}$$

is calculated to provide a series of pseudo random numbers [146]. Using 32-bit integers there are 2^{32} different numbers, hence the period cannot exceed 2^{32}. The method can be improved by taking $\mathcal{Z}$ to be the set of m-tuples of 32-bit integers $r = \{z_1, z_2 \dots z_m\}$ and $f(r)$ a function that converts one m-tuple into another. An m-tuple of successive function values defines the iteration

$$r_i = \{z_i, z_{i-1}, \dots z_{i-m+1}\} \tag{9.56}$$

$$r_{i+1} = \{z_{i+1}, z_i, \dots z_{i-m+2}\} = \{f(z_i, \dots z_{i-m+1}), z_i, \dots z_{i-m+2}\}. \tag{9.57}$$

Using 32-bit integers, (9.57) has a maximum period of 2^{32m} (Example: for $m = 2$ and generating 10^6 numbers per second the period is 584942 years). For the initial seed, here m independent random numbers have to be provided.

The special case of a lagged RNG simply uses

$$r_{i+1} = \{z_{i+1}, z_i, \dots z_{i-m+2}\} = \{f(z_{i-m+1}), z_i, \dots z_{i-m+2}\}. \tag{9.58}$$

Popular kinds of functions $f(r)$ include linear congruent mappings, xorshift, lagged Fibonacci, multiply with carry (MWC), complimentary multiply with carry (CMWC) methods and combinations of these like the famous Mersenne Twister [149] and KISS [146] algorithms. We discuss briefly some important principles

9.2.1 Linear Congruent Mapping (LC)

A simple algorithm, mainly of historical importance due to some well known problems [150], is the linear congruent mapping

$$r_{i+1} = (ar_i + c) \bmod b \tag{9.59}$$

with multiplier a and base b which is usually taken to be $b = 2^{32}$ for 32-bit integers since this can be implemented most easily. The maximum period is given by b.

9.2.2 Xorshift

A 32-Bit integer[2] can be viewed as a vector $\mathbf{r} = (b_0, b_1 \dots b_{31})$ of elements b_i in the field $\mathcal{F}_2 = \{0, 1\}$. Addition of two such vectors (modulo 2) can be implemented with the exclusive or operation as can be seen from comparison with Table 9.1.

An invertible linear transformation of the vector $\mathbf{r}$ can be described by multiplication with a nonsingular 32×32 matrix T

$$f(\mathbf{r}) = \mathbf{r}T. \tag{9.60}$$

Table 9.1 Addition modulo 2

$0 + 0 = 0$
$1 + 0 = 1$
$0 + 1 = 1$
$1 + 1 = 0$

[2] This method can be easily extended to 64-Bit integers.

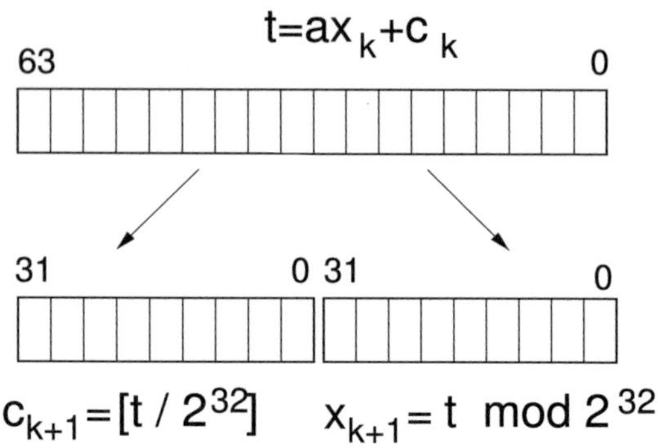

Fig. 9.5 Multiply with carry method using 64-bit integer arithmetic

To simplify the numerical calculation, Marsaglia [146] considers matrices of the special form[3]

$$T = (1 + L^a)(1 + R^b)(1 + L^c) \tag{9.61}$$

where L (R) is a matrix that produces a left (right) shift by one. For properly chosen numbers a, b, c the matrix T is of order $2^{32} - 1$ and the random numbers have the maximum possible period. There are many possible choices, one of them leads to the sequence

$$y = y \text{ xor } (y \ll 13)$$

$$y = y \text{ xor } (\text{y} \gg 17)$$

$$y = y \text{ xor } (y \ll 5). \tag{9.62}$$

9.2.3 Multiply with Carry (MWC)

This method is quite similar to the linear congruent mapping. However, instead of the constant c in (9.59) a varying carry is used.

For base $b = 2^{32}$ and multiplier $a = 698769069$ consider pairs of integers $r = [x, c]$ with $0 \le c < a, \quad 0 \le x < b$ excluding $[0, 0]$and $[a - 1, b - 1]$ and the iteration function[4]

$$f([x, c]) = [ax + c \bmod b, (ax + c)/b]. \tag{9.63}$$

Starting with a random seed $[x_0, c_0]$ the sequence $[x_k, c_k] = f([x_{k-1}, c_{k-1}])$has a period of about 2^{60} [146]. If one calculates $t = ax_k + c_k$ in 64 bits, then for $b = 2^{32}$,$c_{k+1}$is given by the top 32 bits and $x_{k+1}$by the bottom 32 bits (Fig. 9.5).

[3] At least three factors are necessary for 32 and 64-Bit integers.

[4] Using integer arithmetics.

9.2.4 Complementary Multiply with Carry (CMWC)

The simple MWC method has some inherent problems which can be overcome by a slight modification. First, the base is taken to be $b = 2^{32} - 1$ and second the iteration is changed to use the $(b-1)$-complement

$$x_k = (b-1) - (ax_{k-1} + c_{k-1}) \bmod b. \tag{9.64}$$

This method can provide random numbers which pass many tests and have very large periods.

9.2.5 Random Numbers with Given Distribution

Assume we have a program that generates random numbers in the interval [0,1] like in C:

$$rand()/(double)RAND\_MAX.$$

The corresponding cumulative distribution function is

$$F_0(x) = \begin{cases} 0 & for \quad x < 0 \\ x & for \quad 0 \le x \le 1 \\ 1 & for \quad x > 1 \end{cases}. \tag{9.65}$$

Random numbers with cumulative distribution $F(x)$ can be obtained as follows:

choose a RN $r \in [0, 1]$ with $P(r \le x) = F_0(x)$
let $\xi = F^{-1}(r)$.

$F(x)$ increases monotonously and therefore

$$P(\xi \le x) = P(F(\xi) \le F(x)) = P(r \le F(x)) = F_0(F(x)) \tag{9.66}$$

but since $0 \le F(x) \le 1$ we have

$$P(\xi \le x) = F(x). \tag{9.67}$$

This method of course is applicable only if F^{-1} can be expressed analytically.

9.2.6 Examples

Fair Die

A six-sided fair die can be simulated as follows:

choose a random number $r \in [0, 1]$

$$\text{Let } \xi = F^{-1}(r) = \begin{cases} 1 & for \quad 0 \leq r < \frac{1}{6} \\ 2 & for \quad \frac{1}{6} \leq r < \frac{2}{6} \\ 3 & for \quad \frac{2}{6} \leq r < \frac{3}{6} \\ 4 & for \quad \frac{3}{6} \leq r < \frac{4}{6} \\ 5 & for \quad \frac{4}{6} \leq r < \frac{5}{6} \\ 6 & for \quad \frac{5}{6} \leq r < 1. \end{cases}$$

Exponential Distribution

The cumulative distribution function

$$F(x) = 1 - \mathrm{e}^{-x/\lambda} \tag{9.68}$$

which corresponds to the exponential probability density

$$f(x) = \frac{1}{\lambda}\mathrm{e}^{-x/\lambda} \tag{9.69}$$

can be inverted by solving

$$r = 1 - \mathrm{e}^{-x/\lambda} \tag{9.70}$$

for x:

choose a random number $r \in [0, 1]$
Let $x = F^{-1}(r) = -\lambda \ln(1 - r)$.

Random Points on the Unit Sphere

We consider the surface element

$$\frac{1}{4\pi} R^2 d\varphi \sin\theta d\theta. \tag{9.71}$$

Our aim is to generate points on the unit sphere (θ, φ) with the probability density

$$f(\theta, \varphi)d\varphi d\theta = \frac{1}{4\pi}d\varphi \sin\theta d\theta = -\frac{1}{4\pi}d\varphi d\cos\theta. \tag{9.72}$$

The corresponding cumulative distribution is

$$F(\theta, \varphi) = -\frac{1}{4\pi}\int_1^{\cos\theta} d\cos\theta \int_0^{\varphi} d\varphi = \frac{\varphi}{2\pi}\frac{1-\cos\theta}{2} = F_\varphi F_\theta. \tag{9.73}$$

Since this factorizes, the two angles can be determined independently:

choose a first random number $r_1 \in [0, 1]$
Let $\varphi = F_\varphi^{-1}(r_1) = 2\pi r_1$
choose a second random number $r_2 \in [0, 1]$
Let $\theta = F_\theta^{-1}(r_2) = \arccos(1 - 2r_2)$.

Gaussian distribution (Box Muller)

For a Gaussian distribution the inverse F^{-1} has no simple analytical form. The famous Box Muller method [151] is based on a two-dimensional normal distribution with probability density

$$f(x, y) = \frac{1}{2\pi}\exp\left\{-\frac{x^2+y^2}{2}\right\} \tag{9.74}$$

which reads in polar coordinates

$$f(x, y)dxdy = f_p(\rho, \varphi)d\rho d\varphi \frac{1}{2\pi}e^{-\rho^2/2}\rho d\rho d\varphi. \tag{9.75}$$

Hence

$$f_p(\rho, \varphi) = \frac{1}{2\pi}\rho e^{-\rho^2/2} \tag{9.76}$$

and the cumulative distribution factorizes:

$$F_p(\rho, \varphi) = \frac{1}{2\pi}\varphi \cdot \int_0^{\rho} \rho' e^{-\rho'^2/2}d\rho' = \frac{\varphi}{2\pi}(1 - e^{-\rho^2}) = F_\varphi(\varphi)F_\rho(\rho). \tag{9.77}$$

The inverse of F_ρ is

$$\rho = \sqrt{-\ln(1-r)}. \tag{9.78}$$

Algorithm 13 generates Gaussian random numbers:

Algorithm 13: Gaussian random number generator

$r_1 \leftarrow \text{RN} \in [0, 1]$;
$r_2 \leftarrow \text{RN} \in [0, 1]$;
$\rho = \sqrt{-\ln(1 - r_1)}$;
$\varphi = 2\pi r_2$;
$x = \rho \cos\varphi$;

9.3 Monte Carlo Integration

Physical problems often involve high dimensional integrals (for instance path integrals, thermal averages) which cannot be evaluated by standard methods. Here Monte Carlo methods [135] can be very useful. Let us start with a very basic example.

9.3.1 *Numerical Calculation of π*

The area of a unit circle ($r = 1$) is given by $r^2\pi = \pi$. Hence π can be calculated by numerical integration. We use the following algorithm:

Choose N points randomly in the first quadrant, for instance N independent pairs $x, y \in [0, 1]$
Calculate $r^2 = x^2 + y^2$
Count the number of points within the circle, i.e the number of points $Z(r^2 \leq 1)$
$\frac{\pi}{4}$ is approximately given by $\frac{Z(r^2 \leq 1)}{N}$

The result converges rather slowly (Figs. 9.6 and 9.7).

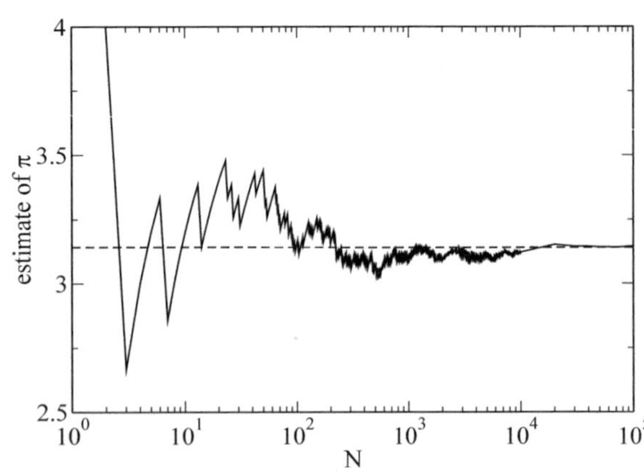

Fig. 9.6 Convergence of the numerical integration

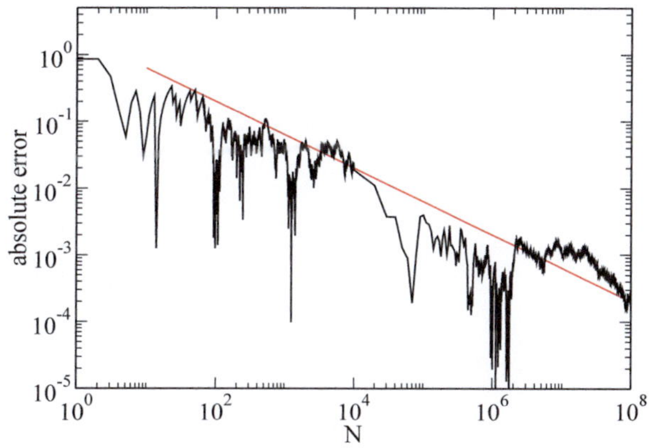

Fig. 9.7 Error of the numerical integration. The red line shows the expected $1/\sqrt{N}$ dependence (9.85) of the integration error. For $N > 10^4$ the black curve shows only one out of 10^4 data points

9.3.2 MC Calculation of an Integral

Let ξ be a random variable in the interval $[a, b]$ with the distribution

$$P(x < \xi \leq x + dx) = f(x)dx = \begin{cases} \frac{1}{b-a} & \text{for} \quad x \in [a, b] \\ 0 & \text{else.} \end{cases} \tag{9.79}$$

The expectation value of a function $g(x)$ is

$$E[g(x)] = \int_{-\infty}^{\infty} g(x) f(x) dx = \int_a^b g(x) dx \tag{9.80}$$

which can be approximated by the average of N randomly taken function values

$$\int_a^b g(x)dx \approx \frac{1}{N} \sum_{i=1}^{N} g(\xi_i). \tag{9.81}$$

To estimate the error we consider the new random variable

$$\gamma = \frac{1}{N} \sum_{i=1}^{N} g(\xi_i). \tag{9.82}$$

Its expectation value is

$$\overline{\gamma} = E[\gamma] = \frac{1}{N} \sum_{i=1}^{N} E[g(x)] = E[g(x)] = \int_a^b g(x)dx \tag{9.83}$$

and the variance follows from

$$\sigma_\gamma^2 = E[(\gamma - \overline{\gamma})^2] = E[(\frac{1}{N}\sum_{i=1}^{N} g(\xi_i) - \overline{\gamma})^2] = E[(\frac{1}{N^2}\sum_{i=1}^{N}(g(\xi_i) - \overline{\gamma}))^2] \tag{9.84}$$

$$= \frac{1}{N^2}E[\sum_{i=1}^{N}(g(\xi_i) - \overline{\gamma})^2] = \frac{1}{N}(\overline{g(\xi)^2} - \overline{g(\xi)}^2) = \frac{1}{N}\sigma^2_{g(\xi)}. \tag{9.85}$$

The width of the distribution and hence the uncertainty falls off as $1/\sqrt{N}$.

9.3.3 MC Integration with More General Random Numbers

Consider now random numbers $\xi \in [a, b]$ with arbitrary (but within $[a, b]$ not vanishing) probability density $f(x)$. The integral is approximated by

$$\frac{1}{N}\sum_{i=1}^{N}\frac{g(\xi_i)}{f(\xi_i)} = E\left[\frac{g(x)}{f(x)}\right] = \int_a^b \frac{g(x)}{f(x)} f(x)dx = \int_a^b g(x)dx. \tag{9.86}$$

The new random variable

$$\tau = \frac{1}{N}\sum_{i=1}^{N}\frac{g(\xi_i)}{f(\xi_i)} \tag{9.87}$$

according to (9.85) has a standard deviation given by

$$\sigma_\tau = \frac{1}{\sqrt{N}}\sigma(\frac{g(\xi)}{f(\xi)}) \tag{9.88}$$

which can be reduced by choosing f similar to g. Then preferentially sample points ξ are generated in regions where the integrand is large (importance sampling).

9.3.4 Configuration Integrals

Consider a system which is described by a *ndim* dimensional configuration space where a certain configuration $\mathbf{q} = (q_1 \ldots q_{ndim})$ has the normalized probability density

$$\varrho(q_1 \ldots q_{ndim}) \tag{9.89}$$

$$\int \ldots \int \varrho(q_1, \ldots q_{ndim}) dq^{ndim} = 1. \tag{9.90}$$

The average of an observable $A(\mathbf{q})$ has the form

$$< A >= \int \ldots \int A(q_1 \ldots q_{ndim}) \varrho(q_1 \ldots q_{ndim}) dq^{ndim} \tag{9.91}$$

which will be calculated by MC integration.

Classical Thermal Averages

An example from classical physics is a N particle system with potential energy

$$V(q_1 \ldots q_{3N}). \tag{9.92}$$

The probability of a certain configuration is given by its normalized Boltzmann-factor[5]

$$\varrho(q_1 \ldots q_{3N}) = \frac{e^{-\beta V(q_1 \ldots q_{3N})}}{\int dq^{3N} e^{-\beta V(q_1 \ldots q_{3N})}} \tag{9.93}$$

and the thermal average of some observable quantity $A(q_1 \ldots q_{3N})$ is given by the configuration integral

$$< A >= \int A(q_1 \ldots q_{3N}) \varrho(q_1 \ldots q_{3N}) dq^{3N}$$

$$= \frac{\int dq^{3N} A(q_1 \ldots q_{3N}) e^{-\beta V(q_1 \ldots q_{3N})}}{\int dq^{3N} e^{-\beta V(q_1 \ldots q_{3N})}}. \tag{9.94}$$

Variational Quantum Monte Carlo method

Consider a quantum mechanical N particle system with Hamiltonian

$$H = T + V(q_1 \ldots q_{3N}). \tag{9.95}$$

According to Ritz's variational principle, the ground state energy is a lower bound to the energy expectation value of any trial wavefunction

$$E_V = \frac{< \Psi_{trial} | H | \Psi_{trial} >}{< \Psi_{trial} | \Psi_{trial} >} \geq E_0. \tag{9.96}$$

[5] $\beta = 1/k_B T$.

Energy and wavefunction of the ground state can be approximated by minimizing the energy of the trial wavefunction, which is rewritten in the form

$$E_V = \frac{\int \Psi^*_{trial}(q_1 \dots q_{3N}) H \Psi_{trial}(q_1 \dots q_{3N}) dq^{3N}}{\int |\Psi_{trial}(q_1 \dots q_{3N})|^2 dq^{3N}}$$

$$= \frac{\int \varrho(q_1 \dots q_{3N}) E_L(q_1 \dots q_{3N}) dq^{3N}}{\int \varrho(q_1 \dots q_{3N}) dq^{3N}} =< E_L > \tag{9.97}$$

with the probability density

$$\varrho(q_1 \dots q_{3N}) = |\Psi_{trial}(q_1 \dots q_{3N})|^2 \tag{9.98}$$

and the so called local energy

$$E_L(q_1^{(t)} \cdots q_{3N}^{(t)}) = \frac{H\Psi_{trial}(q_1 \dots q_{3N})}{\Psi_{trial}(q_1 \dots q_{3N})}. \tag{9.99}$$

9.3.5 Simple Sampling

Let ξ be a random variable which is equally distributed over the range $q_{\min} \cdots q_{\max}$, i.e. a probability distribution

$$P(\xi \in [q, q + dq]) = f(q)dq \tag{9.100}$$

$$f(q) = \begin{cases} \frac{1}{q_{\max} - q_{\min}} & q \in [q_{\min}, q_{\max}] \\ 0 & \text{else} \end{cases} \tag{9.101}$$

$$\int f(q)dq = 1. \tag{9.102}$$

Repeatedly choose *ndim* random numbers $\xi_1^{(m)}, \dots \xi_{ndim}^{(m)}$ and calculate the expectation value

$$E\left(A(\xi_1 \cdots \xi_{ndim})\varrho(\xi_1, \dots \xi_{ndim})\right) = \lim_{M\to\infty} \frac{1}{M} \sum_{m=1}^{M} A(\xi_1^{(m)} \dots \xi_{ndim}^{(m)})\varrho(\xi_1^{(m)} \dots \xi_{ndim}^{(m)})$$

$$= \int A(q_1 \dots q_{ndim})\varrho(q_1 \dots q_{ndim}) f(q_1) \cdots f(q_{ndim}) dq_1 \cdots dq_{ndim}$$

$$= \frac{1}{(q_{\max} - q_{\min})^{ndim}} \int_{q_{\min}}^{q_{\max}} \cdots \int_{q_{\min}}^{q_{\max}} A(q_1 \dots q_{ndim}) \varrho(q_1 \dots q_{ndim}) dq^{ndim}. \tag{9.103}$$

Hence

$$\frac{E\left(A(\xi_1 \cdots \xi_{ndim})\varrho(\xi_1, \dots \xi_{ndim})\right)}{E\left(\varrho(\xi_1, \dots \xi_{ndim})\right)}$$

$$= \frac{\int_{q_{\min}}^{q_{\max}} \cdots \int_{q_{\min}}^{q_{\max}} A(q_1 \dots q_{ndim})\varrho(q_1 \dots q_{ndim}) dq^{ndim}}{\int_{q_{\min}}^{q_{\max}} \cdots \int_{q_{\min}}^{q_{\max}} \varrho(q_1 \dots q_{ndim}) dq^{ndim}} \approx < A > . \tag{9.104}$$

Each set of random numbers $\xi_1 \dots \xi_{ndim}$ defines one sample configuration. The average over a large number M of samples gives an approximation to the average $< A >$, if the range of the q_i is sufficiently large. However, many of the samples will have small weight and contribute only little.

9.3.6 Importance Sampling

Let us try to sample preferentially the most important configurations. Choose the distribution function as

$$f(q_1 \cdots q_{ndim}) = \varrho(q_1 \dots q_{ndim}). \tag{9.105}$$

The expectation value of A now directly approximates the configurational average

$$E\left(A(\xi_1 \cdots \xi_{ndim})\right) = \lim_{M \to \infty} \frac{1}{M} \sum_{m=1}^{M} A(\xi_1^{(m)} \dots \xi_{ndim}^{(m)})$$

$$= \int A(q_1 \dots q_{ndim})\varrho(q_1 \dots q_{ndim}) dq^{ndim} =< A > . \tag{9.106}$$

The fraction of samples within a small integration volume[6] dq^{ndim} around the configuration $(q_1 \dots q_{ndim})$ is given in the limit of large M by

$$\frac{N(q_1 \dots q_{ndim})}{M} = \varrho(q_1 \dots q_{ndim}) dq^{ndim} \tag{9.107}$$

hence the ratio of samples around two different configurations is[7]

---

[6] Obviously, we have to divide the configuration space into small volumes dq^{ndim}, since we can only take samples at a finite number of configurations.

[7] We consider only important configurations with nonzero probability.

$$\frac{N(q_1 \dots q_{ndim})}{N(q_1' \dots q_{ndim}')} = \frac{\varrho(q_1 \dots q_{ndim})}{\varrho(q_1' \dots q_{ndim}')} \frac{dq^{ndim}}{dq^{ndim\,\prime}}. \tag{9.108}$$

Therefore, to generate the important configurations, it is not necessary to know the normalized distribution function ϱ. It is sufficient to know the relative probability (e.g. the unnormalized Boltzmann factors).

9.3.7 Metropolis-Hastings Markov Chain Monte Carlo

This Markov chain–Monte Carlo algorithm which was proposed by Metropolis and coworkers [152] and later generalized by Hastings [153] is often used to draw samples from a given probability distribution.

Starting from an initial configuration $\mathbf{q}^{(0)} = (q_1^{(0)} \cdots q_{ndim}^{(0)})$, a Markov chain of configurations $\mathbf{q}^{(t)} = (q_1^{(t)} \cdots q_{ndim}^{(t)})$ [154, 155] is generated, where each configuration depends only on its predecessor and not on the history of the chain.

If the state space, from which the configurations are to be selected, is continuous, the probability to make a transition from $\mathbf{q}^{(t)}$ to exactly $\mathbf{q}^{(t+1)}$ is zero. Here we have to consider the probability to make a transition into the "surrounding"[8] of $\mathbf{q}^{(t+1)}$ which will be described by a probability density [156].

The transition kernel (Fig. 9.8) is defined as the conditional probability to move from $\mathbf{q}^{(t)}$ to a new configuration within the set A

$$P(\mathbf{q}^{(t+1)} \in A|\mathbf{q}^{(t)}) = K(\mathbf{q}^{(t)}, A) = \int_A K(\mathbf{q}^{(t)}, \mathbf{q}) d^{ndim} q. \tag{9.109}$$

It can be expressed as the sum of an off-diagonal and a diagonal term

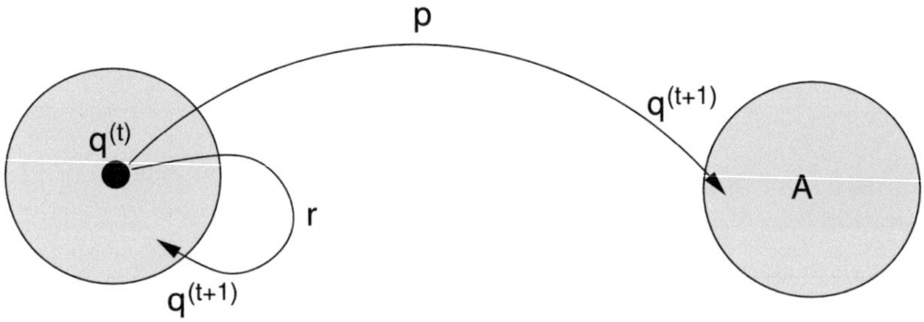

Fig. 9.8 Transition kernel. The chain stays close to its position with probability r and moves to another position with probability p. The probability to find the following configuration $\mathbf{q}^{(t+1)} \in A$, given the configuration $\mathbf{q}^{(t)}$, is known as transition kernel $K(\mathbf{q}^{(t)}, A)$

[8] Strictly speaking, the set A is an element of the Borel σ-field on $\mathbb{R}^{ndim}$.

$$K(\mathbf{q}^{(t)}, A) = \int_A p(\mathbf{q}^{(t)} \to \mathbf{q}^{(t+1)}) d^{ndim} q^{(t+1)} + r(\mathbf{q}^{(t)}) \chi_A(\mathbf{q}^{(t)}) \tag{9.110}$$

where

$$\chi_A(\mathbf{q}^{(t)}) = \begin{cases} 1 & \text{if } \mathbf{q}^{(t)} \in A \\ 0 & \text{else} \end{cases} \tag{9.111}$$

is the indicator function of A

$$p(\mathbf{q}^{(t)} \to \mathbf{q}^{(t+1)}) \tag{9.112}$$

is the probability density of the transition $\mathbf{q}^{(t)} \to \mathbf{q}^{(t+1)}$ and

$$r(\mathbf{q}^{(t)}) = 1 - \int p(\mathbf{q}^{(t)} \to \mathbf{q}') d^{ndim} q' \tag{9.113}$$

the probability that the chain remains at its position

Now assume that p fulfills the condition of detailed balance

$$\varrho(\mathbf{q}^{(t)}) p(\mathbf{q}^{(t)} \to \mathbf{q}^{(t+1)}) = \varrho(\mathbf{q}^{(t+1)}) p(\mathbf{q}^{(t+1)} \to \mathbf{q}^{(t)}). \tag{9.114}$$

Then, if the probability distribution of $\mathbf{q}^{(t)}$ is given by ϱ,

$$P(\mathbf{q}^{(t)} \in A) = \int_A \varrho(\mathbf{q}) d^{ndim} q. \tag{9.115}$$

After the next iteration

$$\begin{aligned} & P(\mathbf{q}^{(i+1)} \in A) \\ &= \int d^{ndim} q \varrho(\mathbf{q}) \int_A p(\mathbf{q} \to \mathbf{q}') d^{ndim} q' \\ &\quad + \int \varrho(\mathbf{q}) d^{ndim} q \, r(\mathbf{q}) \chi_A(\mathbf{q}) \\ &= \int_A d^{ndim} q' \, \varrho(\mathbf{q}') \int d^{ndim} q p(\mathbf{q}' \to \mathbf{q}) \\ &\quad + \int_A r(\mathbf{q}) \varrho(\mathbf{q}) d^{ndim} q \\ &= \int_A d^{ndim} q' \, \varrho(\mathbf{q}') \left[1 - r(\mathbf{q}')\right] + \int_A r(\mathbf{q}) \varrho(\mathbf{q}) d^{ndim} q \end{aligned}$$

$$= \int_A d^{ndim} q' \, \varrho(\mathbf{q}') \tag{9.116}$$

which shows, that ϱ is the invariant density of the kernel [154] if detailed balance is fulfilled. Under certain conditions the Markov chain will converge to this density [157–159]. The Metropolis-Hastings method factorizes the transition probability

$$p(\mathbf{q}^{(t)} \rightarrow \mathbf{q}^{(t+1)}) = p_{trial}(\mathbf{q}^{(t)}, \mathbf{q}^{(t+1)}) p_{accept}(\mathbf{q}^{(t)}, \mathbf{q}^{(t+1)}) \tag{9.117}$$

into a trial step which is generated from a certain distribution function and a probability for acceptance which is tailored to establish the detailed balance condition. If the trial step is not accepted, the Markov chain stays at the old configuration.

Several realizations are discussed in [154, 156].

In the following, we assume a discrete and finite state space

$$\aleph = \{\mathbf{q}_i, i = 1 \cdots \text{nconf}\}. \tag{9.118}$$

[9] We denote the probability of a certain configuration as $p(\mathbf{q})$ and introduce transition probabilities

$$W_{i \rightarrow j} = P(\mathbf{q}_i \rightarrow \mathbf{q}_j) \tag{9.119}$$

which are chosen to fulfill the condition of detailed balance (Fig. 9.9)

$$p(\mathbf{q}_i) W_{i \rightarrow j} = p(\mathbf{q}_j) W_{j \rightarrow i}. \tag{9.120}$$

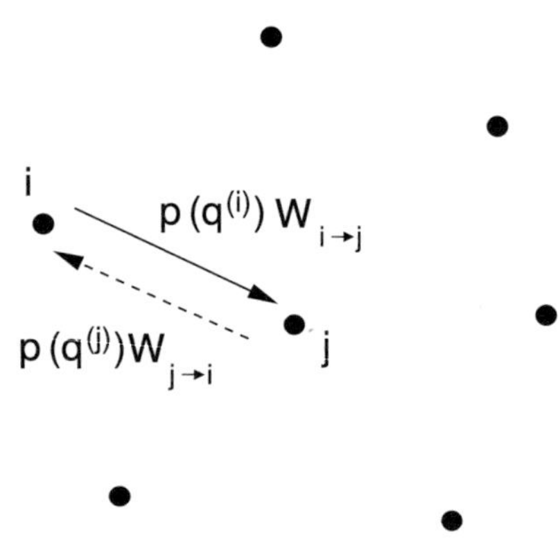

Fig. 9.9 Principle of detailed balance. In equilibrium, transitions $i \rightarrow$j and $j \rightarrow i$ appear with the same frequency

[9] Some authors argue that this is always the case for numerical calculations with floating point numbers but in practice it might be difficult to explore all possible configurations in finite time. For important applications, the configurations are truly discrete, e.g. for Ising models or Boltzmann machines.

Consider an ensemble of such Markov chains and let $N_t(\mathbf{q}_i)$ denote the number of chains which are in the configuration $\mathbf{q}_i$ after t steps. The changes during the following step are

$$\Delta N(\mathbf{q}_i) = N_{t+1}(\mathbf{q}_i) - N_t(\mathbf{q}_i) = \sum_{\mathbf{q}_j \in conf.} N_t(\mathbf{q}_j) W_{j\to i} - N_t(\mathbf{q}_i) W_{i\to j}. \quad (9.121)$$

In equilibrium

$$N_{eq}(\mathbf{q}_i) = N_0 p(\mathbf{q}_i) \quad (9.122)$$

and the changes (9.121) vanish:

$$\begin{aligned} \Delta N(\mathbf{q}_i) &= N_0 \sum_{\mathbf{q}_j} p(\mathbf{q}_j) W_{j\to i} - p(\mathbf{q}_i) W_{i\to j} \\ &= N_0 \sum_{\mathbf{q}_j} p(\mathbf{q}_j) W_{j\to i} - p(\mathbf{q}_i) \left[W_{j\to i} \frac{p(\mathbf{q}_j)}{p(\mathbf{q}_i)} \right] \\ &= 0. \end{aligned} \quad (9.123)$$

A solution of

$$\Delta N(\mathbf{q}_i) = \sum_{\mathbf{q}_j \in conf.} N_t(\mathbf{q}_j) W_{j\to i} - N_t(\mathbf{q}_i) W_{i\to j} = 0 \quad (9.124)$$

corresponds to a zero eigenvalue of the system of equations[10]

$$\sum_{\mathbf{q}_j} N(\mathbf{q}_j) W_{j\to i} - N(\mathbf{q}_i) \sum_{\mathbf{q}_j} W_{i\to j} = \lambda N(\mathbf{q}_i). \quad (9.125)$$

One solution of this eigenvalue equation is given by

$$\frac{N_{eq}(\mathbf{q}_j)}{N_{eq}(\mathbf{q}_i)} = \frac{p(\mathbf{q}_j)}{p(\mathbf{q}_i)}. \quad (9.126)$$

However, there might be other solutions. For instance,if the chain is periodic or if not all states can be reached after a finite number of steps and some isolated configurations are occupied initially. More details can be found in [158, 160, 161] and references therein.

[10] Equivalent to the dominant eigenvalue 1 of the transition matrix.

Symmetric Metropolis Algorithm

The original Metropolis algorithm (Algorithm 14) is symmetric. A trial step is selected with probability

$$T(\mathbf{q}_i \to \mathbf{q}_{trial}) = T(\mathbf{q}_{trial} \to \mathbf{q}_i). \tag{9.127}$$

Algorithm 14:

```
/* trial step                                                    */
/* choose a new configuration q_trial randomly                   */
/* with probability T(q_i → q_trial)                             */
/*                                                               */
if ϱ(q_i) = 0 then
    /* accept trial step                                         */
    q_{i+1} ← q_trial;
else
    /* calculate probability ratio                               */
    R ← p(q_trial) / p(q_i);
    /*                                                           */
    if R ≥ 1 then
        /* accept trial step                                     */
        q_{i+1} ← q_trial;
        /*                                                       */
    else
        /* accept trial step only with probability R             */
        /* choose a random number ξ ∈ [0, 1]                     */
        ξ ← randomnumber();
        if ξ < R then
            q_{i+1} ← q_trial;
        else
            q_{i+1} ← q_i;
        end
    end
end
```

The transition probability is the product

$$W_{i\to j} = T_{i\to j} A_{i\to j} \tag{9.128}$$

of the probability $T_{i\to j}$ to select $i \to j$ as a trial step and the probability $A_{i\to j}$ to accept the trial step. Now we have[11]

[11] For binary distributions it may be necessary to accept the case $R = 1$ only with probability $1/2$ to guarantee convergence [162].

$$\begin{array}{l} \text{for } R \geq 1 \;\rightarrow\; A_{i\rightarrow j} = 1,\; A_{j\rightarrow i} = R^{-1} \\ \text{for } R < 1 \;\rightarrow\; A_{i\rightarrow j} = R,\; A_{j\rightarrow i} = 1 \end{array}. \tag{9.129}$$

Since $T_{i\rightarrow j} = T_{j\rightarrow i}$, in both cases

$$\frac{W_{i\rightarrow j}}{W_{j\rightarrow i}} = \frac{A_{i\rightarrow j}}{A_{j\rightarrow i}} = R = \frac{p(\mathbf{q}_j)}{p(\mathbf{q}_i)}. \tag{9.130}$$

The size of the trial steps has to be adjusted to produce a reasonable acceptance ratio of[12]

$$\frac{N_{\text{accepted}}}{N_{\text{rejected}}} \approx 1. \tag{9.131}$$

Metropolis-Hastings algorithm

If the trial probability is not symmetric,

$$T(\mathbf{q}_i \rightarrow \mathbf{q}_{trial}) \neq T(\mathbf{q}_{trial} \rightarrow \mathbf{q}_i) \tag{9.132}$$

take[13]

$$R = \frac{p(\mathbf{q}_j)}{p(\mathbf{q}_i)} \frac{T_{j\rightarrow i}}{T_{i\rightarrow j}} \tag{9.133}$$

and obtain detailed balance again

$$\frac{W_{i\rightarrow j}}{W_{j\rightarrow i}} = \frac{T_{i\rightarrow j}}{T_{j\rightarrow i}} \frac{A_{i\rightarrow j}}{A_{j\rightarrow i}} = \begin{cases} \frac{T_{i\rightarrow j}}{T_{j\rightarrow i}} \frac{1}{R^{-1}} = \frac{p(\mathbf{q}_j)}{p(\mathbf{q}_i)} \text{ for } R \geq 1 \\ \\ \frac{T_{i\rightarrow j}}{T_{j\rightarrow i}} R = \frac{p(\mathbf{q}_j)}{p(\mathbf{q}_i)} \text{ for } R < 1. \end{cases} \tag{9.134}$$

Gibbs sampling

The Gibbs sampler [163–166] is a special M–H algorithm which has become very popular in the field of machine learning. It does not change the configuration vector in one step but scans the coordinates sequentially one at a time. Starting from a configuration

$$\mathbf{q}^{(t)} = (q_1^t, q_2^t, \cdots q_{ndim}^t) \tag{9.135}$$

the first coordinate q_1 is changed according to its conditional probability given the current values of $q_2, q_3 \cdots q_{ndim}$. This corresponds to a trial step with

[12] The optimum ratio may be dependent on the dimension of the configuration space.

[13] If $\pi(\mathbf{q}_i)T_{i\rightarrow j} = 0$ then the trial step is accepted.

$$T_{(q_1^t,q_2^t,\cdots q_{ndim}^t)\rightarrow(q_1^{t+1},q_2^t,\cdots q_{ndim}^t)} = p(q_1^{t+1}|q_2^t\cdots q_{ndim}^t) \tag{9.136}$$

which is accepted with probability $A = 1$. Then all coordinates are changed one after another with transition probabilities

$$T_{(q_1^{t+1},q_2^t,\cdots q_{ndim}^t)\rightarrow(q_1^{t+1},q_2^{t+1},\cdots q_{ndim}^t)} = p(q_2^{t+1}|q_1^{t+1},q_3^t\cdots q_{ndim}^t) \tag{9.137}$$

$$\vdots$$

$$T_{(q_1^{t+1},q_2^{t+1},\ldots,d_{ndim-1}^{t+1}q_{ndim}^t)\rightarrow(q_1^{t+1},q_2^{t+1},\ldots q_{ndim}^{t+1})} = p(q_{ndim}^{t+1}|q_1^{t+1},q_2^t\cdots q_{ndim-1}^t) \tag{9.138}$$

which gives the next element $\mathbf{q}^{(t+1)}$ of the Markov chain which under rather general conditions [159, 167] converges to the multivariate distribution $p(q_1, q_2, \cdots q_{ndim})$ which is defined by the full set of conditional probabilities. In addition, $q_i^{(t)}$ converges to the marginal probability $p(q_i)$.

For the special case of only two variables $\mathbf{q} = (x, y)$ Gibbs sampling alternates between sampling from the two conditional probabilities $p(x|y)$and $p(y|x)$. Starting from a configuration

$$\mathbf{q}^{(t)} = (x^{(t)}, y^{(t)}) \tag{9.139}$$

a pair of transitions is made with probabilities

$$T_{(x^{(t)},y^{(t)})\rightarrow(x^{(t+1)},y^{(t)})} = p(x^{(t+1)}|y^{(t)}) \tag{9.140}$$

$$T_{(x^{(t+1)},y^{(t)})\rightarrow(x^{(t+1)},y^{(t+1)})} = p(y^{(t+1)}|x^{(t+1)}) \tag{9.141}$$

to obtain the next element of the Markov Chain.

Multiple Walkers

To scan the relevant configurations more completely and reduce correlation between the samples, usually a large number of "walkers" is used (e.g. several hundred) which, starting from different initial conditions, represent independent Markov chains. This also offers a simple possibility for parallelization.

Exercises

Exercise 9.1 Central Limit Theorem

This computer experiment draws a histogram for the random variable τ, which is calculated from N random numbers $\xi_1 \cdots \xi_N$:

$$\tau = \frac{\sum_{i=1}^{N} \xi_i}{\sqrt{N}.} \tag{9.142}$$

The ξ_i are random numbers with zero mean and unit variance and can be chosen as

- $\xi_i = \pm 1$ (coin tossing)
- Gaussian random numbers

Investigate how a Gaussian distribution is approached for large N.

Exercise 9.2 Nonlinear Optimization

MC methods can be used for nonlinear optimization (Traveling salesman problem, structure optimization etc) [168]. Consider an energy function depending on many coordinates

$$E(q_1, q_2 \cdots q_N). \tag{9.143}$$

Introduce a fictitious temperature T and generate configurations with probabilities

$$p(q_1 \cdots q_N) = \frac{1}{Z} \mathrm{e}^{-E(q_1 \cdots q_N)/T}. \tag{9.144}$$

Slow cooling drives the system into a local minimum. By repeated heating and cooling other local minima can be reached (simulated annealing [169]).

In this computer experiment we try to find the shortest path which visits each of N up to 50 given points. The fictitious Boltzmann factor for a path with total length L is

$$P(L) = \mathrm{e}^{-L/T}. \tag{9.145}$$

Starting from an initial path $S = (i_1, i_2, \cdots i_N)$ $n < 5$ and p are chosen randomly and a new path $S' = (i_1, \cdots i_{p-1}, i_{p+n}, \cdots i_p, i_{p+n+1}, \cdots i_N)$ is generated by reverting the sub-path

$$i_p \cdots i_{p+n} \rightarrow i_{p+n} \cdots i_{p.} \tag{9.146}$$

Start at high temperature $T > L$ and cool down slowly.

Chapter 10
Eigenvalue Problems

Eigenvalue problems are omnipresent in physics. Important examples are the time independent Schrödinger equation in a finite orthogonal basis (Chap. 10 in Vol 2)

$$\sum_{j=1}^{M} < \phi_{j'}|H|\phi_j > C_j = EC_{j'} \tag{10.1}$$

or the harmonic motion of a molecule around its equilibrium structure (Sect. 2.4.1 in Vol. 2)

$$\omega^2 m_i(\xi_i - \xi_i^{eq}) = \sum_j \frac{\partial^2 U}{\partial \xi_i \partial \xi_j}(\xi_j - \xi_j^{eq}). \tag{10.2}$$

Most important are ordinary eigenvalue problems,[1] which involve the solution of a homogeneous system of linear equations

$$\sum_{j=1}^{N} a_{ij} x_j = \lambda x_i \tag{10.3}$$

with a Hermitian (or symmetric, if real) matrix [170]

$$a_{ji} = a_{ij}^*. \tag{10.4}$$

[1] We do not consider general eigenvalue problems here.

Supplementary Information The online version contains supplementary material available at https://doi.org/10.1007/978-3-032-07856-8_10.

P. O. J. Scherer, *Computational Physics I*, Graduate Texts in Physics,
https://doi.org/10.1007/978-3-032-07856-8_10

The couple $(\lambda, \mathbf{x})$ consisting of an eigenvector $\mathbf{x}$ and the corresponding eigenvalueλ is called an eigenpair.

Matrices of small dimension can be diagonalized directly by determining the roots of the characteristic polynomial and solving a homogeneous system of linear equations. The Jacobi method uses successive rotations to diagonalize a matrix with a unitary transformation. A very popular method for not too large symmetric matrices reduces the matrix to tridiagonal form which can be diagonalized efficiently with the QL algorithm. Some special tridiagonal matrices can be diagonalized analytically. Krylov-space algorithms are the choice for matrices of very large dimension. We discuss especially the methods by Arnoldi and Lanczos.

10.1 Direct Solution

For matrices of very small dimension (2, 3) the determinant

$$\det \left| a_{ij} - \lambda \delta_{ij} \right| = 0 \tag{10.5}$$

can be written explicitly as a polynomial of λ. The roots of this polynomial are the eigenvalues. The eigenvectors are given by the system of equations

$$\sum_j (a_{ij} - \lambda \delta_{ij}) u_j = 0. \tag{10.6}$$

10.2 Jacobi Method

Any symmetric 2×2 matrix

$$A = \begin{pmatrix} a_{11} & a_{12} \\ a_{12} & a_{22} \end{pmatrix} \tag{10.7}$$

can be diagonalized by a rotation of the coordinate system. Rotation by the angle φ corresponds to an orthogonal transformation with the rotation matrix

$$R_\varphi = \begin{pmatrix} \cos\varphi & -\sin\varphi \\ \sin\varphi & \cos\varphi \end{pmatrix}. \tag{10.8}$$

In the following we use the abbreviations

$$c = \cos\varphi, \quad s = \sin\varphi, \quad t = \tan\varphi. \tag{10.9}$$

The transformed matrix is

$$RAR^{-1} = \begin{pmatrix} c & -s \\ s & c \end{pmatrix} \begin{pmatrix} a_{11} & a_{12} \\ a_{12} & a_{22} \end{pmatrix} \begin{pmatrix} c & s \\ -s & c \end{pmatrix}$$

$$= \begin{pmatrix} c^2 a_{11} + s^2 a_{22} - 2csa_{12} & cs(a_{11} - a_{22}) + (c^2 - s^2)a_{12} \\ cs(a_{11} - a_{22}) + (c^2 - s^2)a_{12} & s^2 a_{11} + c^2 a_{22} + 2csa_{12} \end{pmatrix}. \quad (10.10)$$

It is diagonal if

$$0 = cs(a_{11} - a_{22}) + (c^2 - s^2)a_{12} = \frac{a_{11} - a_{22}}{2} \sin(2\varphi) + a_{12} \cos(2\varphi) \quad (10.11)$$

or

$$\tan(2\varphi) = \frac{2a_{12}}{a_{22} - a_{11}}. \quad (10.12)$$

Calculation of φ is not necessary since only its cosine and sine appear in (10.10). From [170]

$$\frac{1 - t^2}{t} = \frac{c^2 - s^2}{2cs} = \cot(2\varphi) = \frac{a_{22} - a_{11}}{2a_{12}} \quad (10.13)$$

we see that t is a root of

$$t^2 + \frac{a_{22} - a_{11}}{a_{12}} t - 1 = 0 \quad (10.14)$$

hence

$$t = -\frac{a_{22} - a_{11}}{2a_{12}} \pm \sqrt{1 + \left(\frac{a_{22} - a_{11}}{2a_{12}}\right)^2} = \frac{1}{\frac{a_{22}-a_{11}}{2a_{12}} \pm \sqrt{1 + \left(\frac{a_{22}-a_{11}}{2a_{12}}\right)^2}}. \quad (10.15)$$

For reasons of convergence [170] the solution with smaller magnitude is chosen which can be written as

$$t = \frac{\operatorname{sign}\left(\frac{a_{22}-a_{11}}{2a_{12}}\right)}{\left|\frac{a_{22}-a_{11}}{2a_{12}}\right| + \sqrt{1 + \left(\frac{a_{22}-a_{11}}{2a_{12}}\right)^2}}. \quad (10.16)$$

Again for reasons of convergence the smaller solution φ is preferred and therefore we take

$$c = \frac{1}{\sqrt{1 + t^2}} \qquad s = \frac{t}{\sqrt{1 + t^2}}. \quad (10.17)$$

The diagonal elements of the transformed matrix are

$$\tilde{a}_{11} = c^2 a_{11} + s^2 a_{22} - 2csa_{12} \quad (10.18)$$

$$\tilde{a}_{22} = s^2 a_{11} + c^2 a_{22} + 2csa_{12}. \tag{10.19}$$

The trace of the matrix is invariant

$$\tilde{a}_{11} + \tilde{a}_{22} = a_{11} + a_{22} \tag{10.20}$$

whereas the difference of the diagonal elements is

$$\tilde{a}_{11} - \tilde{a}_{22} = (c^2 - s^2)(a_{11} - a_{22}) - 4csa_{12} = \frac{1-t^2}{1+t^2}(a_{11} - a_{22}) - 4\frac{a_{12}t}{1+t^2}$$

$$= (a_{11} - a_{22}) + (-a_{12}\frac{1-t^2}{t})\frac{-2t^2}{1+t^2} - 4\frac{a_{12}t}{1+t^2} = (a_{11} - a_{22}) - 2ta_{12}. \tag{10.21}$$

and the transformed matrix has the simple form

$$\begin{pmatrix} a_{11} - a_{12}t & \\ & a_{22} + a_{12}t \end{pmatrix}. \tag{10.22}$$

For larger dimension $N > 2$ the Jacobi method uses the following algorithm:

(1) look for the dominant non-diagonal element $\max_{i \neq j} |a_{ij}|$
(2) Perform a rotation in the (ij)-plane to cancel the element $\tilde{a}_{ij}$ of the transformed matrix $\tilde{A} = R^{(ij)} \cdot A \cdot R^{(ij)-1}$. The corresponding rotation matrix has the form

$$R^{(ij)} = \begin{pmatrix} 1 & & & & & \\ & \ddots & & & & \\ & & c & & s & \\ & & & \ddots & & \\ & & -s & & c & \\ & & & & & \ddots & \\ & & & & & & 1 \end{pmatrix} \tag{10.23}$$

(3) repeat (1–2) until convergence (if possible).

The sequence of Jacobi rotations gives the over all transformation

$$RAR^{-1} = \cdots R_2 R_1 A R_1^{-1} R_2^{-1} \cdots = \begin{pmatrix} \lambda_1 & & \\ & \ddots & \\ & & \lambda_N \end{pmatrix}. \tag{10.24}$$

Hence

$$AR^{-1} = R^{-1} \begin{pmatrix} \lambda_1 & & \\ & \ddots & \\ & & \lambda_N \end{pmatrix} \tag{10.25}$$

and the column vectors of $R^{-1} = (\mathbf{v}_1, \mathbf{v}_2 \cdots \mathbf{v}_N)$ are the eigenvectors of A:

$$A\,(\mathbf{v}_1, \mathbf{v}_2 \cdots \mathbf{v}_N) = (\lambda_1 \mathbf{v}_1, \lambda_2 \mathbf{v}_2, \cdots \lambda_N \mathbf{v}_N)\,. \tag{10.26}$$

10.3 Tridiagonal Matrices

A tridiagonal matrix has nonzero elements only in the main diagonal and the first diagonal above and below. Many algorithms simplify significantly when applied to tridiagonal matrices.

10.3.1 Characteristic Polynomial of a Tridiagonal Matrix

The characteristic polynomial of a tridiagonal matrix

$$P_A(\lambda) = \det \begin{vmatrix} a_{11}-\lambda & a_{12} & & \\ a_{21} & a_{22}-\lambda & & \\ & & \ddots & a_{N-1N} \\ & & a_{NN-1} & a_{NN}-\lambda \end{vmatrix} \tag{10.27}$$

can be calculated recursively:

$$P_0 = 1$$

$$P_1(\lambda) = a_{11} - \lambda$$

$$P_2(\lambda) = (a_{22} - \lambda) P_1(\lambda) - a_{12} a_{21}$$

$$\vdots$$

$$P_N(\lambda) = (a_{NN} - \lambda) P_{N-1}(\lambda) - a_{N,N-1} a_{N-1,N} P_{N-2}(\lambda). \tag{10.28}$$

10.3.2 Special Tridiagonal Matrices

Certain classes of tridiagonal matrices can be diagonalized exactly [171–173].

Discretized Second Derivatives

Discretization of a second derivative involves, under Dirichlet boundary conditions $f(x_0) = f(x_{N+1}) = 0$, the differentiation matrix (Sect. 7.2 in Vol. 2)

$$M = \begin{pmatrix} -2 & 1 & & & \\ 1 & -2 & 1 & & \\ & \ddots & \ddots & \ddots & \\ & & 1 & -2 & 1 \\ & & & 1 & -2 \end{pmatrix}. \tag{10.29}$$

Its eigenvectors have the form

$$\mathbf{f} = \begin{pmatrix} f_1 \\ \vdots \\ f_n \\ \vdots \\ f_N \end{pmatrix} = \begin{pmatrix} \sin k \\ \vdots \\ \sin(nk) \\ \vdots \\ \sin(Nk) \end{pmatrix}. \tag{10.30}$$

This can be seen by inserting (10.30) into the n-th line of the eigenvalue equation (10.31)

$$M\mathbf{f} = \lambda \mathbf{f} \tag{10.31}$$

$$(M\mathbf{f})_n = (\sin((n-1)k) + \sin((n+1)k) - 2\sin(nk))$$

$$= 2\sin(nk)(\cos(k) - 1) = \lambda(\mathbf{f})_n \tag{10.32}$$

with the eigenvalue

$$\lambda = 2(\cos k - 1) = -4\sin^2\left(\frac{k}{2}\right). \tag{10.33}$$

The first line of the eigenvalue equation (10.31) reads

$$(M\mathbf{f})_1 = (-2\sin(k) + \sin(2k))$$

$$= 2\sin(k)(\cos(k) - 1) = \lambda(\mathbf{f})_n \tag{10.34}$$

and from the last line we have

$$(M\mathbf{f})_N = (-2\sin(Nk) + \sin([N-1]k))$$

$$= \lambda(\mathbf{f})_N = 2(\cos(k) - 1)\sin(Nk) \tag{10.35}$$

which holds if

$$\sin((N-1)k) = 2\sin(Nk)\cos(k). \tag{10.36}$$

This simplifies to

$$\sin(Nk)\cos(k) - \cos(Nk)\sin(k) = 2\sin(Nk)\cos(k)$$

$$\sin(Nk)\cos(k) + \cos(Nk)\sin(k) = 0$$

$$\sin((N+1)k) = 0. \tag{10.37}$$

Hence the possible values of k are

$$k = \frac{\pi}{(N+1)} l \text{ with } l = 1, 2, \cdots N \tag{10.38}$$

and the eigenvectors are explicitly (Fig. 10.1)

$$\mathbf{f} = \begin{pmatrix} \sin\left(\frac{\pi}{N+1} l\right) \\ \vdots \\ \sin\left(\frac{\pi}{N+1} l\, n\right) \\ \vdots \\ \sin\left(\frac{\pi}{N+1} l\, N\right) \end{pmatrix}. \tag{10.39}$$

For Neumann boundary conditions $\frac{\partial f}{\partial x}(x_1) = \frac{\partial f}{\partial x}(x_N) = 0$ the matrix is slightly different (Sect. 7.2 in Vol. 2)

$$M = \begin{pmatrix} -2 & 2 & & & \\ 1 & -2 & 1 & & \\ & \ddots & \ddots & \ddots & \\ & & 1 & -2 & 1 \\ & & & 2 & -2 \end{pmatrix}. \tag{10.40}$$

Its eigenvalues are also given by the expression (10.33). To obtain the eigenvectors, we try a more general ansatz with a phase shift

$$\mathbf{f} = \begin{pmatrix} \sin\Phi_1 \\ \vdots \\ \sin(\Phi_1 + (n-1)k) \\ \vdots \\ \sin(\Phi_1 + (N-1)k) \end{pmatrix}. \tag{10.41}$$

Obviously

$$\sin(\Phi_1 + (n-1)k - k) + \sin(\Phi_1 + (n-1)k + k) - 2\sin(\Phi_1 + (n-1)k)$$
$$= 2(\cos k - 1)\sin(\Phi_1 + (n-1)k). \tag{10.42}$$

The first and last lines of the eigenvalue equation give

$$0 = -2\sin(\Phi_1) + 2\sin(\Phi_1 + k) - 2(\cos k - 1)\sin(\Phi_1)$$
$$= 2\cos\Phi_1 \sin k \tag{10.43}$$

and

$$0 = -2\sin(\Phi_1 + (N-1)k) + 2\sin(\Phi_1 + (N-1)k - k) - 2(\cos k - 1)\sin(\Phi_1 + (N-1)k)$$
$$= 2\cos(\Phi_1 + (N-1)k)\sin k \tag{10.44}$$

which is solved by

$$\Phi_1 = \frac{\pi}{2} \quad k = \frac{\pi}{N-1}l, \quad l = 1, 2 \ldots N \tag{10.45}$$

hence finally the eigenvector is (Fig. 10.1)

$$\mathbf{f} = \begin{pmatrix} 1 \\ \vdots \\ \cos\left(\frac{n-1}{N-1}\pi l\right) \\ \vdots \\ (-1)^l \end{pmatrix}. \tag{10.46}$$

Even simpler is the case of the corresponding cyclic tridiagonal matrix

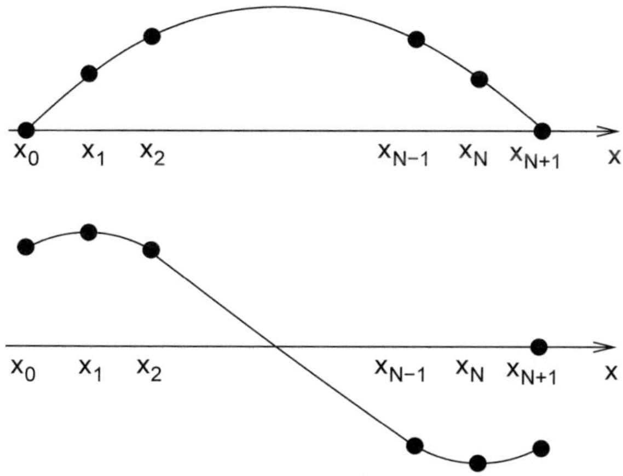

Fig. 10.1 Lowest eigenvector for fixed and open boundaries. **Top:** For fixed boundaries $f_n = \sin(nk)$ which is zero at the additional points x_0, x_{N+1}. **Bottom:** For open boundaries $f_n = \cos((n-1)k)$ with horizontal tangent at x_1, x_N due to the boundary conditions $f_2 = f_0$, $f_{N-1} = f_{N+1}$

$$M = \begin{pmatrix} -2 & 1 & & & 1 \\ 1 & -2 & 1 & & \\ & \ddots & \ddots & \ddots & \\ & & 1 & -2 & 1 \\ 1 & & & 1 & -2 \end{pmatrix} \tag{10.47}$$

which has eigenvectors

$$\mathbf{f} = \begin{pmatrix} e^{ik} \\ \vdots \\ e^{ink} \\ \vdots \\ e^{iNk} \end{pmatrix} \tag{10.48}$$

and eigenvalues

$$\lambda = -2 + e^{-ik} + e^{ik} = 2(\cos(k) - 1) = -4\sin^2\left(\frac{k}{2}\right) \tag{10.49}$$

where the possible $k-$ values again follow from the first and last line

$$-2e^{ik} + e^{i2k} + e^{iNk} = \left(-2 + e^{-ik} + e^{ik}\right) e^{ik} \tag{10.50}$$

$$e^{ik} + e^{i(N-1)k} - 2e^{iNk} = \left(-2 + e^{-ik} + e^{ik}\right) e^{iNk} \tag{10.51}$$

which both lead to

$$e^{iNk} = 1 \tag{10.52}$$

$$k = \frac{2\pi}{N}l, \quad l = 0, 1, \cdots N-1. \tag{10.53}$$

Discretized First Derivatives

Using symmetric differences to discretize a first derivative in one dimension leads to the matrix[2]

$$D = \begin{pmatrix} & 1 & & & & \\ -1 & & 1 & & & \\ & \ddots & & \ddots & & \\ & & \ddots & & \ddots & \\ & & & -1 & & 1 \\ & & & & -1 & \end{pmatrix}. \tag{10.54}$$

The characteristic polynomial of the Hermitian matrix iD is given by the recursion (10.3.1)

$$P_0 = 1$$

$$P_1 = -\lambda$$

$$\vdots$$

$$P_N = -\lambda P_{N-1} - P_{N-2} \tag{10.55}$$

which after the substitution $x = -\lambda/2$ is exactly the recursion for the Chebyshev polynomial of the second kind $U_N(x)$. Hence the eigenvalues of D are given by the roots x_k of $U_N(x)$ as

$$\lambda_D = 2ix_k = 2i\cos\left(\frac{k\pi}{N+1}\right) \quad k = 1, 2 \ldots N. \tag{10.56}$$

The eigenvalues of the corresponding cyclic tridiagonal matrix

[2] This matrix is skew symmetric, hence iD is Hermitian and has real eigenvalues $i\lambda$.

$$D = \begin{pmatrix} & 1 & & & & -1 \\ -1 & & 1 & & & \\ & \ddots & & \ddots & & \\ & & \ddots & & \ddots & \\ & & & -1 & & 1 \\ 1 & & & & -1 & \end{pmatrix} \tag{10.57}$$

are easy to find. Inserting the ansatz for the eigenvector

$$\begin{pmatrix} \exp \mathrm{i}k \\ \vdots \\ \exp \mathrm{i}Nk \end{pmatrix} \tag{10.58}$$

we find the eigenvalues

$$\mathrm{e}^{\mathrm{i}(m+1)k} - \mathrm{e}^{\mathrm{i}(m-1)k} = \lambda \mathrm{e}^{\mathrm{i}mk} \tag{10.59}$$

$$\lambda = 2\mathrm{i} \sin k \tag{10.60}$$

and from the first and last equation

$$1 = \mathrm{e}^{\mathrm{i}Nk} \tag{10.61}$$

$$\mathrm{e}^{\mathrm{i}k} = \mathrm{e}^{\mathrm{i}(N+1)k} \tag{10.62}$$

the possible k-values

$$k = \frac{2\pi}{N} l, \quad l = 0, 1, \ldots N - 1. \tag{10.63}$$

10.4 Reduction to a Tridiagonal Matrix

Eigenproblem algorithms work especially efficient if the matrix is first transformed to tridiagonal form (for real symmetric matrices, upper Hessian form for real non-symmetric matrices) which can be achieved by a series of Householder transformations (5.56)

$$A' = PAP \quad \text{with} \quad P = P^T = 1 - 2\frac{\mathbf{u}\mathbf{u}^T}{|\mathbf{u}|^2}. \tag{10.64}$$

The following orthogonal transformation P_1 brings the first row and column to tridiagonal form. We divide the matrix A according to

$$A = \begin{pmatrix} a_{11} & \boldsymbol{\alpha}^T \\ \boldsymbol{\alpha} & A_{rest} \end{pmatrix} \tag{10.65}$$

with the $(N-1)$-dimensional vector

$$\boldsymbol{\alpha} = \begin{pmatrix} a_{12} \\ \vdots \\ a_{1n} \end{pmatrix}.$$

Now let

$$\mathbf{u} = \begin{pmatrix} 0 \\ a_{12}+\lambda \\ \vdots \\ a_{1N} \end{pmatrix} = \begin{pmatrix} 0 \\ \boldsymbol{\alpha} \end{pmatrix} + \lambda \mathbf{e}^{(2)} \quad \text{with} \quad \mathbf{e}^{(2)} = \begin{pmatrix} 0 \\ 1 \\ 0 \\ \vdots \\ 0 \end{pmatrix}. \tag{10.66}$$

Then

$$|\mathbf{u}|^2 = |\boldsymbol{\alpha}|^2 + \lambda^2 + 2\lambda a_{12} \tag{10.67}$$

and

$$\mathbf{u}^T \begin{pmatrix} a_{11} \\ \boldsymbol{\alpha} \end{pmatrix} = |\alpha|^2 + \lambda a_{12}. \tag{10.68}$$

The first row of A is transformed by multiplication with P_1 according to

$$P_1 \begin{pmatrix} a_{11} \\ \boldsymbol{\alpha} \end{pmatrix} = \begin{pmatrix} a_{11} \\ \boldsymbol{\alpha} \end{pmatrix} - 2\frac{|\alpha|^2 + \lambda a_{12}}{|\boldsymbol{\alpha}|^2 + \lambda^2 + 2\lambda a_{12}} \left[\begin{pmatrix} 0 \\ \boldsymbol{\alpha} \end{pmatrix} + \lambda \mathbf{e}^{(2)} \right]. \tag{10.69}$$

The elements number $3 \ldots N$ are eliminated if we choose[3]

$$\lambda = \pm|\alpha| \tag{10.70}$$

because then

$$2\frac{|\alpha|^2 + \lambda a_{12}}{|\boldsymbol{\alpha}|^2 + \lambda^2 + 2\lambda a_{12}} = 2\frac{|\alpha|^2 \pm |\alpha| a_{12}}{|\alpha|^2 + |\alpha|^2 \pm 2|\alpha| a_{12}} = 1 \tag{10.71}$$

and

$$P_1 \begin{pmatrix} a_{11} \\ \boldsymbol{\alpha} \end{pmatrix} = \begin{pmatrix} a_{11} \\ \boldsymbol{\alpha} \end{pmatrix} - \begin{pmatrix} 0 \\ \boldsymbol{\alpha} \end{pmatrix} - \lambda \mathbf{e}^{(2)} = \begin{pmatrix} a_{11} \\ \mp|\alpha| \\ 0 \\ \vdots \\ 0 \end{pmatrix}. \tag{10.72}$$

[3] To avoid numerical extinction we choose the sign to be that of A_{12}.

Finally we have

$$A^{(2)} = P_1 A P_1 = \begin{pmatrix} a_{11} & a_{12}^{(2)} & 0 & \cdots & 0 \\ a_{12}^{(2)} & a_{22}^{(2)} & a_{23}^{(2)} & \cdots & a_{2N}^{(2)} \\ 0 & a_{23}^{(2)} & \ddots & & a_{3N}^{(2)} \\ \vdots & \vdots & & \ddots & \vdots \\ 0 & a_{2N}^{(2)} & a_{3N}^{(2)} & \cdots & a_{NN}^{(2)} \end{pmatrix} \tag{10.73}$$

as desired.

For the next step we choose

$$\boldsymbol{\alpha} = \begin{pmatrix} a_{22}^{(2)} \\ \vdots \\ a_{2N}^{(2)} \end{pmatrix}, \quad \mathbf{u} = \begin{pmatrix} 0 \\ 0 \\ \alpha \end{pmatrix} \pm |\alpha| \mathbf{e}^{(3)} \tag{10.74}$$

to eliminate the elements $a_{24} \ldots a_{2N}$. Note that P_2 does not change the first row and column of $A^{(2)}$ and therefore

$$A^{(3)} = P_2 A^{(2)} P_2 = \begin{pmatrix} a_{11} & a_{12}^{(2)} & 0 & \cdots & \cdots & 0 \\ a_{12}^{(2)} & a_{22}^{(2)} & a_{23}^{(3)} & 0 & \cdots & 0 \\ 0 & a_{23}^{(3)} & a_{33}^{(3)} & \cdots & \cdots & a_{3N}^{(3)} \\ \vdots & 0 & \vdots & & & \vdots \\ \vdots & \vdots & \vdots & & & \vdots \\ 0 & 0 & a_{3N}^{(3)} & \cdots & \cdots & a_{NN}^{(3)} \end{pmatrix}. \tag{10.75}$$

After $N - 1$ transformations finally a tridiagonal matrix is obtained.

10.5 The Power Iteration Method

A real symmetric $N \times N$ matrix with (orthonormal) eigenvectors and eigenvalues[4]

$$A\mathbf{u}_i = \lambda_i \mathbf{u}_i \tag{10.76}$$

can be expanded as

$$A = \sum_i \mathbf{u}_i \lambda_i \mathbf{u}_i^T. \tag{10.77}$$

The sequence of powers

[4] We do not consider degenerate eigenvalues explicitly here.

$$A^n = \sum_i \mathbf{u}_i \lambda_i^n \mathbf{u}_i^T \tag{10.78}$$

converges to

$$A^n \to \mathbf{u}_{\max} \lambda_{\max}^n \mathbf{u}_{\max}^T$$

where[5]

$$|\lambda_{\max}| = \max. \tag{10.79}$$

Hence for any initial vector $\mathbf{v}_1$ (which is arbitrary but not perpendicular to $\mathbf{u}_{\max}$) the sequence

$$\mathbf{v}_{n+1} = A\mathbf{v}_n \tag{10.80}$$

converges to a multiple of $\mathbf{u}_{\max}$. To obtain all eigenvectors simultaneously, we could use a set of independent start vectors, e.g the N unit vectors and iterate simultaneously for all of them

$$(\mathbf{v}_1^{(1)}, \ldots \mathbf{v}_1^{(N)}) = (\mathbf{e}_1, \ldots \mathbf{e}_N) = \begin{pmatrix} 1 & & \\ & \ddots & \\ & & 1 \end{pmatrix} \tag{10.81}$$

$$(\mathbf{v}_2^{(1)}, \ldots \mathbf{v}_2^{(N)}) = A(\mathbf{v}_1^{(1)}, \ldots \mathbf{v}_1^{(N)}) = A. \tag{10.82}$$

Most probably, all column vectors of A then converge to multiples of the same eigenvector. To assure linear independence, an orthogonalization step has to follow each iteration. This can be done (QR decomposition, Sect. 5.2) by decomposing the matrix into the product of an upper triangular[6] matrix R and an orthogonal matrix $Q^T = Q^{-1}$ (Sect. 5.2)

$$A = QR. \tag{10.83}$$

For symmetric tridiagonal matrices this factorization can be efficiently realized by multiplication with a sequence of Givens rotation matrices which eliminate the off-diagonal elements in the lower part one by one[7]

$$Q = R_{\alpha_{N-1}}^{(N-1,N)} \ldots R_{\alpha_2}^{(2,3)} R_{\alpha_1}^{(1,2)} \tag{10.84}$$

beginning with

---

[5] For simplicity we do not consider eigenvalues which are different but have the same absolute value.

[6] The equivalent QL method uses a lower triangular matrix.

[7] This is quite different from the Jacobi method since it is not an orthogonal transformation.

$$R_{\alpha_1}^{(1,2)} A = \begin{pmatrix} c & s & & & \\ -s & c & & & \\ & & 1 & & \\ & & & \ddots & \\ & & & & 1 \end{pmatrix} \begin{pmatrix} a_{11} & a_{12} & & & \\ a_{12} & a_{22} & a_{23} & & \\ & \ddots & \ddots & \ddots & \\ & & a_{N-2,N-1} & a_{N-1,N-1} & a_{N-1,N} \\ & & & a_{N-1,N} & a_{N,N} \end{pmatrix}$$

$$= \begin{pmatrix} ca_{11}+sa_{12} & ca_{12}+sa_{22} & sa_{23} & & \\ 0 & -sa_{12}+ca_{22} & ca_{23} & & \\ & a_{34} & a_{33} & a_{34} & \\ & & \ddots & \ddots & \ddots \\ & & & a_{N-1,N} & a_{N,N} \end{pmatrix} \tag{10.85}$$

where the rotation angle α_1 has to be chosen such that

$$\tan\alpha_1 = \frac{s}{c} = \frac{a_{12}}{a_{11}}. \tag{10.86}$$

Finally, this leads to a method known as orthogonal simultaneous power iteration

$$W^{(1)} = A = Q^{(1)} R^{(1)} \tag{10.87}$$

$$W^{(n+1)} = A Q^{(n)} \tag{10.88}$$

$$Q^{(n+1)} R^{(n+1)} = W^{(n+1)}. \tag{10.89}$$

This method calculates a sequence of orthogonal matrices $Q^{(n)}$ which converge to a set of independent eigenvectors. Moreover, from (10.88), (10.89)

$$A = W^{(n+1)} Q^{(n)T} = Q^{(n+1)} R^{(n+1)} Q^{(n)T} \tag{10.90}$$

and therefore powers of A are given by

$$A^n = \left(Q^{(n)} R^{(n)} Q^{(n-1)T}\right)\left(Q^{(n-1)} R^{(n-1)} Q^{(n-2)T}\right)\ldots\left(Q^{(2)} R^{(2)} Q^{(1)T}\right)\left(Q^{(1)} R^{(1)}\right)$$

$$= Q^{(n)} R^{(n)} R^{(n-1)} \ldots R^{(1)}. \tag{10.91}$$

The product of two upper triangular matrices is upper triangular again which can be seen from

$$(R^{(m)} R^{(n)})_{i,k} = \sum_{j:i\le j\le k} R_{i,j}^{(m)} R_{jk}^{(n)} = 0 \text{ if } i > k. \tag{10.92}$$

Therefore the QR decomposition of A^n is

$$A^n = Q^{(n)} \overline{R}^{(n)} \tag{10.93}$$

with

$$\overline{R}^{(n)} = R^{(n)} \dots R^{(1)}. \tag{10.94}$$

To obtain other than the dominant eigenvalues, the inverse power iteration method with shift is useful. Consider the matrix

$$\tilde{A} = (A - \sigma)^{-1} \tag{10.95}$$

where μ is not an eigenvalue $\lambda_i$of A. Obviously its has the same eigenvectors as A and eigenvalues given by

$$\tilde{A}\mathbf{u}_i = \tilde{\lambda}_i \mathbf{u}_i = \frac{1}{\lambda_i - \sigma}\mathbf{u}_i. \tag{10.96}$$

Hence, if σ is close to λ_i, the power iteration method will converge to a multiple of $\mathbf{u}_i$.For practical calculations, an equivalent formulation of the power iteration method is used which is known as the QR (or QL) method.

10.6 The QR Algorithm

The QR algorithm [174] is an iterative algorithm. It uses a series of orthogonal transformations which conserve the eigenvalues. Starting from the decomposition of A

$$A = Q_1 R_1 \tag{10.97}$$

$$A_2 = R_1 Q_1 = Q_1^T A Q_1 \tag{10.98}$$

we iterate

$$\vdots$$

$$A_n = Q_n R_n \tag{10.99}$$

$$A_{n+1} = R_n Q_n = Q_n^T A_n Q_n. \tag{10.100}$$

From (10.99), (10.100)

$$Q_{n+1} R_{n+1} = R_n Q_n$$

and the n-th power of A is

$$A^n = AA \ldots A = Q_1 R_1 Q_1 R_1 \ldots Q_1 R_1 = Q_1 (Q_2 R_2 \ldots Q_2 R_2) R_1$$

$$= Q_1 Q_2 (Q_3 R_3 \ldots Q_3 R_3) R_2 R_1 \cdots = \overline{Q}_n \overline{R}_n \tag{10.101}$$

$$\overline{Q}_n = Q_1 \ldots Q_n \quad \overline{R}_n = R_n \ldots R_1. \tag{10.102}$$

But since QR decomposition is unique, comparison of (10.93, 10.101) shows

$$\overline{Q}_n = Q^{(n)} \quad \overline{R}_n = \overline{R}^{(n)} \tag{10.103}$$

i.e. the column vectors of $\overline{Q}_n$ converge to a set of eigenvectors and the transformed matrix

$$A_{n+1} = Q_n^T A_n Q_n = Q_n^T Q_{n-1}^T A_{n-1} Q_{n-1} Q_n = \cdots = \overline{Q}_n^T A \overline{Q}_n \tag{10.104}$$

converges to a diagonal matrix. Now consider the inverse power

$$A^{-n} = \overline{R}_n^{-1} \overline{Q}_n^T. \tag{10.105}$$

The inverse of a symmetric matrix is also symmetric and

$$A^{-n} = \overline{Q}_n \left(\overline{R}_n^{-1} \right)^T \tag{10.106}$$

shows, that the QR algorithm uses the same orthogonal transformations as ordinary and also inverse power iteration. The inverse of an upper triangular matrix is also upper triangular but the transpose is lower triangular. Therefore we modify (10.106) by multiplying with a permutation matrix

$$P = \begin{pmatrix} & & 1 \\ & \ddots & \\ 1 & & \end{pmatrix} \quad P^2 = 1 \tag{10.107}$$

from the right side, which reverses the order of the columns and

$$A^{-n} P = \overline{Q}_n P P \left(\overline{R}_n^{-1} \right)^T P = \tilde{Q} \tilde{R} \tag{10.108}$$

is the QR decomposition of $A^{-n} P$. This shows the close relationship between the QR algorithm[8] and the inverse power iteration method

To improve convergence, a shift σ is introduced and the QR factorization applied to $A_n - \sigma$. The modified iteration then reads

[8] Or the equivalent QL algorithm.

$$A_n - \sigma = Q_n R_n \tag{10.109}$$

$$A_{n+1} = R_n Q_n + \sigma = Q_n^T (A_n - \sigma) Q_n + \sigma = Q_n^T A_n Q_n. \tag{10.110}$$

Symmetry and tridiagonal form are conserved by this algorithm. The simplest choice for the shift is to take a diagonal element $\sigma_m = a_{m,m} = (A_n)_{m,m}$ corresponding to the Rayleigh quotient method. An even more robust and very popular choice [170] is Wilkinson's shift

$$\sigma_m = a_{m,m} + \delta - \text{sign}(\delta)\sqrt{\delta^2 + a_{m,m-1}^2} \quad \delta = \frac{a_{m-1,m-1} - a_{m,m}}{2} \tag{10.111}$$

which is that eigenvalue of the matrix $\begin{pmatrix} a_{m-1,m-1} & a_{m-1,m} \\ a_{m-1,m} & a_{m,m} \end{pmatrix}$ which is closer to $a_{n,n}$. The calculation starts with σ_N and iterates until the off-diagonal element $a_{N-1,N}$ becomes sufficiently small.[9] Then the transformed matrix has the form

$$\begin{pmatrix} a_{11} & a_{12} & & & \\ a_{12} & a_{22} & \ddots & & \\ & \ddots & \ddots & a_{N-2,N-1} & \\ & & a_{N-2,N-1} & a_{N-1,N-1} & 0 \\ & & & 0 & a_{NN} \end{pmatrix}. \tag{10.112}$$

Now the last column and row can be discarded (deflation method) and the next iteration performed with the shift σ_{N-1} on a tridiagonal matrix of dimension N$-$1. This procedure has to be repeated N times to obtain all eigenvalues. Convergence is usually cubic (or at least quadratic if there are degenerate eigenvalues).

10.7 Hermitian Matrices

In quantum mechanics often Hermitian matrices have to be diagonalized (which have real valued eigenvalues). To avoid complex arithmetics, an Hermitian eigenproblem can be replaced by a symmetric real valued problem of double dimension by introducing

$$B = \Re(A) \quad C = \Im(A) \quad \mathbf{x} = \mathbf{u} + \mathrm{i}\mathbf{v} \tag{10.113}$$

where, for Hermitian A

$$A = B + \mathrm{i}C = A^H = B^T - \mathrm{i}C^T \tag{10.114}$$

[9] For the QL method, it is numerically more efficient to start at the upper left corner of the matrix.

hence

$$B = B^T \quad C = -C^T \tag{10.115}$$

and the eigenvalue problem can be rewritten as

$$0 = (A\mathbf{x} - \lambda\mathbf{x}) = (B\mathbf{u} - C\mathbf{v} - \lambda\mathbf{u}) + \mathrm{i}(B\mathbf{v} + C\mathbf{u} - \lambda\mathbf{v}) \tag{10.116}$$

or finally in the real symmetric form

$$\begin{pmatrix} B & C^T \\ C & B \end{pmatrix} \begin{pmatrix} \mathbf{u} \\ \mathbf{v} \end{pmatrix} = \lambda \begin{pmatrix} \mathbf{u} \\ \mathbf{v} \end{pmatrix}. \tag{10.117}$$

Each eigenvalue of the N-dimensional Hermitian problem corresponds to two eigenvectors of the 2N-dimensional problem since for any solution of (10.117)

$$\begin{pmatrix} B & -C \\ C & B \end{pmatrix} \begin{pmatrix} -\mathbf{v} \\ \mathbf{u} \end{pmatrix} = \begin{pmatrix} -B\mathbf{v} - C\mathbf{u} \\ -C\mathbf{v} + B\mathbf{u} \end{pmatrix} = \lambda \begin{pmatrix} -\mathbf{v} \\ \mathbf{u} \end{pmatrix} \tag{10.118}$$

provides a different solution, while the complex vectors $\mathbf{u} + \mathrm{i}\mathbf{v}$ and $\mathrm{i}(\mathbf{u} + \mathrm{i}\mathbf{v}) = -\mathbf{v} + \mathrm{i}\mathbf{u}$ only differ by a phase factor.

10.8 Large Matrices

Many problems in computational physics involve very large matrices, for which standard methods are not applicable. It might be even difficult or impossible to keep the full matrix in memory. Here methods are used which only involve the product of the matrix with a vector which can be computed on the fly. Krylov methods are very similar to power iteration but diagonalize only the projection of the matrix onto a Krylov space of much smaller dimension $n \ll N$ which is constructed by multiplying a normalized start vector $\mathbf{q}_1$ repeatedly with A

$$K_n(A, \mathbf{q}_1) = \mathrm{span}\{\mathbf{q}_1, A\mathbf{q}_1, A^2\mathbf{q}_1, \dots A^{n-1}\mathbf{q}_1\}. \tag{10.119}$$

We use the Arnoldi method Sect. 5.6.5 to construct an orthonormalized basis of this space. For a symmetric matrix this simplifies to a three-term recursion also known as symmetric Lanczos algorithm [175]. Applying the Arnoldi method

$$h_{j,n} = (\mathbf{q}_j^T A\mathbf{q}_n) \quad j \le n \tag{10.120}$$

$$\tilde{\mathbf{q}}_{n+1} = A\mathbf{q}_n - \sum_{j=1}^{n} h_{jn}\mathbf{q}_j \tag{10.121}$$

$$h_{n+1,n} = |\tilde{\mathbf{q}}_{n+1}| \quad \mathbf{q}_{n+1} = \frac{\tilde{\mathbf{q}}_{n+1}}{h_{n+1,n}} \tag{10.122}$$

to a symmetric matrix A, we find

$$h_{n+1,n}^2 = \mathbf{q}_n^T A^2 \mathbf{q}_n - 2\sum_j (\mathbf{q}_n^T A \mathbf{q}_j)(\mathbf{q}_j^T A \mathbf{q}_n) + \sum_{jj'} (\mathbf{q}_j^T A \mathbf{q}_n)(\mathbf{q}_{j'}^T A \mathbf{q}_n)\delta_{jj'}$$

$$= \mathbf{q}_n^T A \left[h_{n+1,n}\mathbf{q}_{n+1} + \sum_j h_{jn}\mathbf{q}_j \right] - \sum_j h_{jn}^2 = h_{n+1,n}h_{n,n+1} \tag{10.123}$$

hence

$$h_{n+1,n} = h_{n,n+1}. \tag{10.124}$$

Furthermore,

$$h_{n-2,n} = \mathbf{q}_{n-2}^T A \mathbf{q}_n = \mathbf{q}_{n-2}^T A \frac{1}{h_{n,n-1}} \left[A\mathbf{q}_{n-1} - \sum_{j=1}^{n-1} h_{jn-1}\mathbf{q}_j \right]$$

$$= \frac{1}{h_{n.n-1}} \left[-\sum_{j=1}^{n-2} h_{jn-1}h_{j,n-2} - h_{n-1,n-1}h_{n-2,n-1} + \mathbf{q}_{n-1}^T A \left(h_{n-1,n-2}\mathbf{q}_{n-1} + \sum_{j=1}^{n-2} h_{jn-2}\mathbf{q}_j \right) \right]$$

$$= \frac{1}{h_{n.n-1}} \left[-\sum_{j=1}^{n-2} h_{jn-1}h_{j,n-2} - h_{n-1,n-1}h_{n-2,n-1} + h_{n-1,n-1}h_{n-1,n-2} + \sum_{j=1}^{n-2} h_{jn-1}h_{jn-2} \right] = 0 \tag{10.125}$$

and similar for $s > 2$

$$h_{n-s,n} = \mathbf{q}_{n-s}^T A \mathbf{q}_n = \mathbf{q}_{n-s}^T A \frac{1}{h_{n,n-1}} \left[A\mathbf{q}_{n-1} - \sum_{j=1}^{n-1} h_{jn-1}\mathbf{q}_j \right]$$

$$= \frac{1}{h_{n.n-1}} \left[-\sum_{j=1}^{n-2} h_{jn-1}h_{j,n-s} - h_{n-1,n-1}h_{n-s,n-1} + \mathbf{q}_{n-1}^T A (h_{n-s+1,n-s}\mathbf{q}_{n-s+1} + \sum_{j=1}^{n-s} h_{jn-s}\mathbf{q}_j) \right]$$

$$= \frac{1}{h_{n.n-1}} \left[-\sum_{j=1}^{n-2} h_{jn-1}h_{j,n-s} - h_{n-1,n-1}h_{n-s,n-1} + h_{n-s+1,n-s}h_{n-1,n-s+1} + \sum_{j=1}^{n-s} h_{jn-1}h_{jn-s} \right]$$

$$= \frac{1}{h_{n,n-1}} \left[- \sum_{j=n-s+1}^{n-2} h_{j,n-1} h_{j,n-s} - h_{n-1,n-1} h_{n-s,n-1} + h_{n-s,n-s+1} h_{n-s+1,n-1} \right] \tag{10.126}$$

$$h_{n-s,n} = \frac{1}{h_{n,n-1}} \left[- \sum_{j=n-s+2}^{n-2} h_{j,n-1} h_{j,n-s} - h_{n-1,n-1} h_{n-s,n-1} \right]. \tag{10.127}$$

Starting from (10.125) for $s = 2$ we increment s repeatedly and find

$$h_{n-2,n} = h_{n-3,n} = \ldots h_{1,n} = 0 \tag{10.128}$$

since (10.127) only involves smaller values of s, for which (10.128) already has been shown. The Arnoldi decomposition produces an upper Hessenberg matrix Sect. (5.6.5),

$$U_n = (\mathbf{q}_1, \ldots \mathbf{q}_n) \tag{10.129}$$

$$AU_n = U_{n+1} H = (U_n, \mathbf{q}_{n+1}) \begin{pmatrix} H_n \\ h_{n+1,n} \mathbf{e}_n^T \end{pmatrix} = U_n H_n + h_{n+1} \mathbf{q}_{n+1} \mathbf{e}_n^T \tag{10.130}$$

which for symmetric A becomes tridiagonal

$$H = \begin{pmatrix} h_{11} & h_{12} & \cdots & h_{1n} \\ h_{21} & h_{22} & \cdots & h_{2n} \\ & h_{32} & \ddots & \\ & & \ddots & h_{nn} \\ & & & h_{n+1,n} \end{pmatrix} = \begin{pmatrix} a_1 & b_1 & & & \\ b_1 & a_2 & b_2 & & \\ & \ddots & \ddots & \ddots & \\ & & b_{n-2} & a_{n-1} & b_{n-1} \\ & & & b_{n-1} & a_n \\ & & & & b_n \end{pmatrix} = \begin{pmatrix} T \\ b_n \mathbf{e}_n^T \end{pmatrix} \tag{10.131}$$

with a symmetric tridiagonal matrix T, which is the desired projection of A into the Krylov space K_n

$$U_n^T A U_n = U_n^T U_{n+1} \, H = U_n^T \; (U_n, \mathbf{q}_{n+1}) \begin{pmatrix} T \\ b_n \mathbf{e}_n^T \end{pmatrix}$$

$$= (E_n, 0) \begin{pmatrix} T \\ b_n \mathbf{e}_n^T \end{pmatrix} = T. \tag{10.132}$$

For an eigenpair $(\lambda, \mathbf{v})$ of T

$$A(U_n \mathbf{v}) = U_{n+1} H \mathbf{v} = (U_n, \mathbf{q}_{n+1}) \begin{pmatrix} T \\ b_n \mathbf{e}_n^T \end{pmatrix} \mathbf{v}$$

$$= \left(U_n T + b_n \mathbf{q}_{n+1} \mathbf{e}_n^T\right) \mathbf{v} = \lambda(U_n \mathbf{v}) + b_n \mathbf{q}_{n+1} \mathbf{e}_n^T \mathbf{v}. \quad (10.133)$$

Hence, an approximate eigenpair of A is given by the Ritz pair $(\lambda, U_n \mathbf{v})$ and the error can be estimated from the residual norm

$$\frac{|(A-\lambda)U_n \mathbf{v}|}{|\mathbf{v}|} = |b_n||\mathbf{e}_n^T \mathbf{v}|. \quad (10.134)$$

Due to numerical errors, orthogonality of the Lanczos vectors $\mathbf{q}_n$ can get lost and reorthogonalization is necessary [176, 177]. If this takes to much time or if memory limits do not allow to store enough Lanczos vectors, the procedure has to be restarted with a new initial vector which is usually taken as a linear combination of selected eigenvectors which have already been found [178, 179]. Furthermore, special care has to be taken to determine possible degeneracies of the eigenvalues.

10.9 Non-symmetric Matrices

Eigenvalue problems with non-symmetric matrices are more complicated. Left and right eigenvectors have to be distinguished and the eigenvalues can be complex valued even if the matrix is real. The QR method [174] is applicable also to a non-symmetric matrix but very expensive unless the matrix is first brought to upper triangular (instead of tridiagonal) form, which can be achieved by a series of similarity transformations with Householder reflections (5.2.2). The implicit QR method with double shift avoids complex arithmetics by treating pairs of complex conjugated eigenvalues simultaneously. For very large matrices the Arnoldi method brings a non-symmetric matrix to upper Hessenberg form, which provides the projection onto the Krylov space as an upper triangular matrix.

Exercises

Exercise 10.1 Disorder in a Tight-binding Model
We consider a two-dimensional lattice of interacting particles. Pairs of nearest neighbors have an interaction V and the diagonal energies are chosen from a Gaussian distribution

$$P(E) = \frac{1}{\Delta\sqrt{2\pi}} e^{-E^2/2\Delta^2}. \quad (10.135)$$

The wave function of the system is given by a linear combination

$$\psi = \sum_{ij} C_{ij} \psi_{ij} \quad (10.136)$$

where on each particle (i, j) one basis function ψ_{ij} is located. The nonzero elements of the interaction matrix are given by

$$H(ij|ij) = E_{ij} \tag{10.137}$$

$$H(ij|i \pm 1, j) = H(ij|i, j \pm 1) = V. \tag{10.138}$$

The Matrix H is numerically diagonalized and the amplitudes C_{ij} of the lowest state are shown as circles located at the grid points. As a measure of the degree of localization the quantity

$$\sum_{ij} |C_{ij}|^4 \tag{10.139}$$

is evaluated. Explore the influence of coupling V and disorder Δ.

Chapter 11
Data Fitting

Often a set of data points has to be fitted by a continuous function, either to obtain approximate function values in between the data points or to describe a functional relationship between two or more variables by a smooth curve, i.e. to fit a certain model to the data. If uncertainties of the data are negligibly small, an exact fit is possible, for instance with polynomials, spline functions or trigonometric functions (Chap. 2). If the uncertainties are considerable, a curve has to be constructed that fits the data points approximately. Consider a two-dimensional data set

$$(x_i, y_i) \quad i = 1 \ldots m \tag{11.1}$$

and a model function

$$f(x, a_1 \ldots a_n) \quad m \geq n \tag{11.2}$$

which depends on the variable x and $n \leq m$ additional parameters a_j. The errors of the fitting procedure are given by the residuals

$$r_i = y_i - f(x_i, a_1 \ldots a_n). \tag{11.3}$$

Supplementary Information The online version contains supplementary material available at https://doi.org/10.1007/978-3-032-07856-8_11.

P. O. J. Scherer, *Computational Physics I*, Graduate Texts in Physics,
https://doi.org/10.1007/978-3-032-07856-8_11

The parameters a_j have to be determined such, that the overall error is minimized, which in most practical cases is measured by the mean squared difference[1]

$$S_{sd}(a_1 \dots a_n) = \frac{1}{m}\sum_{i=1}^{m} r_i^2. \tag{11.4}$$

The optimal parameters are determined by solving the system of normal equations. If the model function depends linearly on the parameters, orthogonalization offers a numerically more stable method. The dimensionality of a data matrix can be reduced with the help of singular value decomposition, which allows to approximate a matrix by another matrix of lower rank and is also useful for linear regression, especially if the columns of the data matrix are linearly dependent.

11.1 Least Squares Fit

A (local) minimum of (11.4) corresponds to a stationary point with zero gradient. For n model parameters there are n, generally nonlinear, equations which have to be solved [180]. From the general condition

$$\frac{\partial S_{sd}}{\partial a_j} = 0 \quad j = 1 \dots n \tag{11.5}$$

we find

$$\sum_{i=1}^{m} r_i \frac{\partial f(x_i, a_1 \dots a_n)}{\partial a_j} = 0 \tag{11.6}$$

which can be solved with the methods discussed in Chap. 6. For instance, the Newton-Raphson method starts from a suitable initial guess of parameters

$$(a_1^0 \dots a_n^0) \tag{11.7}$$

and tries to improve the fit iteratively by making small changes to the parameters

$$a_j^{s+1} = a_j^s + \Delta a_j^s. \tag{11.8}$$

The changes Δa_j^s are determined approximately by expanding the model function

$$f(x_i, a_1^{s+1} \dots a_n^{s+1}) = f(x_i, a_1^s \dots a_n^s) + \sum_{j=1}^{n} \frac{\partial f(x_i, a_1^s \dots a_n^s)}{\partial a_j} \Delta a_j^s + \dots \tag{11.9}$$

[1] Minimization of the sum of absolute errors $\sum |r_i|$ is much more complicated.

to approximate the new residuals

$$r_i^{s+1} = r_i^s - \sum_{j=1}^{n} \frac{\partial f(x_i, a_1^s \dots a_m^s)}{\partial a_j} \Delta a_j^s \tag{11.10}$$

and the derivatives

$$\frac{\partial r_i^s}{\partial a_j} = -\frac{\partial f(x_i, a_1^s \dots a_m^s)}{\partial a_j}. \tag{11.11}$$

Equation (11.6) now becomes

$$\sum_{i=1}^{m} \left(r_i^s - \sum_{j=1}^{n} \frac{\partial f(x_i)}{\partial a_j} \Delta a_j^s \right) \frac{\partial f(x_i)}{\partial a_k} \tag{11.12}$$

which is a system of n (usually overdetermined) linear equations for the Δa_j, the so-called normal equations:

$$\sum_{i=1}^{m} \sum_{j=1}^{n} \frac{\partial f(x_i)}{\partial a_j} \frac{\partial f(x_i)}{\partial a_k} \Delta a_j^s = \sum_{i=1}^{m} r_i^s \frac{\partial f(x_i)}{\partial a_k}. \tag{11.13}$$

With the definition

$$A_{kj} = \frac{1}{m} \sum_{i=1}^{m} \frac{\partial f(x_i)}{\partial a_k} \frac{\partial f(x_i)}{\partial a_j} \tag{11.14}$$

$$b_k = \frac{1}{m} \sum_{i=1}^{m} y_i \frac{\partial f(x_i)}{\partial a_k} \tag{11.15}$$

the normal equations can be written as

$$\sum_{j=1}^{n} A_{kj} \Delta a_j = b_k. \tag{11.16}$$

11.1.1 Linear Least Squares Fit

Especially important are model functions which depend linearly on all parameters (Fig. 11.1 shows an example which is discussed in Exercise 11.1)

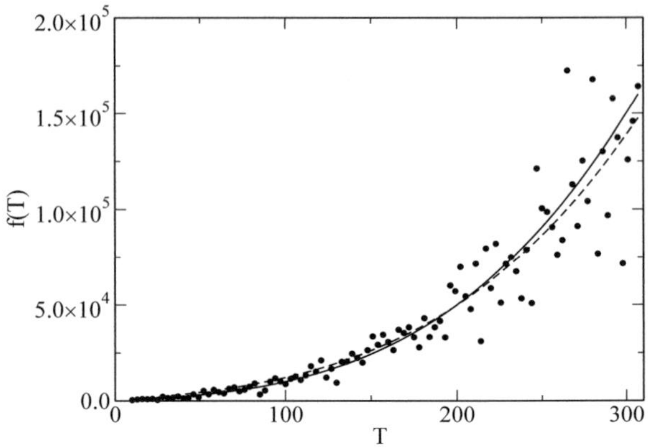

Fig. 11.1 Least squares fit The polynomial $C(T) = aT + bT^3$ (full curve) is fitted to a set of data points which are distributed randomly around the "exact" values $C(T) = a_0T + b_0T^3$ (dashed curve). For more details see Exercise 11.1

$$f(x, a_1 \ldots a_n) = \sum_{j=1}^{n} a_j f_j(x). \tag{11.17}$$

The derivatives are

$$\frac{\partial f(x_i)}{\partial a_j} = f_j(x_i) \tag{11.18}$$

and the minimum of (11.4) is given by the solution of the normal equations

$$\frac{1}{m} \sum_{j=1}^{n} \sum_{i=1}^{m} f_k(x_i) f_j(x_i) a_j = \frac{1}{m} \sum_{i=1}^{m} y_i f_k(x_i) \tag{11.19}$$

which for a linear fit problem become

$$\sum_{j=1}^{n} A_{kj} a_j = b_k \tag{11.20}$$

with

$$A_{kj} = \frac{1}{m} \sum_{i=1}^{m} f_k(x_i) f_j(x_i) \tag{11.21}$$

$$b_k = \frac{1}{m} \sum_{i=1}^{m} y_i f_k(x_i). \tag{11.22}$$

Example: Linear Regression

For a linear fit function

$$f(x) = a_0 + a_1 x \tag{11.23}$$

the mean square difference is

$$S_{sd} = \frac{1}{m}\sum_{i=1}^{m}(y_i - a_0 - a_1 x_i)^2 \tag{11.24}$$

and we have to solve the equations

$$0 = \frac{\partial S_{sd}}{\partial a_0} = \frac{1}{m}\sum_{i=1}^{m}(y_i - a_0 - a_1 x_i) = \overline{y} - a_0 - a_1\overline{x}$$

$$0 = \frac{\partial S_{sd}}{\partial a_1} = \frac{1}{m}\sum_{i=1}^{m}(y_i - a_0 - a_1 x_i)x_i = \overline{xy} - a_0\overline{x} - a_1\overline{x^2} \tag{11.25}$$

which can be done here with determinants

$$a_0 = \frac{\begin{vmatrix} \overline{y} & \overline{x} \\ \overline{xy} & \overline{x^2} \end{vmatrix}}{\begin{vmatrix} 1 & \overline{x} \\ \overline{x} & \overline{x^2} \end{vmatrix}} = \frac{\overline{y}\,\overline{x^2} - \overline{x}\,\overline{xy}}{\overline{x^2} - \overline{x}^2} \tag{11.26}$$

$$a_1 = \frac{\begin{vmatrix} 1 & \overline{y} \\ \overline{x} & \overline{xy} \end{vmatrix}}{\begin{vmatrix} 1 & \overline{x} \\ \overline{x} & \overline{x^2} \end{vmatrix}} = \frac{\overline{xy} - \overline{x}\,\overline{y}}{\overline{x^2} - \overline{x}^2}. \tag{11.27}$$

11.1.2 Linear Least Squares Fit with Orthogonalization

With the definitions

$$\mathbf{x} = \begin{pmatrix} a_1 \\ \vdots \\ a_n \end{pmatrix} \quad \mathbf{b} = \begin{pmatrix} y_1 \\ \vdots \\ y_m \end{pmatrix} \tag{11.28}$$

and the $m \times n$ matrix

$$A = \begin{pmatrix} a_{11} & \cdots & a_{1n} \\ \vdots & \ddots & \vdots \\ a_{m1} & \cdots & a_{mn} \end{pmatrix} = \begin{pmatrix} f_1(x_1) & \cdots & f_n(x_1) \\ \vdots & \ddots & \vdots \\ f_1(x_m) & \cdots & f_n(x_m) \end{pmatrix} \tag{11.29}$$

the linear least squares fit problem (11.20) can be formulated as a search for the minimum of

$$|A\mathbf{x}-\mathbf{b}| = \sqrt{(A\mathbf{x}-\mathbf{b})^T(A\mathbf{x}-\mathbf{b})}. \tag{11.30}$$

In the last section we calculated the gradient

$$\frac{\partial |A\mathbf{x}-\mathbf{b}|^2}{\partial \mathbf{x}} = A^T(A\mathbf{x}-\mathbf{b}) + (A\mathbf{x}-\mathbf{b})^T A = 2A^T A\mathbf{x} - 2A^T\mathbf{b} \tag{11.31}$$

and solved the normal equations

$$A^T A\mathbf{x} = A^T\mathbf{b}. \tag{11.32}$$

This method can become numerically unstable. Alternatively we use orthogonalization of the n column vectors $\mathbf{a}_k$ of A to have

$$A = (\mathbf{a}_1 \cdots \mathbf{a}_n) = (\mathbf{q}_1 \cdots \mathbf{q}_n)\begin{pmatrix} r_{11} & r_{12} & \cdots & r_{1n} \\ & r_{22} & \cdots & r_{2n} \\ & & \ddots & \vdots \\ & & & r_{nn} \end{pmatrix} \tag{11.33}$$

where $\mathbf{a}_k$ and $\mathbf{q}_k$ are now vectors of dimension m. Since the $\mathbf{q}_k$ are orthonormal $\mathbf{q}_i^T\mathbf{q}_k = \delta_{ik}$ we have

$$\begin{pmatrix} \mathbf{q}_1^T \\ \vdots \\ \mathbf{q}_n^T \end{pmatrix} A = \begin{pmatrix} r_{11} & r_{12} & \cdots & r_{1n} \\ & r_{22} & \cdots & r_{2n} \\ & & \ddots & \vdots \\ & & & r_{nn} \end{pmatrix}. \tag{11.34}$$

The $\mathbf{q}_k$ can be augmented by another $(m-n)$ vectors to provide an orthonormal basis of $\mathbb{R}^m$. These will not be needed explicitly. They are orthogonal to the first n vectors and hence to the column vectors of A. All vectors $\mathbf{q}_k$ together form an orthogonal matrix

$$Q = \begin{pmatrix} \mathbf{q}_1 & \cdots & \mathbf{q}_n & \mathbf{q}_{n+1} & \cdots & \mathbf{q}_m \end{pmatrix} \tag{11.35}$$

and we can define the transformation of the matrix A:

$$\widetilde{A} = \begin{pmatrix} \mathbf{q}_1^T \\ \vdots \\ \mathbf{q}_n^T \\ \mathbf{q}_{n+1}^T \\ \vdots \\ \mathbf{q}_m^T \end{pmatrix} (\mathbf{a}_1 \cdots \mathbf{a}_n) = Q^T A = \begin{pmatrix} R \\ 0 \end{pmatrix} \quad R = \begin{pmatrix} r_{11} & \cdots & r_{1n} \\ & \ddots & \vdots \\ & & r_{nn} \end{pmatrix}. \tag{11.36}$$

The vector $\mathbf{b}$ transforms as

$$\tilde{\mathbf{b}} = Q^T\mathbf{b} = \begin{pmatrix} \mathbf{b}_u \\ \mathbf{b}_l \end{pmatrix} \quad \mathbf{b}_u = \begin{pmatrix} \mathbf{q}_1^T \\ \vdots \\ \mathbf{q}_n^T \end{pmatrix} \mathbf{b} \quad \mathbf{b}_l = \begin{pmatrix} \mathbf{q}_{n+1}^T \\ \vdots \\ \mathbf{q}_m^T \end{pmatrix} \mathbf{b}. \tag{11.37}$$

Since the norm of a vector is not changed by unitary transformations

$$|\mathbf{b} - A\mathbf{x}| = \sqrt{(\mathbf{b}_u - R\mathbf{x})^2 + \mathbf{b}_l^2} \tag{11.38}$$

which is minimized if

$$R\mathbf{x} = \mathbf{b}_u. \tag{11.39}$$

The error of the fit is given by

$$|\mathbf{b} - A\mathbf{x}| = |\mathbf{b}_l|. \tag{11.40}$$

Example: Linear Regression

Consider again the fit function

$$f(x) = a_0 + a_1 x \tag{11.41}$$

for the measured data (x_i, y_i). The fit problem is to determine

$$\left| \begin{pmatrix} 1 & x_1 \\ \vdots & \vdots \\ 1 & x_m \end{pmatrix} \begin{pmatrix} a_0 \\ a_1 \end{pmatrix} - \begin{pmatrix} y_1 \\ \vdots \\ y_m \end{pmatrix} \right| = \min. \tag{11.42}$$

Orthogonalization of the column vectors

$$\mathbf{a}_1 = \begin{pmatrix} 1 \\ \vdots \\ 1 \end{pmatrix} \quad \mathbf{a}_2 = \begin{pmatrix} x_1 \\ \vdots \\ x_m \end{pmatrix} \tag{11.43}$$

with the Schmidt method gives:

$$r_{11} = \sqrt{m} \tag{11.44}$$

$$\mathbf{q}_1 = \begin{pmatrix} \frac{1}{\sqrt{m}} \\ \vdots \\ \frac{1}{\sqrt{m}} \end{pmatrix} \tag{11.45}$$

$$r_{12} = \frac{1}{\sqrt{m}} \sum_{i=1}^{m} x_i = \sqrt{m}\,\overline{x} \tag{11.46}$$

$$\mathbf{b}_2 = (x_i - \overline{x}) \tag{11.47}$$

$$r_{22} = \sqrt{\sum (x_i - \overline{x})^2} = \sqrt{m}\sigma_x \tag{11.48}$$

$$\mathbf{q}_2 = (\frac{x_i - \overline{x}}{\sqrt{m}\sigma_x}). \tag{11.49}$$

Transformation of the right hand side gives

$$\begin{pmatrix} \mathbf{q}_1^T \\ \mathbf{q}_2^T \end{pmatrix} \begin{pmatrix} y_1 \\ \vdots \\ y_m \end{pmatrix} = \begin{pmatrix} \sqrt{m}\,\overline{y} \\ \sqrt{m}\frac{\overline{xy} - \overline{x}\,\overline{y}}{\sigma_x} \end{pmatrix} \tag{11.50}$$

and we have to solve the system of linear equations

$$R\mathbf{x} = \begin{pmatrix} \sqrt{m} & \sqrt{m}\,\overline{x} \\ 0 & \sqrt{m}\sigma \end{pmatrix} \begin{pmatrix} a_0 \\ a_1 \end{pmatrix} = \begin{pmatrix} \sqrt{m}\,\overline{y} \\ \sqrt{m}\frac{\overline{xy} - \overline{x}\,\overline{y}}{\sigma_x} \end{pmatrix}. \tag{11.51}$$

The solution

$$a_1 = \frac{\overline{xy} - \overline{x}\,\overline{y}}{\overline{(x - \overline{x})^2}} \tag{11.52}$$

$$a_0 = \overline{y} - \overline{x}a_1 = \frac{\overline{y}\,\overline{x^2} - \overline{x}\,\overline{xy}}{\overline{(x - \overline{x})^2}} \tag{11.53}$$

coincides with the earlier results since

$$\overline{(x - \overline{x})^2} = \overline{x^2} - \overline{x}^2. \tag{11.54}$$

11.2 Singular Value Decomposition

Computational physics often has to deal with large amounts of data. Singular value decomposition is a very useful tool to reduce redundancies and to extract the most important information from data. It has been used for instance for image compression

[181], it is very useful to extract the essential dynamics from molecular dynamics simulations [182, 183] and it is an essential tool of Bio-informatics [184].

11.2.1 Full Singular Value Decomposition

For $m \geq n$,[2] any real[3] $m \times n$ matrix A of rank $r \leq n$ can be decomposed into a product

$$A = U\,\Sigma\,V^T \tag{11.55}$$

$$\begin{pmatrix} a_{11} & \dots & a_{1n} \\ \vdots & \ddots & \vdots \\ a_{m1} & \dots & a_{mn} \end{pmatrix} = \begin{pmatrix} u_{11} & \dots & u_{1m} \\ \vdots & \ddots & \vdots \\ u_{m1} & \dots & u_{mm} \end{pmatrix} \begin{pmatrix} s_1 & & \\ & \ddots & \\ & & s_n \\ 0 & \dots & 0 \end{pmatrix} \begin{pmatrix} v_{11} & \dots & v_{n1} \\ \vdots & \ddots & \vdots \\ v_{1n} & \dots & v_{nn} \end{pmatrix} \tag{11.56}$$

where U is a $m \times m$ orthogonal matrix, Σ is a $m \times n$ matrix, in which the upper part is a $n \times n$ diagonal matrix and V is an orthogonal $n \times n$ matrix.

The diagonal elements s_i are called singular values. Conventionally, they are sorted in descending order and the last $n - r$ of them are zero. For a square $n \times n$ matrix singular value decomposition (11.56) is equivalent to diagonalization

$$A = USU^T. \tag{11.57}$$

11.2.2 Reduced Singular Value Decomposition

We write

$$U = (U_n, U_{m-n}) \tag{11.58}$$

with the $m \times n$ matrix U_n and the $m \times (m - n)$ matrix U_{m-n} and

$$\Sigma = \begin{pmatrix} S \\ 0 \end{pmatrix} \tag{11.59}$$

with the diagonal $n \times n$ matrix S. The singular value decomposition then becomes

$$A = (U_n, U_{m-n}) \begin{pmatrix} S \\ 0 \end{pmatrix} V^T = U_n S V^T \tag{11.60}$$

---

[2] Otherwise consider the transpose matrix.

[3] Generalization to complex matrices is straightforward.

which is known as reduced singular value decomposition. U_n (usually simply denoted by U) is not unitary but its column vectors, called the left singular vectors, are orthonormal

$$\sum_{i=1}^{m} u_{i,r} u_{i,s} = \delta_{r,s} \tag{11.61}$$

as well as the column vectors of V which are called the right singular vectors

$$\sum_{i=1}^{n} v_{i,r} v_{i,s} = \delta_{r,s}. \tag{11.62}$$

Hence the products

$$U_n^T U_n = V^T V = E_n \tag{11.63}$$

give the $n \times n$ unit matrix.

In principle, U and V can be obtained from diagonalization of $A^T A$ and AA^T, since

$$A^T A = (V \Sigma^T U^T)(U \Sigma V^T) = V(S, 0) \begin{pmatrix} S \\ 0 \end{pmatrix} V^T = V S^2 V^T \tag{11.64}$$

$$AA^T = (U \Sigma V^T)(V \Sigma^T U^T) = U \begin{pmatrix} S \\ 0 \end{pmatrix} (S, 0) U^T = U_n S^2 U_n^T. \tag{11.65}$$

However, calculation of U by diagonalization is very inefficient, since usually only the first n rows are needed (i.e. U_n). To perform a reduced singular value decomposition, we first diagonalize

$$A^T A = V D V^T \tag{11.66}$$

which has positive eigenvalues $d_i \geq 0$, sorted in descending order and obtain the singular values

$$S = D^{1/2} = \begin{pmatrix} \sqrt{d_1} & & \\ & \ddots & \\ & & \sqrt{d_n} \end{pmatrix}. \tag{11.67}$$

Now we determine a matrix U such, that

$$A = U S V^T \tag{11.68}$$

or, since V is unitary

$$Y = AV = US. \tag{11.69}$$

The last $n - r$ singular values are zero if $r < n$. Therefore we partition the matrices (indices denote the number of rows)

$$\left(Y_r \; 0_{n-r} \right) = \left(U_r \; U_{n-r} \right) \begin{pmatrix} S_r & \\ & 0_{n-r} \end{pmatrix} = \left(U_r S_r \; 0 \right). \tag{11.70}$$

We retain only the first r columns and obtain a system of equations

$$\begin{pmatrix} y_{11} & \dots & y_{1r} \\ \vdots & \ddots & \vdots \\ y_{m1} & \dots & y_{mr} \end{pmatrix} = \begin{pmatrix} u_{11} & \dots & u_{1r} \\ \vdots & \ddots & \vdots \\ u_{m1} & \dots & u_{mr} \end{pmatrix} \begin{pmatrix} s_1 & & \\ & \ddots & \\ & & s_r \end{pmatrix} \tag{11.71}$$

which can be easily solved to give the first r rows of U

$$\begin{pmatrix} u_{11} & \dots & u_{1r} \\ \vdots & \ddots & \vdots \\ u_{m1} & \dots & u_{mr} \end{pmatrix} = \begin{pmatrix} y_{11} & \dots & y_{1n} \\ \vdots & \ddots & \vdots \\ y_{m1} & \dots & y_{mn} \end{pmatrix} \begin{pmatrix} s_1^{-1} & & \\ & \ddots & \\ & & s_r^{-1} \end{pmatrix}. \tag{11.72}$$

The remaining $n - r$ column vectors of U have to be orthogonal to the first r columns but are otherwise arbitrary. They can be obtained for instance by the Gram Schmidt method.

For larger matrices direct decomposition algorithms are available, for instance [185], which is based on a reduction to bidiagonal form and a variant of the QL algorithm as first introduced by Golub and Kahan [186].

11.2.3 Low Rank Matrix Approximation

Component-wise (11.60) reads

$$a_{i,j} = \sum_{k=1}^{r} u_{i,k} s_k v_{j,k}. \tag{11.73}$$

Approximations to A of lower rank are obtained by reducing the sum to only the largest singular values (the smaller singular values are replaced by zero). It can be shown [187] that the matrix of rank $l \leq r$

$$a_{i,j}^{(l)} = \sum_{k=1}^{l} u_{i,k} s_k v_{j,k} \tag{11.74}$$

is the rank-l matrix which minimizes

$$\sum_{i,j} |a_{i,j} - a_{i,j}^{(l)}|^2. \tag{11.75}$$

If only the largest singular value is taken into account, A is approximated by the rank−1 matrix

$$a_{i,j}^{(1)} = s_1 u_{i,1} v_{j,1}. \tag{11.76}$$

As an example, consider a $m \times n$ matrix

$$A = \begin{pmatrix} x_1(t_1) & \dots & x_n(t_1) \\ \vdots & & \vdots \\ x_1(t_m) & \dots & x_n(t_m) \end{pmatrix} \tag{11.77}$$

which contains the values of certain quantities $x_1 \dots x_n$observed at different times $t_1 \dots t_m$. For convenience, we assume that the average values have been subtracted, such that $\sum_{j=1}^{m} x_i = 0$. Approximation (11.76) reduces the dimensionality to 1, i.e. a linear relation between the data. The i-th row of A,

$$\begin{pmatrix} x_1(t_i) & \dots & x_n(t_i) \end{pmatrix} \tag{11.78}$$

is approximated by

$$s_1 \, u_{i,1} \begin{pmatrix} v_{1,1} & \dots & v_{n,1} \end{pmatrix} \tag{11.79}$$

which describes a direct proportionality of different observables

$$\frac{1}{v_{j,1}} x_j(t_i) = \frac{1}{v_{k,1}} x_k(t_i). \tag{11.80}$$

According to (11.75) this linear relation minimizes the mean square distance between the data points (11.78) and their approximation (11.79).

Example: Linear approximation [188]

Consider the data matrix

$$A^T = \begin{pmatrix} 1 & 2 & 3 & 4 & 5 \\ 1 & 2.5 & 3.9 & 3.5 & 4.0 \end{pmatrix}. \tag{11.81}$$

First subtract the row averages

$$\overline{x} = 3 \quad \overline{y} = 2.98 \tag{11.82}$$

to obtain

$$A^T = \begin{pmatrix} -2 & -1 & 0 & 1 & 2 \\ -1.98 & -0.48 & 0.92 & 0.52 & 1.02 \end{pmatrix}. \tag{11.83}$$

Diagonalization of

$$A^T A = \begin{pmatrix} 10.00 & 7.00 \\ 7.00 & 6.308 \end{pmatrix} \tag{11.84}$$

gives the eigenvalues

$$d_1 = 15.393 \quad d_2 = 0.915 \tag{11.85}$$

and the eigenvectors

$$V = \begin{pmatrix} 0.792 & -0.610 \\ 0.610 & -0.792 \end{pmatrix}. \tag{11.86}$$

Since there are no zero singular values we find

$$U = AVS^{-1} = \begin{pmatrix} -0.181 & -0.380 \\ -0.070 & 0.252 \\ 0.036 & 0.797 \\ 0.072 & -0.217 \\ 0.143 & -0.451 \end{pmatrix}. \tag{11.87}$$

This gives the decomposition[4]

$$A = \begin{pmatrix} \mathbf{u}_1 & \mathbf{u}_2 \end{pmatrix} \begin{pmatrix} s_1 & \\ & s_2 \end{pmatrix} \begin{pmatrix} \mathbf{v}_1^T \\ \mathbf{v}_2^T \end{pmatrix} = s_1 \mathbf{u}_1 \mathbf{v}_1^T + s_2 \mathbf{u}_2 \mathbf{v}_2^T$$

$$= \begin{pmatrix} -2.212 & -1.704 \\ -0.860 & -0.662 \\ 0.445 & 0.343 \\ 0.879 & 0.677 \\ 1.748 & 1.347 \end{pmatrix} + \begin{pmatrix} 0.212 & -0.276 \\ -0.140 & 0.182 \\ -0.445 & 0.577 \\ 0.121 & -0.157 \\ 0.252 & -0.327 \end{pmatrix}. \tag{11.88}$$

If we neglect the second contribution corresponding to the small singular value s_2 we have an approximation of the data matrix by a rank $-$ 1 matrix. The column vectors of the data matrix, denoted as $\mathbf{x}$ and $\mathbf{y}$, are approximated by

$$\mathbf{x} = s_1 v_{11} \mathbf{u}_1 \quad \mathbf{y} = s_1 v_{21} \mathbf{u}_1 \tag{11.89}$$

which describes a proportionality between $\mathbf{x}$ and $\mathbf{y}$ (Fig. 11.2).

---

[4] $\mathbf{u}_i \mathbf{v}_i^T$ is the outer or matrix product of two vectors.

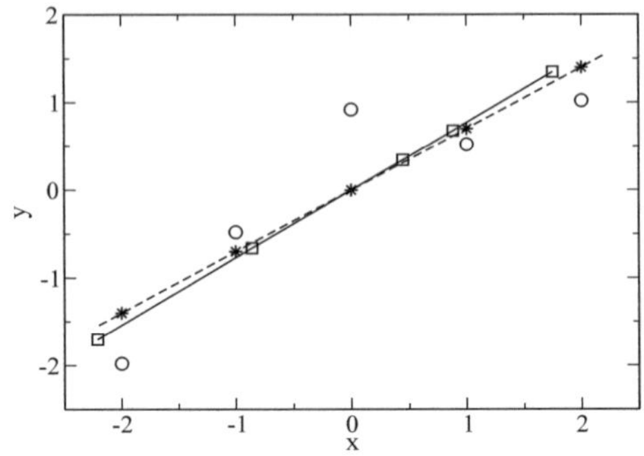

Fig. 11.2 Linear approximation by singular value decomposition
The data set (11.81) is shown as circles. The linear approximation which is obtained by retaining only the dominant singular value is shown by the squares and the full line. It minimizes the mean square distance to the data points. Stars and the dashed line show the approximation by linear regression, which minimizes the mean square distance in vertical direction

11.2.4 Linear Least Squares Fit with Singular Value Decomposition

The singular value decomposition can be used for linear regression [188]. Consider a set of data, which have to be fitted to a linear function

$$y = c_0 + c_1 x_1 \cdots + c_n x_n \tag{11.90}$$

with the residual

$$r_i = c_0 + c_1 x_{i,1} \cdots + c_n x_{i,n} - y_i. \tag{11.91}$$

Let us subtract the averages

$$r_i - \overline{r} = c_1(x_{i,1} - \overline{x}_1) \cdots + c_n(x_{i,n} - \overline{x}_n) - (y_i - \overline{y}) \tag{11.92}$$

which we write in matrix notation as

$$\begin{pmatrix} r_1 - \overline{r} \\ \vdots \\ r_m - \overline{r} \end{pmatrix} = \begin{pmatrix} x_{1,1} - \overline{x}_1 & \dots & x_{1,n} - \overline{x}_n \\ \vdots & & \\ x_{m,1} - \overline{x}_1 & \dots & x_{m,n} - \overline{x}_n \end{pmatrix} \begin{pmatrix} c_1 \\ \vdots \\ c_n \end{pmatrix} - \begin{pmatrix} y_1 - \overline{y} \\ \vdots \\ y_m - \overline{y} \end{pmatrix} \tag{11.93}$$

or shorter

$$\mathbf{r} = X\mathbf{c} - \mathbf{y}. \tag{11.94}$$

Now let us insert the full decomposition of X

$$\mathbf{r} = U \Sigma V^T \mathbf{c} - \mathbf{y}. \tag{11.95}$$

Since U is orthogonal

$$U^T \mathbf{r} = \Sigma V^T \mathbf{c} - U^T \mathbf{y} = \Sigma \mathbf{a} - \mathbf{b} \tag{11.96}$$

where we introduce the abbreviations

$$\mathbf{a} = V^T \mathbf{c} \quad \mathbf{b} = U^T \mathbf{y}. \tag{11.97}$$

The sum of squared residuals has the form

$$|\mathbf{r}|^2 = |U^T \mathbf{r}|^2 = \left| \begin{pmatrix} S_r & 0 \\ 0 & 0 \end{pmatrix} \begin{pmatrix} \mathbf{a}_r \\ \mathbf{a}_{n-r} \end{pmatrix} - \begin{pmatrix} \mathbf{b}_r \\ \mathbf{b}_{n-r} \end{pmatrix} \right|^2$$

$$= |S_r \mathbf{a}_r - \mathbf{b}_r|^2 + \mathbf{b}_{n-r}^2 \leq |S_r \mathbf{a}_r - \mathbf{b}_r|^2. \tag{11.98}$$

Hence $\mathbf{a}_{n-r}$ is arbitrary and one minimum of S_D is given by

$$\mathbf{a}_r = S_r^{-1} \mathbf{b}_r \quad \mathbf{a}_{n-r} = 0 \tag{11.99}$$

which can be written more compactly as

$$\mathbf{a} = \Sigma^+ \mathbf{b} \tag{11.100}$$

with the Moore-Penrose pseudoinverse [189] of Σ

$$\Sigma^+ = \begin{pmatrix} s_1^{-1} & & & \\ & \ddots & & \\ & & s_r^{-1} & \\ & & & 0 \end{pmatrix}. \tag{11.101}$$

Finally we have

$$\mathbf{c} = V \Sigma^+ U^T \mathbf{y} = X^+ \mathbf{y} \tag{11.102}$$

where

$$X^+ = V \Sigma^+ U^T \tag{11.103}$$

is the Moore-Penrose pseudoinverse of X.

Example

The following data matrix has rank 2

$$X = \begin{pmatrix} -3 & -4 & -5 \\ -2 & -3 & -4 \\ 0 & 0 & 0 \\ 2 & 3 & 4 \\ 3 & 4 & 5 \end{pmatrix} \quad \mathbf{y} = \begin{pmatrix} 1.0 \\ 1.1 \\ 0 \\ -1.0 \\ -1.1 \end{pmatrix}. \tag{11.104}$$

A solution to the linear fit problem is given by

$$\mathbf{c} = X^+\mathbf{y} = \begin{pmatrix} -0.917 & 1.167 & 0 & -1.167 & 0.917 \\ -0.167 & 0.167 & 0 & -0.167 & 0.167 \\ 0.583 & -0.833 & 0 & 0.833 & -0.583 \end{pmatrix} \begin{pmatrix} 1.0 \\ 1.1 \\ 0 \\ -1.0 \\ -1.1 \end{pmatrix} = \begin{pmatrix} 0.525 \\ 0.000 \\ -0.525 \end{pmatrix}. \tag{11.105}$$

The fit function is

$$y = 0.525(x_1 - x_3) \tag{11.106}$$

and the residuals are

$$X\mathbf{c} - \mathbf{y} = \begin{pmatrix} 0.05 \\ -0.05 \\ 0 \\ -0.05 \\ 0.05 \end{pmatrix}. \tag{11.107}$$

11.2.5 Singular and Underdetermined Linear Systems of Equations

SVD is also very useful to solve linear systems with a singular or almost singular matrix. Consider a system

$$\begin{pmatrix} a_{11} & \dots & a_{1n} \\ \vdots & \ddots & \vdots \\ a_{m1} & \dots & a_{mn} \end{pmatrix} \begin{pmatrix} x_1 \\ \vdots \\ x_n \end{pmatrix} = \begin{pmatrix} b_1 \\ \vdots \\ b_m \end{pmatrix} \tag{11.108}$$

with $n > m$, i.e. more unknowns than equations. SVD transforms this system into

$$\begin{pmatrix} u_{11} & \dots & u_{1m} \\ \vdots & \ddots & \vdots \\ u_{m1} & \dots & u_{mm} \end{pmatrix} \begin{pmatrix} s_1 & & & 0 \dots 0 \\ & \ddots & & \vdots \\ & & s_m & 0 \dots 0 \end{pmatrix} \begin{pmatrix} v_{11} & \dots & v_{n1} \\ \vdots & \ddots & \vdots \\ v_{1n} & \dots & v_{nn} \end{pmatrix} \begin{pmatrix} x_1 \\ \vdots \\ x_n \end{pmatrix} = \begin{pmatrix} b_1 \\ \vdots \\ b_m \end{pmatrix} \tag{11.109}$$

Substituting

$$\begin{pmatrix} y_1 \\ \vdots \\ y_n \end{pmatrix} = \begin{pmatrix} v_{11} & \dots & v_{n1} \\ \vdots & \ddots & \vdots \\ v_{1n} & \dots & v_{nn} \end{pmatrix} \begin{pmatrix} x_1 \\ \vdots \\ x_n \end{pmatrix} \text{ and } \begin{pmatrix} c_1 \\ \vdots \\ c_m \end{pmatrix} = \begin{pmatrix} u_{11} & \dots & u_{1m} \\ \vdots & \ddots & \vdots \\ u_{m1} & \dots & u_{mm} \end{pmatrix}^{-1} \begin{pmatrix} b_1 \\ \vdots \\ b_m \end{pmatrix} \tag{11.110}$$

it remains to solve

$$\begin{pmatrix} s_1 & & & 0 \dots 0 \\ & \ddots & & \vdots \\ & & s_m & 0 \dots 0 \end{pmatrix} \begin{pmatrix} y_1 \\ \vdots \\ y_n \end{pmatrix} = \begin{pmatrix} c_1 \\ \vdots \\ c_m \end{pmatrix}. \tag{11.111}$$

For $y_1 \dots y_m$ the solution is

$$y_i = s_i^{-1} c_i \quad i = 1, \dots m \tag{11.112}$$

whereas $y_{m+1} \dots y_n$ are arbitrary and parameterize the solution manifold. Back substitution gives

$$\mathbf{x} = \mathbf{x}_p + \mathbf{z} \tag{11.113}$$

with the particular solution

$$\mathbf{x}_p = \begin{pmatrix} v_{11} & \dots & v_{1m} \\ \vdots & \ddots & \vdots \\ v_{n1} & \dots & v_{nm} \end{pmatrix} \begin{pmatrix} s_1^{-1} & & \\ & \ddots & \\ & & s_m^{-1} \end{pmatrix} \begin{pmatrix} u_{11} & \dots & u_{1m} \\ \vdots & \ddots & \vdots \\ u_{m1} & \dots & u_{mm} \end{pmatrix}^{-1} \begin{pmatrix} b_1 \\ \vdots \\ b_m \end{pmatrix} \tag{11.114}$$

and

$$\mathbf{z} = \begin{pmatrix} v_{1,m+1} & \dots & v_{1n} \\ \vdots & \ddots & \vdots \\ v_{n,m+11} & \dots & v_{nn} \end{pmatrix} \begin{pmatrix} y_{m+1} \\ \vdots \\ y_n \end{pmatrix} \tag{11.115}$$

which is in the nullspace of A

$$A\mathbf{z} = U \Sigma V^T V \begin{pmatrix} 0 \\ y_{m+1} \\ \vdots \\ y_n \end{pmatrix} = U \left(S \; 0 \right) \begin{pmatrix} 0 \\ y_{m+1} \\ \vdots \\ y_n \end{pmatrix} = 0.$$

If $m - r$ singular values are zero (or if the smallest singular values are set to zero) (11.111) becomes

$$\begin{pmatrix} s_1 & & & 0\dots 0 \\ & \ddots & & \vdots \\ & & s_r & 0\dots 0 \\ 0 & \dots & 0 & 0\dots 0 \\ \vdots & & \vdots & \vdots \\ 0 & \dots & 0 & 0\dots 0 \end{pmatrix} \begin{pmatrix} y_1 \\ \vdots \\ y_n \end{pmatrix} = \begin{pmatrix} c_1 \\ \vdots \\ c_m \end{pmatrix}$$

which gives on the one hand

$$y_i = s_i^{-1} c_i \quad i = 1 \dots r$$

$$y_i = \text{arbitrary} \quad i = r + 1 \dots m$$

but also requires

$$c_i = 0 \quad i = r + 1 \dots m$$

If this condition is not fulfilled, the equations are contradictory and no solution exists (e.g. if two rows of A are the same but the corresponding elements of **b** are different).

Exercises

Exercise 11.1 Least Squares Fit

At temperatures far below Debye and Fermi temperatures the specific heat of a metal contains contributions from electrons and lattice vibrations and can be described by

$$C(T) = aT + bT^3. \tag{11.116}$$

The computer experiment generates data

$$T_j = T_0 + j\Delta t \tag{11.117}$$

$$C_j = (a_0 T_j + b_0 T_j^3)(1 + \varepsilon_j) \tag{11.118}$$

with relative error

$$\varepsilon_j = \varepsilon\, \xi_j. \tag{11.119}$$

Random numbers ξ_j are taken from a Gaussian normal distribution function (Sect. 9.2.6).

The fit parameters a, b are determined from minimization of the sum of squares

$$S = \frac{1}{n} \sum_{j=1}^{n} (C_j - aT_i - bT_i^3)^2. \tag{11.120}$$

Compare the "true values" a_0, b_0 with the fitted values a, b.

Chapter 12
Data Analysis with Artificial Neural Networks

Artificial neural networks (ANN) [190] have become a very efficient tool for the analysis of large data sets. In this chapter we consider the two basic tasks of classification and regression which are elementary in the field of physics informed machine learning. We discuss the simple perceptron and the support vector machine as linear classifiers and an extension to nonlinear classification. Simple multilayer feed forward networks solve the xor-problem and are applied to interpolation and regression of a data set. A recursive network is used for time series analysis. The necessary equations including the gradients are discussed in detail and implemented in a series of computer exercises. Finally one-layer networks are used to find the principal components of large data sets. In a series of computer experiments the reader can watch a simple ANN while solving the xor-problem, approximating a given data set by interpolation and regression or analyzing a time series.

12.1 Linear Classifiers

We want to classify a set of data which are given as a set of N-dimensional vectors $\mathbf{x}_i = (x_1^{(i)}, \cdots x_N^{(i)}) \in \mathbb{R}^N$ and corresponding labels $y_i = (y_1^{(i)}, \cdots y_N^{(i)}) \in \{-1, 1\}^N$ which distinguish between two subsets[1] (Fig. 12.1). The data set is called linearly separable if the two subsets can be divided by a (N-1) dimensional hyperplane.

---

[1] For binary classifiers often the polar definition $y = \pm 1$ is more convenient than the binary definition $y \in \{0, 1\}$.

Supplementary Information The online version contains supplementary material available at https://doi.org/10.1007/978-3-032-07856-8_12.

P. O. J. Scherer, *Computational Physics I*, Graduate Texts in Physics,
https://doi.org/10.1007/978-3-032-07856-8_12

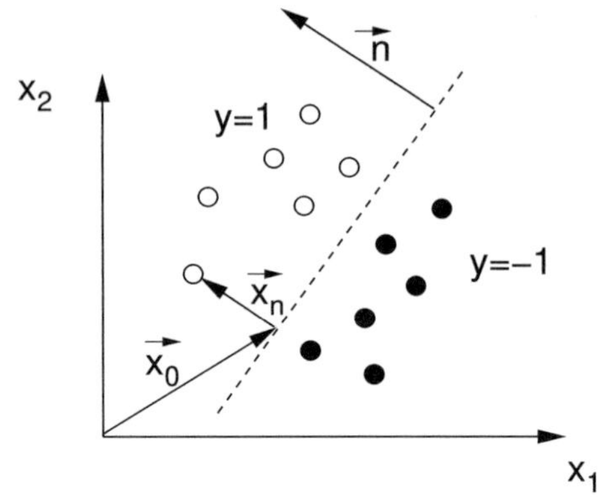

Fig. 12.1 Linear separation of a two-dimensional data set. The data points $(x_1^{(i)}, x_2^{(i)})$ are divided into two subsets (open and filled circles) by the dashed line

The hyperplane can be described by a linear equation

$$\mathbf{n}^T \mathbf{x} = b \tag{12.1}$$

where $\mathbf{n}$ is a vector normal to the hyperplane pointing to the subset labeled as "1". Dividing the data point $\mathbf{x}$ into its orthogonal projection $\mathbf{x}_0$ onto the hyperplane and its normal component $\mathbf{x}_n$ we have

$$\mathbf{n}^T (\mathbf{x}_0 + \mathbf{x}_n) - b = \mathbf{n}^T \mathbf{x}_n. \tag{12.2}$$

The sign of this scalar product distinguishes between the two half spaces which are separated by the hyperplane and the proper classification is achieved by

$$y = \text{sign}(\mathbf{n}^T \mathbf{x} - b). \tag{12.3}$$

12.2 Perceptron

Artificial neural networks try to mimic the processing of information in a real neural network. The perceptron model was first proposed in 1958 by Frank Rosenblatt [95] and later turned into a computational unit by Minsky and Papert [191]. A Perceptron (Fig. 12.2) calculates the weighted sum of N input values $x_1 \cdots x_N$ and "fires" if this sum exceeds a certain threshold value. It produces an output value

$$y = \begin{cases} 1 \text{ if } \sum_{j=1}^{N} w_i x_i > b \\ 0 \text{ else} \end{cases} \tag{12.4}$$

which is similar to the linear classifier (12.3) with the normal vector replaced by the vector of weights $\mathbf{w}$ and the polar labels ± 1 by binary values 0 and 1. In the following

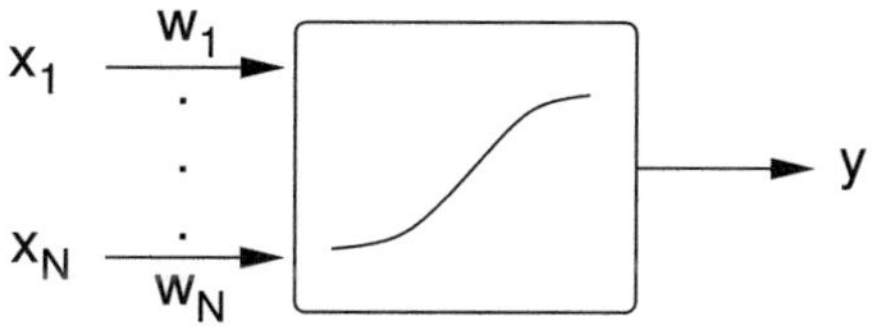

Fig. 12.2 Perceptron. The perceptron calculates the weighted sum of the input data and fires if it is above the threshold value

we prefer to change the sign of b and write more generally

$$s = \mathbf{w}^T \mathbf{x} + b \tag{12.5}$$

$$y = \sigma(s) \tag{12.6}$$

where b is called the bias and σ the activation function.[2]

Addition of the bias is often replaced by using instead augmented vectors $(w_1 \cdots w_N, b)$ and $(x_1 \cdots x_N, 1)$ to simplify the description.

The weights are determined by a supervised learning process which tries to minimize the error of the classification which is measured by a suitable loss function $L(\mathbf{w}, b)$ (Fig. 12.3). Mostly a gradient descent method (Sect. 6.3.1) is applied making small steps along the negative gradient which leads to an iterative algorithm (Algorithm 1),

---

Algorithm 1: Supervised learning

---

initialize weight and bias parameters randomly or otherwise;
repeat
  $\mathbf{w} = \mathbf{w} - \eta \frac{\partial}{\partial \mathbf{w}} L$;
  $b = b - \eta \frac{\partial}{\partial b} L$;
until *converged*;

The stepsize η, called "learning rate" in this context is constant in the simplest case. More sophisticated methods introducing momentum and adaptive stepsizes are discussed in Sect. 6.3.1.[3]

In the following we use again the polar labels $y = \pm 1$. The simplest loss function [194, 195] for binary classifiers is the 0-1 loss. If the classifier made a correct prediction for $\mathbf{x}_i$ then label and score have the same sign and $s_i y_i > 0$. The 0-1 loss function

$$\frac{1}{N} \sum_{i=1}^{N} L_{0/1}(s_i, y_i) \tag{12.7}$$

---

[2] For networks with more than one layer instead of step functions like $\text{sign}(x)$ or $\Theta(x)$ differentiable activation functions have to be used like the sigmoid function $1/(1 + e^{-x})$ or the hyperbolic tangent $(e^x - e^{-x})/(e^x + e^{-x})$ to make error backpropagation possible.

[3] For very large data sets the gradient is often approximated by a randomly chosen subset or "minibatch" of data (stochastic gradient descent method SGD) [192, 193].

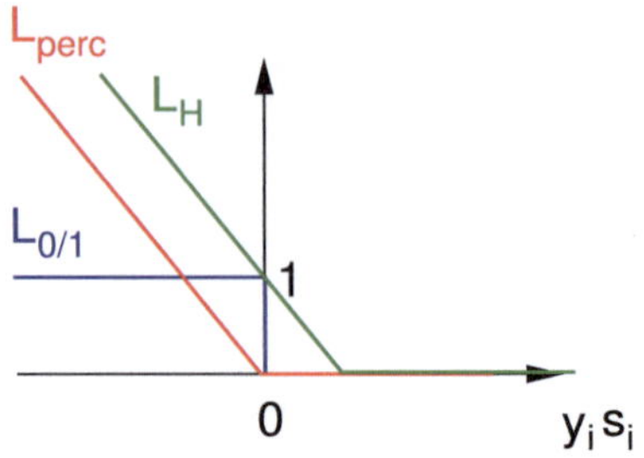

Fig. 12.3 Perceptron loss functions. The simple 0-1 loss $L_{0/1}$ is not a convex function and not suitable for gradient descent methods. The perceptron loss L_{perc} increases with the score if it has the wrong sign. The Hinge loss L_H also increases linearly if the score has the proper sign but the margin is too small

with

$$L_{0/1}(s_i, y_i) = \Theta(-y_i s_i) \tag{12.8}$$

counts the number of errors and returns the average error probability. The 0-1 loss is not convex and not everywhere differentiable. It is not suitable for optimization by gradient descent methods, but it can be substituted by other differentiable loss functions which provide an upper boundary to it.

The perceptron loss function takes into account the extent of the classification error according to

$$L_{perc}(s_i, y_i) = -y_i s_i \Theta(-s_i y_i) = \max(0, -y_i s_i). \tag{12.9}$$

This function is convex but not differentiable for $s_i = 0$. Strictly speaking, we have to apply the concept of subgradients at this point. For the max function we have for $z \neq 0$

$$\frac{\partial}{\partial z}\max(0, z) = \frac{\partial}{\partial z}(\frac{z + |z|}{2}) = \frac{1 + \text{sign}(z)}{2} = \Theta(z). \tag{12.10}$$

For $z = 0$ the subgradients[4] of the max function are given by the interval

$$\frac{\partial}{\partial z}\max(0, z = 0) = [0, 1]. \tag{12.11}$$

We find the (sub)gradients

$$\nabla_w L_{perc} = \Theta(-y_i s_i)(-y_i \nabla_w s_i) = (-y_i \mathbf{x}_i)\Theta(-y_i s_i) \tag{12.12}$$

[4] If a function $f(x)$ is convex but not differentiable in x_0, subgradients are given by all linear functions $l(x)$with $l(x_0) = f(x_0)$ and $l(x) \leq f(x) \quad \forall x$. If $f(x)$ is differentiable then the subgradient is the ordinary gradient.

$$\frac{\partial L_{perc}}{\partial b} = \Theta(-y_i s_i)(-y_i \frac{\partial}{\partial b} s_i) = -y_i \Theta(-y_i s_i). \tag{12.13}$$

12.3 Support Vector Machine (SVM)

The classifying hyperplane is not uniquely defined (Fig. 12.4).

Two parallel hyperplanes are defined by the equations.

$$s = \mathbf{w}^T \mathbf{x} + b = \pm 1. \tag{12.14}$$

We distinguish the following cases:

$$\begin{array}{rl} sy = 1 & \text{very correct} \\ sy > 1 & \text{more than correct} \\ sy < -1 & \text{very wrong} \\ |sy| < 1 & \text{inside margins.} \end{array} \tag{12.15}$$

The Hinge loss

$$L_H = \max(0, 1 - sy) \tag{12.16}$$

penalizes whenever the classification is wrong or inside the margin. Its (sub) gradients are

$$\frac{\partial}{\partial \mathbf{w}} L_H = -\Theta(1 - sy) y \mathbf{x} \tag{12.17}$$

$$\frac{\partial}{\partial b} L_H = -\Theta(1 - sy) y. \tag{12.18}$$

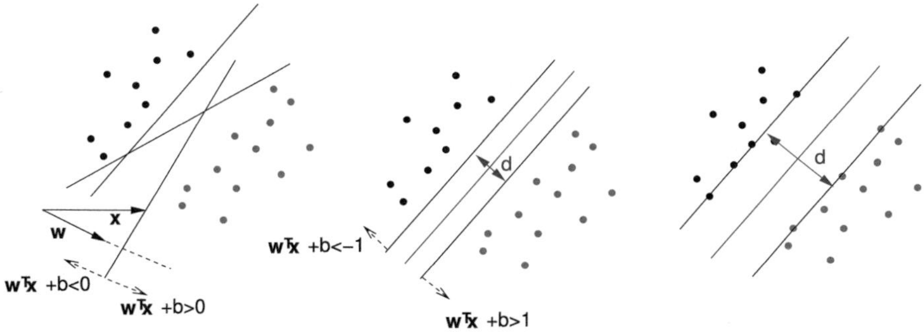

Fig. 12.4 Decision boundaries. **Left:** The decision boundary of the simple perceptron is not uniquely defined since infinitely many hyperplanes classify the data properly. **Middle:** The Hinge loss introduces two boundaries parallel to the classifying hyperplane. **Right:** Maximization of the margin d

The distance of a point $\mathbf{x}$ from the hyperplane is the absolute value of

$$\frac{\mathbf{w}^T\mathbf{x}_n}{|\mathbf{w}|} = \frac{\mathbf{w}^T(\mathbf{x}-\mathbf{x}_0)}{|\mathbf{w}|} = \frac{\mathbf{w}^T\mathbf{x}+b}{|\mathbf{w}|} = \frac{s}{|\mathbf{w}|} \tag{12.19}$$

therefore the width of the margin is

$$d = \frac{2}{|w|}. \tag{12.20}$$

The "support vector machine" (SVM) method [196, 197] maximizes the margin. For linearly separable data this can be formulated as a quadratic optimization problem with the constraint of zero total loss. Introducing a Lagrange parameter λ we have to minimize the objective function[5, 6]

$$|\mathbf{w}|^2 + \frac{\lambda}{N}\sum_{i=1}^{N} \max(0, 1 - s_i y_i). \tag{12.21}$$

This method is also applicable if the data are not linearly separable. Then the parameter λ determines the trade-off between margin width and classification errors. Since the objective function is convex, gradient descent methods can be applied. The "Pegasos" algorithm [198] combines stochastic gradient descent and a special sequence of step lengths.

Let us consider a very simple example of only two data points $(1, 1)$and $(-1, -1)$. The objective function is[7]

$$f(w, b) = \frac{1}{2}w^2 + \lambda\left[\max(0, 1-(w+b)) + \max(0, 1-(w-b))\right]. \tag{12.22}$$

The constraint requires

$$w \geq 1 + b. \tag{12.23}$$

The gradients are

$$\frac{\partial f}{\partial w} = w - \lambda\Theta(1-b-w) - \lambda\Theta(1-w+b) \tag{12.24}$$

$$\frac{\partial f}{\partial b} = -\lambda\Theta(1-b-w) + \lambda\Theta(1-w+b). \tag{12.25}$$

---

[5] A much more general treatment can be found in [197].

[6] Some authors prefer to multiply the first summand by a parameter. Then the method can be alternatively understood as minimization of the loss function with a regularization term [195].

[7] We use more convenient prefactors.

Stationarity requires

$$w \leq 1 - b \text{ or } w \geq 1 + b \tag{12.26}$$

$$\frac{w}{\lambda} = [\Theta(1 - b - w) + \Theta(1 - w + b)]. \tag{12.27}$$

For $b \neq 0$ there is no solution of the second equation with $w > 1 + b$ since the right hand side is zero for $w > 1 + b$. For $b = 0$, $w = 1$

$$\frac{\partial f}{\partial w} = 1 - 2\lambda\xi, \quad \xi \in [0, 1] \tag{12.28}$$

which has the subgradient 0 if $\lambda > 1/2$.

12.4 Nonlinear Classification

The SVM method can be applied to linearly nonseparable data if the dimension of the data is increased by adding nonlinear functions of the original coordinates as new ones such that the data become linearly separable in this space of much higher dimension. Formally, a nonlinear mapping is applied to the data before the classification is done:

$$\mathbf{x} \in \mathbb{R}^{N} \rightarrow \phi(\mathbf{x}) \in \mathbb{R}^{M}. \tag{12.29}$$

This involves for instance quadratic terms $x_1^2 \dots x_N^2, x_1x_2, \cdots$ [197]. The separating hyperplane in the high dimensional space corresponds to a nonlinear (in this case polynomial) decision boundary in the original space. In practice the dimension of the feature space can be very high and the numerical effort can be reduced by the Kernel method [199–201] which works in dual space and involves numerically efficient dot-products.

As a very simple example we consider the well known xor-problem [191] (Table 12.1). A single perceptron can not perform a logical xor operation since the corresponding data are not linearly separable. They become separable by introducing the product $x_3 = x_1x_2$ as a new coordinate.[8] In fact, this product alone is sufficient since it coincides with the result of the xor operation (Fig. 12.5).

12.5 Multilayer Networks

ANN usually have a huge number of elements which are mostly arranged in several layers. The simplest example is the feed forward architecture where the outputs of one layer are the inputs of the next layer. We consider a very simple system consisting

[8] The numerical values –1 and +1 represent Boolean values false and true.

Table 12.1 Truth table of the xor- gate

x_1	x_2	$x_3 = x_1x_2$	$x_1$xor x_2
False = –1	False = –1	1	True
False = –1	True = 1	–1	False
True = 1	False = –1	–1	False
True = 1	True = 1	1	True

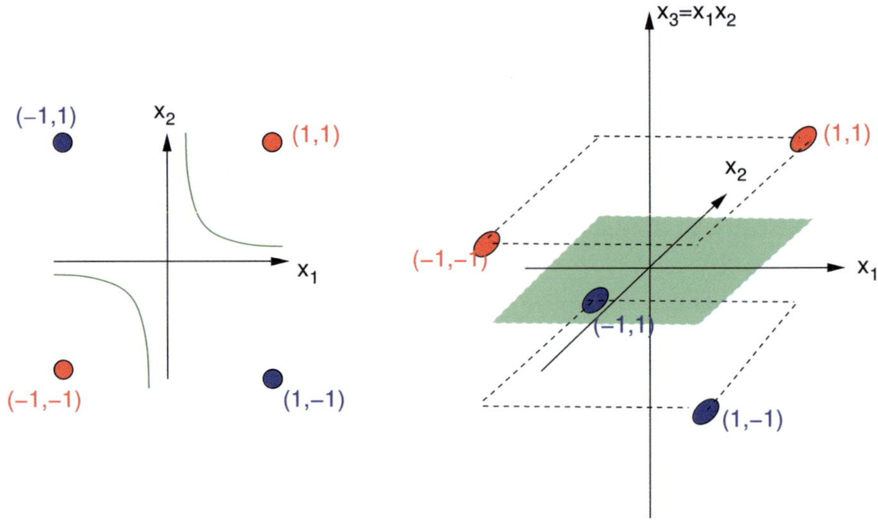

Fig. 12.5 Logical xor operation by nonlinear classification. The data from Table 12.1 are not linearly separable. They become so if the product $x_3 = x_1x_2$ is added as a new coordinate. Any horizontal plane (green) between $-1 < x_3 < 1$ separates the data properly. In the two-dimensional original space this corresponds to nonlinear hyperbolic decision curves with $x_1x_2 = c \in [-1, 1]$ (green)

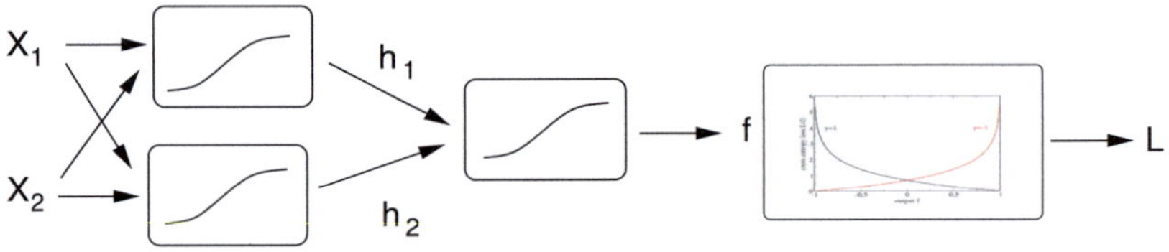

Fig. 12.6 Very simple feed forward network

of two input and one output elements (Fig. 12.6). As we will see, this system can resolve the xor-problem.

The input data $x_{1,2}$ are processed in parallel by two units to produce the intermediate data $h_{1,2}$ which are then processed by the third unit to provide the output f. Finally the cross entropy loss function (Fig. 12.7) is calculated which is commonly used for binary classification.

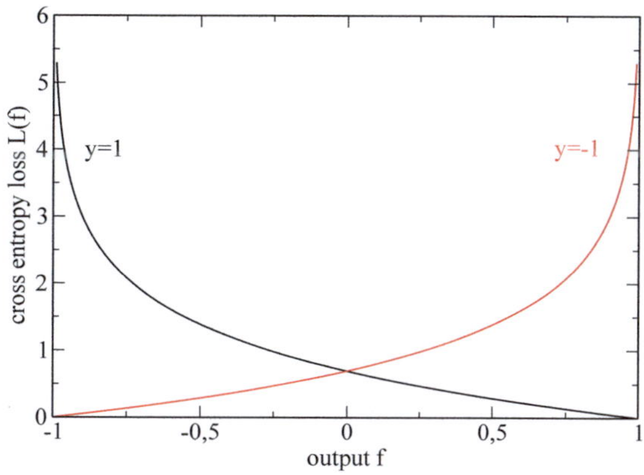

Fig. 12.7 Cross entropy loss function (12.36)

We divide the algorithm into a sequence of steps producing intermediate values as the results of elementary operations and functions like addition, multiplication or the tanh function. This approach is very useful for automatic differentiation by computer programs since it can be applied to any differentiable function [202].

First both units apply a linear function to the input data $x_{1,2}$:

$$s_1 = w_{11}x_1 + w_{21}x_2 + b_1 \tag{12.30}$$

$$s_2 = w_{12}x_1 + w_{22}x_2 + b_2. \tag{12.31}$$

Now the activation function σ is applied. We chose the tanh function which switches smoothly between –1 and 1 and is differentiable.

$$h_1 = \sigma(s_1) \tag{12.32}$$

$$h_2 = \sigma(s_2). \tag{12.33}$$

Then the two results of the hidden layer are combined by another linear function

$$s_3 = w_{13}h_1 + w_{23}h_2 + b_3. \tag{12.34}$$

The activation function is applied once more to obtain the prediction

$$f = \sigma(s_3) \tag{12.35}$$

and finally the loss function is calculated as

$$L = \begin{cases} -\ln\left(\frac{1+f}{2}\right) & \text{if } y = 1 \\ -\ln\left(\frac{1-f}{2}\right) & \text{if } y = -1 \end{cases}. \tag{12.36}$$

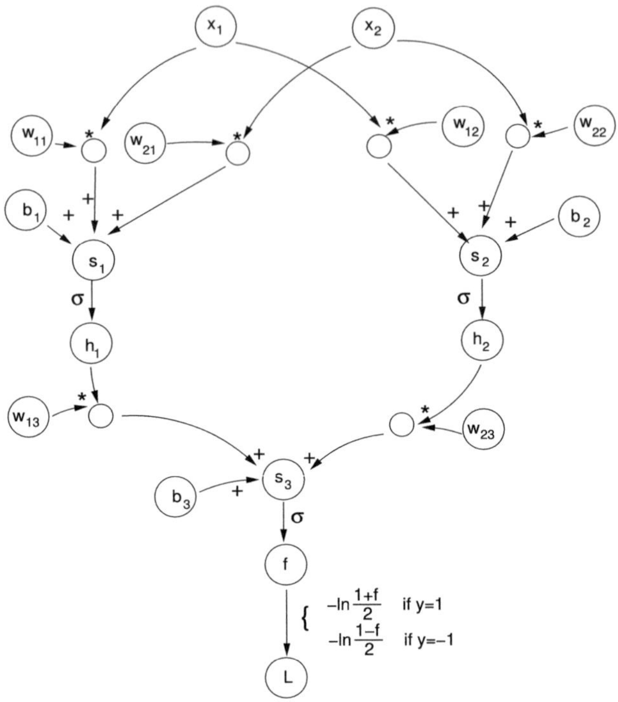

Fig. 12.8 Computational graph for the calculation of the loss function L

This sequence of operations can be visualized as a computational graph [203] (Fig. 12.8), which is helpful for the calculation of gradients which are needed for gradient descent based optimization methods (Sect. 6.3).

12.5.1 Solving the Xor-problem

In computer experiment 12.1 we apply the gradient descent method to minimize the total loss function for the four data points. The first layer makes two different classifications which can be separated by the second layer (Fig. 12.9).

12.5.2 Forward and Backward Propagation

To optimize the 9 parameters, we use the backpropagation method [204] (essentially we apply the chain rule to a sequence of nested functions). The variation of the loss function can be written as

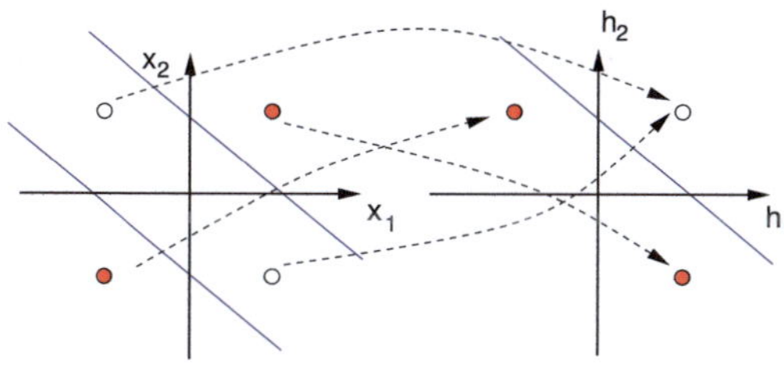

Fig. 12.9 Xor-operation of a 2-layer network. **Left:** The first layer makes two different classifications. **Right:** The neuron in the second layer separates them properly. Decision boundaries (blue lines) are visualized in computer experiment 12.1.

$$\delta L = \begin{cases} -\frac{1}{1+f}\delta f \text{ if } y = 1 \\ \frac{1}{1-f}\delta f \text{ if } y = -1 \end{cases} \tag{12.37}$$

with

$$\delta f = \sigma'(s_3)\delta s_3 \tag{12.38}$$

$$\delta s_3 = \delta w_{13}h_1 + w_{13}\delta h_1 + \delta w_{23}h_2 + w_{23}\delta h_2 + \delta b_3. \tag{12.39}$$

The variations of the intermediate output values are given by

$$\delta h_1 = \sigma'(s_1)\delta s_1 \tag{12.40}$$

$$\delta h_2 = \sigma'(s_2)\delta s_2 \tag{12.41}$$

$$\delta s_1 = \delta w_{11}x_1 + \delta w_{21}x_2 + \delta b_1 \tag{12.42}$$

$$\delta s_2 = \delta w_{12}x_1 + \delta w_{22}x_2 + \delta b_2. \tag{12.43}$$

To obtain,e.g. the partial derivative $\frac{\partial L}{\partial w_{11}}$ we could start with variation of only δw_{11}, neglect all other parameter variations and propagate the perturbation forward which finally gives the result (Fig. 12.10)

$$\frac{\partial L}{\partial w_{11}} = \begin{cases} -\frac{1}{1+f}\sigma'(s_3)w_{13}\sigma'(s_1)x_1 \text{ if } y = 1 \\ \frac{1}{1-f}\sigma'(s_3)w_{13}\sigma'(s_1)x_1 \text{ if } y = -1 \end{cases}. \tag{12.44}$$

This works but we have to start a separate chain of calculations for each gradient. Since only products of variables and function values are required, we can revert the order of multiplications and follow the graph from its end (Fig. 12.11). Since many common products appear for different gradients, backward propagation is more efficient.

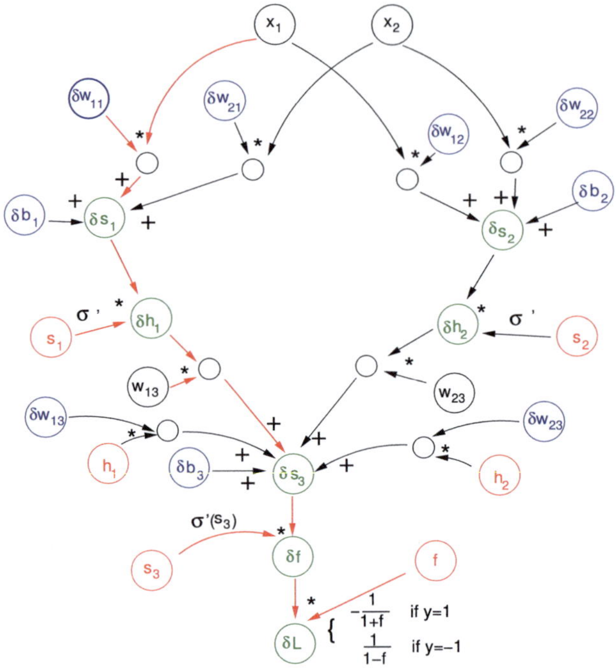

$$\delta s_1 = x_1 \delta w_{11}$$

$$\delta h_1 = \sigma'(s_1)\delta s_1$$

$$\delta s_3 = w_{13}\delta h_1$$

$$\delta f = \sigma'(s_3)\delta s_3$$

$$\delta L = \begin{cases} -\frac{1}{1+f}\delta f \text{ if } y = 1 \\ \frac{1}{1-f}\delta f \text{ if } y = -1 \end{cases}$$

Fig. 12.10 Forward propagation of the variation δw_{11}. Values in red are calculated and stored during calculation of L, blue values are the parameter variations, green are the variations of intermediate values, small circles denote additional intermediate products. To obtain e.g. the gradient $\frac{\partial L}{\partial w_{11}}$, we set $\delta w_{11} = 1$ and all other parameter variations equal to zero. The desired gradient is then given by the resulting variation δL

12.6 Regression and Interpolation

Multilayer feed forward networks can be used to approximate a function quite well [205, 206]. In this example we apply a network with one hidden layer (Fig. 12.12) to the elementary tasks of interpolation and regression.

To approximate a set of data points $\left[x^{(i)}, y^{(i)}\right]$ by a function $f(x)$,[9] we consider a network with continuous output values. A simple one-dimensional example is very instructive. The data values $x^{(i)}, i = 1 \cdots \mathrm{N_{data}}$ are input to N elements with hyperbolic tangent activation [207] to calculate N intermediate values h_n which then are combined by one element in the second layer with linear activation to produce the prediction f.

[9] This can be generalized to vector functions $f_j(x_1, \cdots x_n), \quad j = 1 \cdots m$.

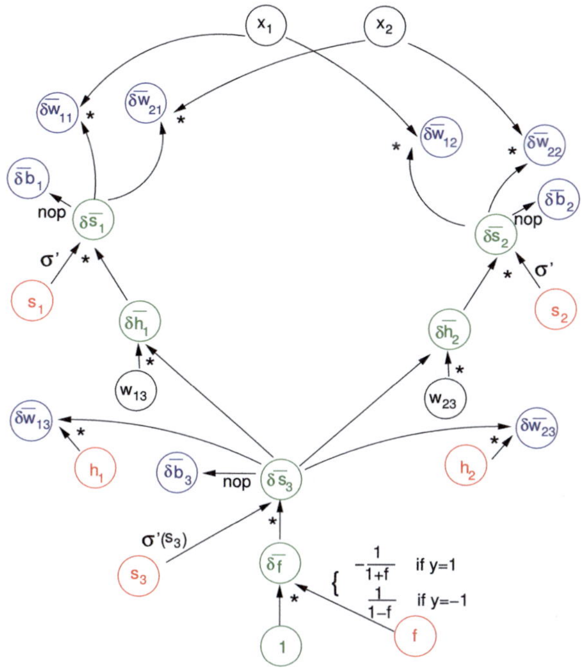

$$\delta\overline{f} = \begin{cases} -\frac{1}{1+f} & \text{if } y = 1 \\ \frac{1}{1-f} & \text{if } y = -1 \end{cases}$$

$$\delta\overline{s}_3 = \sigma'(s_3)\delta\overline{f}$$

$$\delta\overline{h}_1 = w_{13}\delta\overline{s}_3$$

$$\delta\overline{s}_1 = \sigma'(s_1)\delta\overline{h}_1$$

$$\delta\overline{w}_{11} = x_1\delta\overline{s}_1$$

Fig. 12.11 Backward propagation of δw_{11}. Values in red have already been calculated and stored during calculation of L, green are new intermediate values and blue values are the resulting gradients

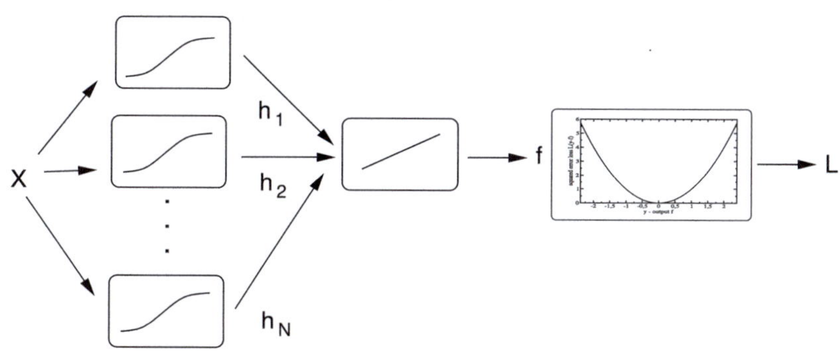

Fig. 12.12 Simple network for regression

First each node applies a linear function

$$s_n = w_n x + b_n \tag{12.45}$$

which is followed by the sigmoid activation function (in our example the tanh)

$$h_n = \sigma(s_n). \tag{12.46}$$

Finally the results of all nodes are combined by another linear function to provide the approximation

$$f = \sum_{n=1}^{N} w_{on} h_n + b_o. \tag{12.47}$$

These steps are repeated for all sample points x_i and the accuracy of the approximations is measured by the squared deviations

$$L^{(i)} = \frac{1}{2}(f(x_i) - y_i)^2 \tag{12.48}$$

which finally are combined to obtain the mean squares loss function

$$L = \frac{1}{\mathrm{N_{data}}} \sum_{i=1}^{\mathrm{N_{data}}} L^{(i)} \tag{12.49}$$

which then has to be optimized by varying the parameters w_{on}, b_o, w_n, b_n.

Gradients are found from

$$\delta L = \frac{1}{\mathrm{N_{data}}} \sum_{i=1}^{\mathrm{N_{data}}} \delta L^{(i)} \tag{12.50}$$

$$\delta L^{(i)} = (f^{(i)} - y^{(i)})\delta f^{(i)} \tag{12.51}$$

$$\delta f^{(i)} = \delta w_{on} h_n^{(i)} + w_{on} \delta h_n^{(i)} + \delta b_o \tag{12.52}$$

$$\delta h_n^{(i)} = \sigma'(w_n x^{(i)} + b_n)(\delta w_n x^{(i)} + \delta b_n) \tag{12.53}$$

by variation of only one parameter at a time and keeping all others fixed

$$\frac{\partial L^{(i)}}{\partial b_o} = (f^{(i)} - y^{(i)}) \tag{12.54}$$

$$\frac{\partial L^{(i)}}{\partial w_{on}} = (f^{(i)} - y^{(i)}) h_n^{(i)} \tag{12.55}$$

$$\frac{\partial L^{(i)}}{\partial b_n} = (f^{(i)} - y^{(i)}) w_{on} \sigma'(w_n x^{(i)} + b_n) \tag{12.56}$$

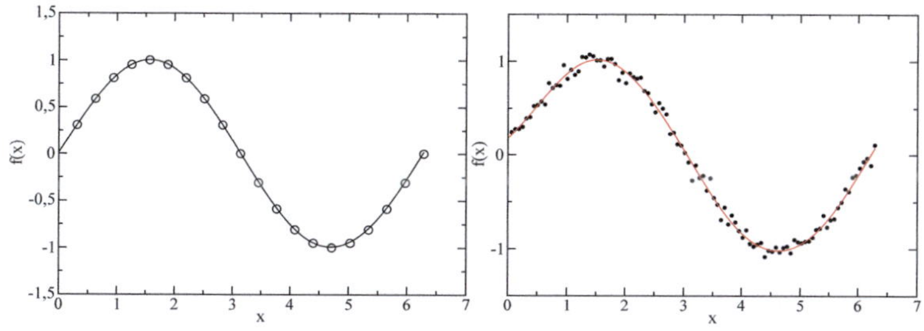

Fig. 12.13 Regression and interpolation. A simple network with 10 units is used to approximate a set of data points. **Left:** For 20 data points of a sine function this corresponds to interpolation. **Right:** For the 100 data points of a noisy sine function this is similar to a regression task

$$\frac{\partial L^{(i)}}{\partial w_n} = (f^{(i)} - y^{(i)}) w_{on} \sigma'(w_n x^{(i)} + b_n) x^{(i)}. \tag{12.57}$$

Figure 12.13 shows examples from Exercise 12.2.

12.7 General Feed Forward Networks

So far we discussed examples with rather small networks. For real ai applications much larger arrays are used which follow the same principles. Often used are dense feed forward networks (Fig. 12.14) containing several layers $l = 1 \cdots M$ which may have different numbers of elements $k = 1 \cdots N_l$. Each element of one layer is connected to all elements of the preceding and following but no other layers. The activation function is often a sigmoid function like tanh or ReLU. The input data $X_1 \cdots X_p$ are fed to all elements of the first layer. The output data of the last layer are combined to one or more prediction values $f_1 \cdots f_q$ with sigmoid or linear activation function. These are finally compared to the training values $Y_1 \cdots Y_q$ corresponding to the input data X and a suitable loss function L measures the overall error.[10]

During an initial stage of training the loss function is minimized by applying a set of training data and varying the bias and weight parameters. Afterwards the parameters are kept fixed and the network can then make predictions for arbitrary new input data. To validate the predictions usually a subset of the training data is used which has not yet been used before for training.

The first layer calculates the weighted sum of the input data

$$s_{1,j} = b_{1,j} + \sum_{k=1}^{p} w_{1,j,k} x_k \tag{12.58}$$

[10] E.g. the squared difference.

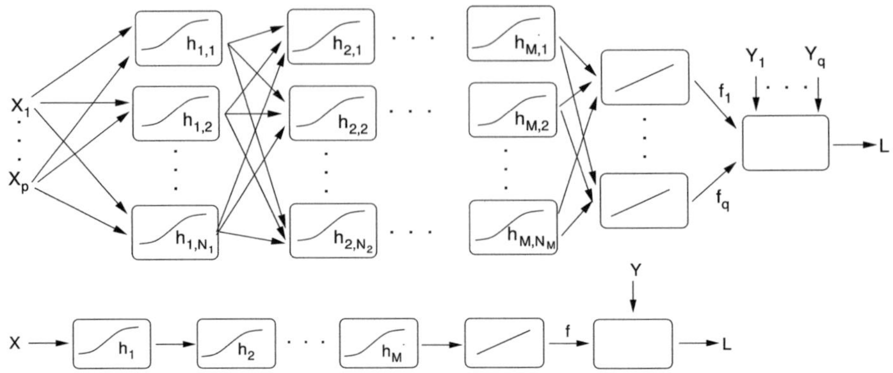

Fig. 12.14 General feed forward network. The p-dimensional vector of input data **X** is fed to a network of M layers. Layer l contains N_l elements, all of them are connected to all elements of the previous $(l-1)$ and the next $(l+1)$ layer. The output data of the last layer M are combined to provide a q-dimensional vector of predictions **f**. Its deviation from the desired training values **Y** enters the loss function L. The lower part shows a simplified scheme, where a whole layer is represented by a single vectorial element

and applies the activation function σ to obtain the output

$$h_{1,j} = \sigma\left(s_{1,j}\right) \tag{12.59}$$

which becomes the input of the next layer. The output of the hidden layer l is calculated according to

$$s_{l,j} = b_{l,j} + \sum_{k=1}^{N_{l-1}} w_{l,j,k} h_{l-1,k} \tag{12.60}$$

$$h_{l,j} = \sigma\left(s_{l,j}\right) \tag{12.61}$$

and finally the output layer calculates the predictions

$$s_{o,j} = b_{o,j} + \sum_{k=1}^{N_M} w_{o,j,k} h_{M,k} \tag{12.62}$$

$$f_j = \sigma_o\left(s_{o,j}\right) \tag{12.63}$$

probably with a different activation function σ_o. Finally the loss function is calculated, e.g. using squared differences

$$L = \frac{1}{2}\sum_{j=1}^{q}(f_j - Y_j)^2. \tag{12.64}$$

The equations look simpler if indices are avoided by using matrix-vector notation. Here and in the following a function of a vector acts componentwise

$$\sigma\left(\begin{bmatrix} s_1 \\ \vdots \\ s_n \end{bmatrix}\right) = \begin{bmatrix} \sigma(s_1) \\ \vdots \\ \sigma(s_n) \end{bmatrix} \tag{12.65}$$

and the symbol $\odot$ denotes the so called Hadamard product where two vectors are multiplied componentwise[11]

$$\begin{bmatrix} a_1 \\ \vdots \\ a_n \end{bmatrix} \odot \begin{bmatrix} b_1 \\ \vdots \\ b_n \end{bmatrix} = \begin{bmatrix} a_1 b_1 \\ \vdots \\ a_n b_n \end{bmatrix}. \tag{12.66}$$

Equations (12.58)–(12.64) then read

$$\mathbf{s}_1 = \mathbf{b}_1 + W_1 \mathbf{x} \tag{12.67}$$

$$\mathbf{h}_1 = \sigma(\mathbf{s}_1) \tag{12.68}$$

$$\mathbf{s}_l = \mathbf{b}_l + W_l \mathbf{h}_{l-1} \quad l = 2 \cdots M \tag{12.69}$$

$$\mathbf{h}_l = \sigma(\mathbf{s}_l) \quad l = 2 \cdots M \tag{12.70}$$

$$\mathbf{s}_o = \mathbf{b}_o + W_o \mathbf{h}_M \tag{12.71}$$

$$\mathbf{f} = \sigma_o(\mathbf{s}_o) \tag{12.72}$$

$$L = \frac{1}{2}|\mathbf{f} - \mathbf{Y}|^2. \tag{12.73}$$

To obtain the gradients with respect to the weight and bias parameters, we apply again backpropagation. Starting from the variation of the loss function

$$\delta L = \nabla_{\mathbf{f}} L \, \delta \mathbf{f} = (\mathbf{f} - \mathbf{Y}) \delta \mathbf{f} \tag{12.74}$$

we go backwards

$$\delta \mathbf{f} = \sigma_o'(\mathbf{s}_o) \odot \delta \mathbf{s}_o \tag{12.75}$$

$$\delta \mathbf{s}_o = \delta \mathbf{b}_o + \boldsymbol{\delta} W_o \mathbf{h}_M + W_o \delta h_M \tag{12.76}$$

$$\boldsymbol{\delta} h_l = \sigma'(\mathbf{s}_l) \odot \boldsymbol{\delta} \mathbf{s}_l \quad l = M \cdots 2 \tag{12.77}$$

[11] Alternatively we could use a diagonal matrix for the left factor.

$$\boldsymbol{\delta s}_l = \boldsymbol{\delta}\mathbf{b_l} + \delta W_l \mathbf{h}_{l-1} + W_l \boldsymbol{\delta} h_{l-1} \quad l = M \cdots 2 \tag{12.78}$$

$$\boldsymbol{\delta}\mathbf{h}_1 = \sigma'(\mathbf{s}_1) \odot \boldsymbol{\delta} s_1 \tag{12.79}$$

$$\boldsymbol{\delta} s_1 = \boldsymbol{\delta} b_1 + \delta W_1 \mathbf{x}. \tag{12.80}$$

We make use of the recursion

$$\mathbf{v_o} = \sigma_o'(s_o) \tag{12.81}$$

$$\mathbf{v}_M = \mathbf{v}_o \odot W_o \sigma'(\mathbf{s}_M) \tag{12.82}$$

$$\vdots$$

$$\mathbf{v}_{l-1} = \mathbf{v}_l \odot W_l \sigma'(s_{l-1}) \quad l = M, \cdots 2 \tag{12.83}$$

to write the variation of the loss function in the form

$$\delta L = (\nabla_{\mathbf{f}} L) \left[\mathbf{v}_o \odot (\delta \mathbf{b}_o + \boldsymbol{\delta} W_o \mathbf{h}_M) + \mathbf{v}_M \odot (\mathbf{b_l} + \delta W_M \mathbf{h}_{M-1}) \cdots + \mathbf{v}_1 \odot (\delta b_1 + \delta W_1 \mathbf{x})\right]. \tag{12.84}$$

The gradients with respect to the bias vectors are then

$$\frac{\partial f_j}{\partial b_{oj}} = v_{o,j} \tag{12.85}$$

$$\frac{\partial f_j}{\partial b_{l,j}} = v_{l,j} \tag{12.86}$$

and with respect to the weight parameters

$$\frac{\partial f_j}{\partial W_{o,j,k}} = v_{o,j} h_{M,k} \tag{12.87}$$

$$\frac{\partial f_j}{\partial W_{l,j,k}} = v_{l,j} h_{l-1,k} \tag{12.88}$$

$$\frac{\partial f_j}{\partial W_{1,j,k}} = v_{1,j} x_k. \tag{12.89}$$

12.8 Recurrent Neural Networks (RNN)

Experimental data often occur as time series, i.e. as a sequence of values at different times[12] $x_1, x_2 \cdots$. Here we assume a constant time step Δt and denote the time series as

$$x^{(t)} = x(t_0 + t\Delta t). \tag{12.90}$$

Recurrent neural networks (Fig. 12.15) are specially designed for time series. They treat the input data iteratively and are useful to discover temporal patterns. From the input data at time t an intermediate output h_t (called the "hidden state") is formed according to the equations

$$\mathbf{s}^{(t)} = \mathbf{w_x} x^{(t)} + W_h \mathbf{h}^{(t-1)} + \mathbf{b} \tag{12.91}$$

$$\mathbf{h}^{(t)} = \sigma(\mathbf{s}^{(t)}) \tag{12.92}$$

where $\mathbf{h}^{(t-1)}$is the result from the previous step or zero for the initial step. $\mathbf{s}$, $\mathbf{w}_x$, $\mathbf{b}$ and $\mathbf{h}$ are vectorial quantities where different components correspond to different units, W_h is a matrix since the output of one unit is fed back to all units. The activation function σ is usually chosen as the tanh or ReLU function.

In principle, the length of the sequence can be arbitrary but for practical reasons we assume a fixed length of n_l. After all n_l input data are processed, the final output[13] is obtained from

$$s_o = \mathbf{w}_o^T \mathbf{h}^{(t)} + b_o \tag{12.93}$$

$$f^{(t)} = \sigma_o(s_o) \tag{12.94}$$

where the activation function σ_o could be simply linear if the range is not restricted. The recurrent network can be "unrolled" and is equivalent to a dense network with $n_l$layers (Fig. 12.16) leading to the system of equations

$$\mathbf{h}^{(1)} = \sigma(\mathbf{w_x} x^{(1)} + \mathbf{b}) \tag{12.95}$$

$$f^{(1)} = \sigma_o(\mathbf{w}_o^T \mathbf{h}^{(1)} + b_o) \tag{12.96}$$

$$\mathbf{h}^{(2)} = \sigma(\mathbf{w_x} x^{(2)} + W_h \mathbf{h}^{(1)} + \mathbf{b}) \tag{12.97}$$

$$f^{(2)} = \sigma_o(\mathbf{w}_o^T \mathbf{h}^{(2)} + b_o) \tag{12.98}$$

$$\vdots$$

[12] This can be easily generalized to a vectorial time series $\mathbf{x}(t)$ with more than one feature.

[13] In the following we only make use of the final output f^{n_l}. Certain applications make also use of the intermediate output values $f^{(1)} \cdots f^{(n_l-1)}$.

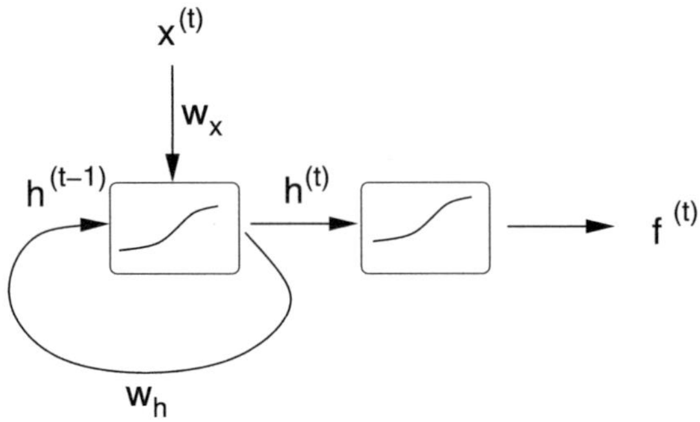

Fig. 12.15 RNN structure. The RNN works iteratively. The results $\mathbf{h}^{(t-1)}$ from the previous step are fed back to the network. The square on the left symbolizes a whole layer of units as in Fig. 12.14

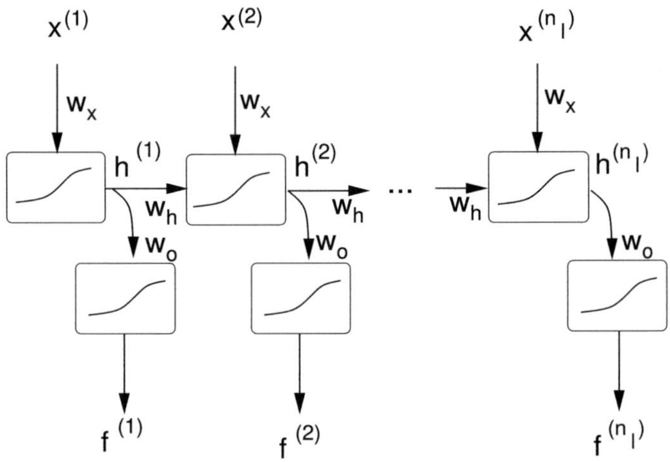

Fig. 12.16 Unrolled RNN. The RNN is equivalent to a dense network with n_l layers. The weight parameters are identical for all layers. The upper row of squares symbolizes whole layers of units

$$\mathbf{h}^{(n_l)} = \sigma(\mathbf{w_x} x^{(n_l)} + W_h \mathbf{h}^{(n_l-1)} + \mathbf{b}) \tag{12.99}$$

$$f^{(n_l)} = \sigma_o(\mathbf{w}_o^T \mathbf{h}^{(n_l)} + b_o). \tag{12.100}$$

Note that the weight parameters are identical for all layers.

The gradients with respect to the weight parameters and biases can be found by applying back propagation iteratively[14]

$$\delta f^{(t)} = \sigma_o^{'}(s_o^{(t)}) \delta s_o^{(t)} \tag{12.101}$$

$$\delta s_o^{(t)} = (\delta \mathbf{w}_o^T)\mathbf{h}^{(t)} + \mathbf{w}_o^T \delta \mathbf{h}^{(t)} + \delta b_o \tag{12.102}$$

---

[14] This is known as backpropagation in time.

$$\delta\mathbf{h}^{(t)} = \sigma'(\mathbf{s}^{(t)}) \odot \left[(\delta\mathbf{w}_x)x^{(t)} + \delta W\mathbf{h}^{(t-1)} + \delta\mathbf{b}_h\right] \tag{12.103}$$

$$+\sigma'(\mathbf{s}^{(t)}) \odot W\sigma'(\mathbf{s}^{(t-1)}) \odot \left[(\delta\mathbf{w}_x)x^{(t-1)} + \delta W\mathbf{h}^{(t-2)} + \delta\mathbf{b}_h\right] + \cdots . \tag{12.104}$$

Similar to (12.81)–(12.83) we introduce

$$v^{(t)} = \sigma'(\mathbf{s}^{(t)}) \tag{12.105}$$

$$v^{(t-1)} = v^{(t)} \odot W\sigma'(\mathbf{s}^{(t-1)}) \tag{12.106}$$

and expand

$$\delta\mathbf{h}^{(t)} = v^{(t)} \odot (\delta\mathbf{w}_x)x^{(t)} + v^{(t-1)} \odot (\delta\mathbf{w}_x)x^{(t-1)} + \cdots \tag{12.107}$$

$$+ v^{(t)} \odot \delta W\mathbf{h}^{(t-1)} + v^{(t-1)} \odot \delta W\mathbf{h}^{(t-2)} + \cdots \tag{12.108}$$

$$+ v^{(t)} \odot \delta\mathbf{b}_h + v^{(t-1)} \odot \delta\mathbf{b}_h + \cdots . \tag{12.109}$$

This set of equations allows an efficient calculation of the gradients. In the computer exercises we try to predict a new function value $x^{(t)}$ from the previous values $x^{(t-1)} \cdots x^{(t-n_l)}$. Therefore our target is the actual new value $y = x^{(t)}$. We use again the mean squares loss function

$$L = \frac{1}{N_{\text{data}}} \sum_{i=1}^{N_{\text{data}}} L_i \tag{12.110}$$

$$L_i = \frac{1}{2}(f(x_{i-n_l}, \cdots x_{i-1}) - x_i)^2 \tag{12.111}$$

with

$$\delta L = \frac{1}{N_{\text{data}}} \sum_{i=1}^{N_{\text{data}}} \delta L_i \tag{12.112}$$

$$\delta L_i = (f_i - y_i)\delta f_i . \tag{12.113}$$

The following computer experiments (Figs. 12.17 and 12.18, Exercise 12.3) perform three steps. First the model is trained on a number of data points, i.e. we try to predict a new function value on the basis of n_l preceding ones and minimize the total loss. Then we validate the model by applying it to a further set of data points, which are not contained in the training set. Finally we try to predict the values of the validation set only on the basis of the last n_l points of the training set.

The simple RNN structure has several drawbacks. From (12.106) we see that with increasing sequence length higher powers of the weight matrix W and the derivatives

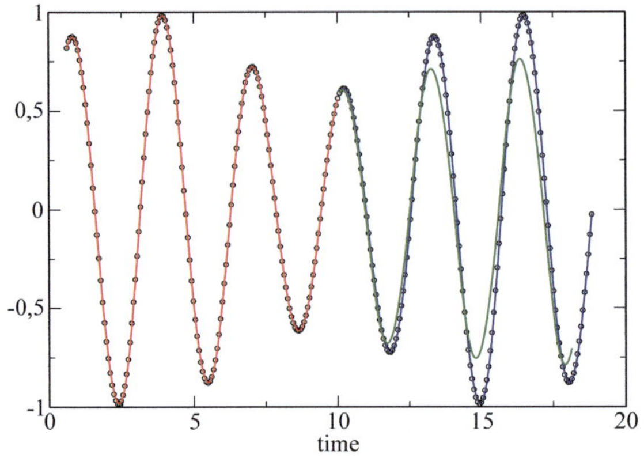

Fig. 12.17 Simple RNN. A simple RNN with 35 elements is trained with 150 data points (black circles) to predict a new point from 10 previous ones. The results (red curve) agree quite well with the input data. With a learning rate of 0.002 the total square loss $\frac{1}{2}\sum_i (f_i - y_i)^2 = 10^{-3}$ after 20000 steps of full gradient optimization. After training, the model is validated with further 140 data points (blue curve). Finally the model is used to predict these points starting from the last 10 points of the training set (green curve). From computer experiment 12.3

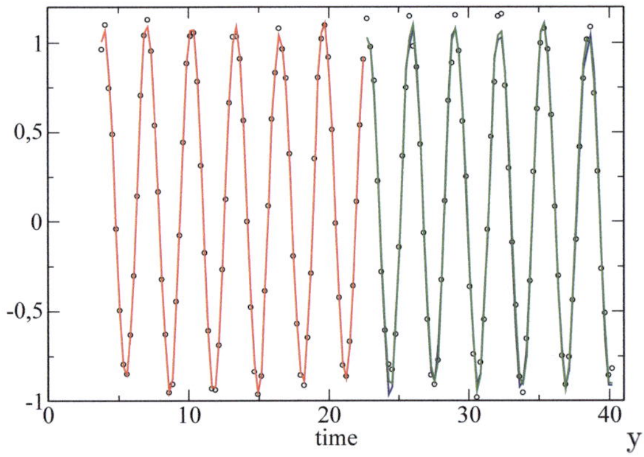

Fig. 12.18 Simple RNN. A simple RNN with 35 elements is trained with 75 noisy data points (black circles) to predict a new point from 15 previous ones. The results (red) agree quite well with the input data. With a learning rate of 0.002 the total square loss $\frac{1}{2}\sum_i (f_i - y_i)^2 = 0.05$ after 10000 steps of full gradient optimization. After training, the model is validated with further 70 data points (blue). Finally the model is used to predict these points starting from the last 15 points of the training set (green). From Exercise 12.3

σ' appear. This can lead to the well known phenomenon of vanishing or exploding gradients. Such problems can be reduced by initializing the weight matrix to an identity matrix [208]. The introduction of special gates which control the flow of information and help to keep only the relevant part, resolves the gradient problem and allows for much longer sequences. The minimal gated RNN [209] has only one gate whereas more complex structures involve two or three [210]. The related LSTM model [211, 212] has three gates and an additional memory variable.

12.8.1 Minimal Gated Unit

The equations of the simple RNN without gate

$$\mathbf{h}^{(t)} = \sigma_h(W_{hx}\mathbf{x}^{(t)} + W_{hh}\mathbf{h}^{(t-1)} + \mathbf{b}_h) \tag{12.114}$$

are extended (Fig. 12.19) for the MGU [209, 213] by first calculating the gate variables which switch between 0 (off) and 1 (on)

$$\mathbf{g}^{(t)} = \sigma_g(W_{gx}\mathbf{x}^{(t)} + W_{gh}\mathbf{h}^{(t-1)} + \mathbf{b}_g). \tag{12.115}$$

Next a proposal for the hidden state is obtained similar to the simple RNN equation

$$\hat{h}^{(t)} = \sigma_h(W_{hx}\mathbf{x}^{(t)} + W_{hh}\left[\mathbf{g}^{(t)} \odot \mathbf{h}^{(t-1)}\right] + \mathbf{b}_h) \tag{12.116}$$

and finally the new hidden state is a mixture of proposed and old values

$$\mathbf{h}^{(t)} = \left[(1-\mathbf{g}^{(t)}) \odot \mathbf{h}^{(t-1)}\right] + \left[\mathbf{g}^{(t)} \odot \hat{h}^{(t)}\right]. \tag{12.117}$$

Hence the gate determines how much the hidden state is changed

12.9 Principal Component Analysis

Principal component analysis [214] (PCA) is a dimensionality reduction method for large data sets which is directly related to eigenvalue decomposition (EVD, Chap. 10) and singular value decomposition (SVD, Sect. 11.2). Neural networks can perform the PCA analysis online from the input data and are especially useful for data sets with very high dimension.

PCA uses a linear transformation to a new coordinate system where the essential variation of the data is concentrated in a subspace of reduced dimension. Along the first principal direction the variation has the maximum value. Consider a data matrix (see also Sect. 11.2.3) of dimension $m \times n$

$$A = \begin{pmatrix} x_1(t_1) & \dots & x_n(t_1) \\ \vdots & & \vdots \\ x_1(t_m) & \dots & x_n(t_m) \end{pmatrix}$$

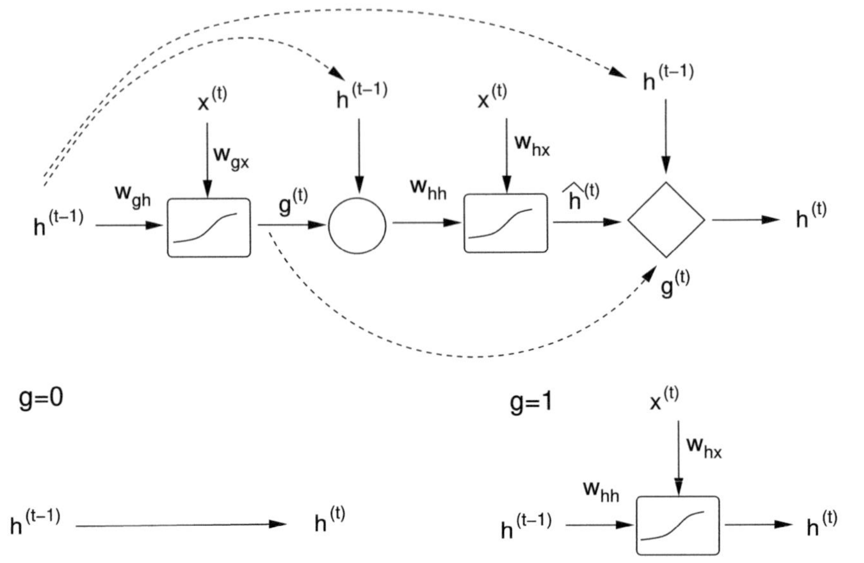

Fig. 12.19 Minimal gated unit. **Top:** The minimal gated unit (MGU) is the simplest gated unit. It has one gate which determines how much the hidden state is changed. **Bottom left:** If the gate is closed ($g = 0$) the hidden state is unchanged. **Bottom right:** If the gate is open ($g = 1$) the unit works like usual

where each m-dimensional row vector represents the results from a repeatedly performed experiment, i.e. the value of the j-th observable measured at time t_i is

$$x_j^{(i)} = A_{ij} \quad i = 1 \cdots m, \quad j = 1 \cdots n. \tag{12.118}$$

For convenience, we assume that the average values have been subtracted, such that $\sum_{j=1}^{m} x_i^{(j)} = 0$. Figure 12.20 shows the principle for a two-dimensional data set. The coordinate system is rotated in the $x_1 - x_2$ plane such that the data variance has the maximum value along the principal axis w_1.

The distance of the data point $(x_1^{(j)}, \cdots x_n^{(j)})$ at time $t^{(j)}$ from a plane perpendicular to the unit vector $\mathbf{w}$ is given by

$$d^{(i)} = \sum_{j=1}^{n} A_{ij} w_j \tag{12.119}$$

and the sum of the squared distances by

$$\sum_{i=1}^{m} (d^{(i)})^2 = \sum_{i=1}^{m} \sum_{j=1}^{n} \sum_{j'=1}^{n} A_{ij} w_j A_{ij'} w_{j'} = \mathbf{w}^T A^T A \mathbf{w}. \tag{12.120}$$

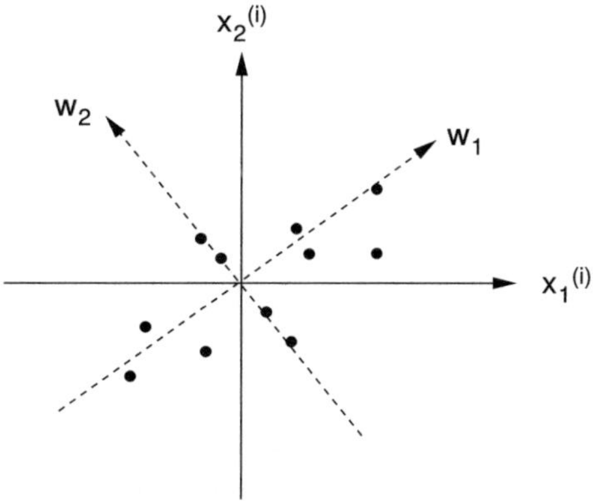

Fig. 12.20 Principle components. The dots show data points in the $x_1 - x_2$ plane. PCA determines a new coordinate system for which the variation of the data becomes maximal along the first principal direction $\mathbf{w}_1$

The first principal direction is given by the maximum variance

$$\mathbf{w}_{(1)} = \underset{|\mathbf{w}|=1}{\arg\max}\,(\mathbf{w}^T A^T A\mathbf{w}). \tag{12.121}$$

The covariance matrix $A^T A$ is symmetric and can therefore be diagonalized. Consider an eigenpair $(\mathbf{u}, \lambda)$ for which

$$A^T A\mathbf{u} = \lambda\mathbf{u} \tag{12.122}$$

$$\mathbf{u}^T A^T A\mathbf{u} = \lambda|\mathbf{u}|^2 \geq 0. \tag{12.123}$$

All eigenvalues are nonnegative and $(A^T A)$ is semipositive definite. To find the maximum variance we introduce a Lagrange parameter α to establish normalization of $\mathbf{w}$. Stationarity then requires

$$0 = \frac{\partial}{\partial w_j}\left[\sum_{j'j''} w_{j'}w_{j''}(A^T A)_{j'j''} - \alpha\sum_{j'} w_{j'}^2\right]$$

$$= \sum_{j''} w_{j''}(A^T A)_{jj''} + \sum_{j'} w_{j'}(A^T A)_{j'j} - 2\alpha w_j \tag{12.124}$$

and since $A^T A$ is symmetric

$$0 = \sum_{j'}(A^T A)_{jj'}w_{j'} - \alpha w_j \tag{12.125}$$

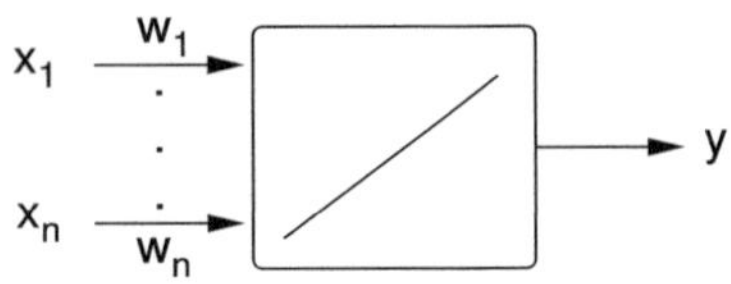

Fig. 12.21 Simple linear one-layer network for PCA

i.e. stationarity is obtained for the eigenvectors of $A^T A$, hence the first principal direction is given by the eigenvector corresponding to the largest eigenvalue. As is discussed in Sect. 11.2.2, diagonalization of $A^T A$ also corresponds to a singular value decomposition of A.

The simplest ANN approach (Fig. 12.21) to find the first principal direction uses a linear one-layer network which calculates the scalar product

$$y^{(i)} = \sum_{j=1}^{n} w_j x_j^{(i)}. \tag{12.126}$$

The weight parameters are updated according to the Hebbian learning rule [215, 216]

$$w_j^{(i+1)} = w_j^{(i)} + \eta x_j^{(i)} y^{(i)} = w_j^{(i)} + \eta x_j^{(i)} \sum w_{j'}^{(i)} x_{j'}^{(i)} = (\delta_{jj'} + \eta \sum_{j'} A_{ij} A_{ij'}) w_{j'}^{(i)}. \tag{12.127}$$

If we feed the network with the whole data set during one epoch, then for small η the changes add up in leading order. After one epoch the total change of the weight vector is given by[15]

$$\mathbf{w} + \Delta\mathbf{w} = (1 + \eta A^T A)\mathbf{w} \tag{12.128}$$

and after many epochs

$$\mathbf{w}(\mathrm{N_{epochs}}) = (1 + \eta A^T A)^{\mathrm{N_{epochs}}} \mathbf{w}(0) \tag{12.129}$$

which will converge to a multiple of the dominant eigenvector (Sect. 10.5). Since $A^T A$ is positive semidefinite, its largest eigenvalue $\lambda_{max} > 0$ [16] and $1 + \eta\lambda_{max} > 0$, which shows that $|\mathbf{w}|$ will continuously increase. This can be avoided by reorthogonalization [217] which changes the learning rule to

$$w_j^{(i+1)} = \frac{w_j^{(i)} + \eta x_j^{(i)} y^{(i)}}{|w_j^{(i)} + \eta x_j^{(i)} y^{(i)}|}. \tag{12.130}$$

[15] If the network is fed online with stochastic data [218], $(A^T A)_{ij}$is replaced by the correlation matrix $C_{ij} =< x_i x_j >$.

[16] The case of only zero eigenvalues is trivial.

For small η, series expansion (with $|w_j^{(i)}| = 1$) gives Oja's famous update rule [217–219]

$$w_j^{(i+1)} = \left(w_j^{(i)} + \eta x_j^{(i)} y^{(i)}\right)\left(1 - \eta \sum_{j'} w_{j'}^{(i)} x_{j'}^{(i)} y^{(j)} + \cdots\right)$$

$$= w_j^{(i)} + \eta\left(x_j^{(i)} y^{(i)} - w_j^{(i)} y^2\right) + \cdots \tag{12.131}$$

which for linear y becomes

$$w_j^{(i+1)} = w_j^{(i)}\left(1 - \eta\left(\sum_{j'=1}^{n} w_{j'} x_{j'}^{(i)}\right)^2\right) + \eta x_j^{(i)} \sum_{j'=1}^{n} w_{j'} x_{j'}^{(i)} \tag{12.132}$$

$$= w_j^{(i)}\left(1 - \sum_{j'=1}^{n} w_{j'} A_{ij'} \sum_{j''=1}^{n} w'_{j''} A_{ij''}\right) + \eta A_{ij} \sum_{j'=1}^{n} w_{j'} A_{ij'}. \tag{12.133}$$

After one epoch we end up with

$$\mathbf{w} + \Delta\mathbf{w} = \left[1 - \eta \mathbf{w}^T A^T A \mathbf{w} + \eta A^T A\right]\mathbf{w}. \tag{12.134}$$

Stationarity is reached when

$$0 = \Delta w = \left[\eta \mathbf{w}^T A^T A \mathbf{w} - \eta A^T A\right]\mathbf{w} = \eta\left[\lambda - A^T A\right]\mathbf{w}. \tag{12.135}$$

Hence $\mathbf{w}$ must be eigenvector of $A^T A$ and the eigenvalue

$$\lambda = \mathbf{w}^T A^T A \mathbf{w} = \lambda |\mathbf{w}|^2. \tag{12.136}$$

Therefore stationarity requires that $\mathbf{w}$ is a normalized eigenvector of $A^T A$. To analyze the stability of this method, we consider a small perturbation and expand the weight vector in the basis of (orthonormal) eigenvectors. We insert

$$\mathbf{w} = \mathbf{u}_\lambda + \sum_{\lambda'} \epsilon_\lambda \mathbf{u}_{\lambda'} \tag{12.137}$$

into (12.134) and obtain

$$\Delta\mathbf{w} = \eta\left(\lambda \mathbf{u}_\lambda + \sum \epsilon_{\lambda'} \lambda' \mathbf{u}_{\lambda'}\right)$$

$$- \eta(\mathbf{u}_\lambda^T + \sum \epsilon_{\lambda'} \mathbf{u}_{\lambda'}^T)(\lambda \mathbf{u}_\lambda + \sum \lambda'' \epsilon_{\lambda''} \mathbf{u}_{\lambda''})(\mathbf{u}_\lambda + \sum \epsilon_{\lambda'''} \mathbf{u}_{\lambda'''}). \tag{12.138}$$

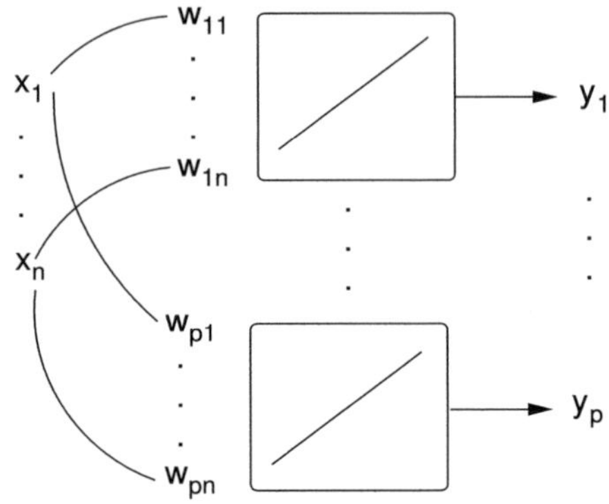

Fig. 12.22 One-layer network for determination of several principal components

Taking only terms linear in the perturbation

$$\Delta \mathbf{w} = \eta \left(\lambda \mathbf{u}_\lambda + \sum \epsilon_{\lambda'} \lambda' \mathbf{u}_{\lambda'} \right) \tag{12.139}$$
$$- \eta \left[\lambda \mathbf{u}_\lambda + \lambda \sum \epsilon_{\lambda'''} \mathbf{u}_{\lambda'''} + \sum \lambda'' \epsilon_{\lambda''} \mathbf{u}_\lambda^T \mathbf{u}_{\lambda''} \mathbf{u}_\lambda + \lambda \sum \epsilon_{\lambda'} \mathbf{u}_{\lambda'}^T \mathbf{u}_\lambda \mathbf{u}_\lambda \right]$$

and due to orthonormality

$$\Delta \mathbf{w} = \eta \left(\lambda \mathbf{u}_\lambda + \sum \epsilon_{\lambda'} \lambda' \mathbf{u}_{\lambda'} \right) - \eta \left[\lambda \mathbf{u}_\lambda + \lambda \sum \epsilon_{\lambda'''} \mathbf{u}_{\lambda'''} + 2\lambda \epsilon_\lambda \mathbf{u}_\lambda \right]$$
$$= \eta \left(\sum \epsilon_{\lambda'} (\lambda' - \lambda) \mathbf{u}_{\lambda'} - 2\lambda \epsilon_\lambda \mathbf{u}_\lambda \right). \tag{12.140}$$

This shows that a perturbation parallel to $\mathbf{u}_\lambda$ decays, also all perturbations with $\lambda' < \lambda$. However perturbation components with $\lambda' > \lambda$ would increase. We conclude that only the dominant eigenvector (with maximum eigenvalue) can be stable. To guarantee convergence, it is however necessary that the learning rate $\eta^{(t)}$ decreases as a function of time [217].

If more principle vectors besides the dominant are required, the Hebbian learning rule can be generalized for a network with several output variables (Fig. 12.22) [216].

$$w_{ij}^{(t+1)} = w_{ij}^{(t)} + \eta^{(t)} \left[y_i^{(t)} x_j^{(t)} - y_i^{(t)} \sum_{k \le i} w_{kj}^{(t)} y_k^{(t)} \right]. \tag{12.141}$$

Other extensions like subspace PCA are discussed in [219–222].

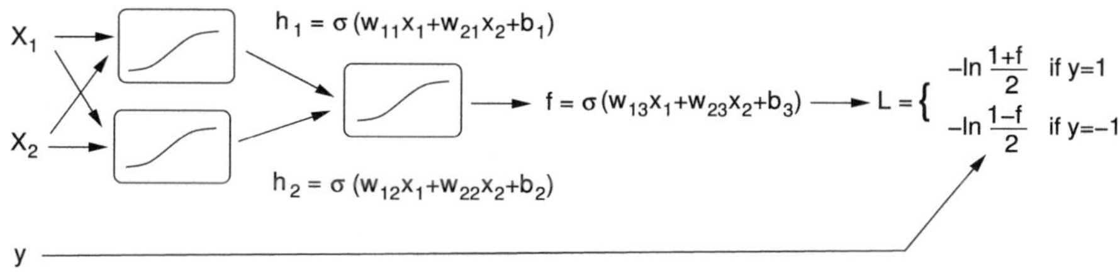

Fig. 12.23 Feed forward network for the xor-problem

Exercises

Exercise 12.1 Xor-problem

A very simple feed forward network with three units (Fig. 12.23) is used to solve the xor-problem. The input data $x_{1,2}$ from Table 12.1 are fed repeatedly to the two input nodes and the loss function is calculated according to (12.30–12.36). The mean square deviation is minimized by a simple gradient descent algorithm with constant learning rate η. Gradients are calculated by backpropagation (Sect. 12.5.2). Initially the weight parameters are chosen randomly around some reasonable values. The program visualizes the evolution of the decision boundaries in the $x_1 - x_2$ and $h_1 - h_2$ planes (Fig. 12.9) and the network parameters. After the training period the predicted truth table is shown. Depending on the initial conditions, convergence is not guaranteed.

Exercise 12.2 Interpolation and Regression

A simple feed forward network with one hidden layer (Fig. 12.12) is used to fit a set of data which are taken from a sine function with variable noise. The loss function

$$L = \frac{1}{2} \sum_{i=1}^{\mathrm{N_{data}}} (y_i - f(x_i))^2 \tag{12.142}$$

is minimized by a simple gradient descent algorithm with constant learning rate. You can change the number of data points and units. For zero noise the network interpolates the sine function. See also Fig. 12.13.

Exercise 12.3 Time Series Analysis

A simple recurrent network (Figs. 12.15 and 12.16) with 35 elements is used to learn a periodic signal. The loss function

$$L = \frac{1}{2\text{N}_{\text{data}}} \sum_{i=1}^{\text{N}_{\text{data}}} (f(x_{i-n_l}, \cdots x_{i-1}) - x_i)^2 \tag{12.143}$$

is minimized by a simple gradient descent algorithm with constant learning rate.

You can choose a pure sine function or switch on amplitude or frequency modulation and add some random noise. From 75 sample points the network learns to predict a new value based on the previous 15 ones. Then 70 further sample points are used to validate the model. Finally starting from the last 15 points of the training set, further 70 values are predicted. See also Figs. 12.17 and 12.18.

Chapter 13
Discretization of Differential Equations

Many processes in science and technology can be described by differential equations involving the rate of changes in time or space of a continuous variable, the unknown function. While the simplest differential equations can be solved exactly, a numerical treatment is necessary in most cases and the equations have to be discretized to turn them into a finite system of equations which can be solved by computers [223–225]. In this chapter we discuss different methods to discretize differential equations. The simplest approach is the method of finite differences, which replaces the differential quotients by difference quotients (Chap. 3). It is often used for the discretization of time. Finite difference methods for the space variables work best on a regular grid. Finite volume methods are very popular in computational fluid dynamics. They take averages over small control volumes and can be easily used with irregular grids. Finite differences and finite volumes belong to the general class of finite element methods which are prominent in the engineering sciences and use an expansion in piecewise polynomials with small support. Spectral methods, on the other hand, expand the solution as a linear combination of global basis functions like polynomials or trigonometric functions. A general concept for the discretization of differential equations is the method of weighted residuals which minimizes the weighted residual of a numerical solution. Most popular is Galerkin's method which uses the expansion functions also as weight functions. Simpler are the point collocation and subdomain collocation methods which fulfill the differential equation only at certain points or averaged over certain control volumes. More demanding is the least-squares method which has become popular in computational fluid dynamics and computational electrodynamics. The least-square integral provides a measure for the quality of the solution which can be used for adaptive grid size control.

Supplementary Information The online version contains supplementary material available at https://doi.org/10.1007/978-3-032-07856-8_13.

P. O. J. Scherer, *Computational Physics I*, Graduate Texts in Physics,
https://doi.org/10.1007/978-3-032-07856-8_13

If the Green's function is available for a problem, the method of boundary elements is an interesting alternative. It reduces the dimensionality and is, for instance, very popular in chemical physics to solve the Poisson-Boltzmann equation.

Artificial neural networks (ANN) can be used to provide an analytic and differentiable expression as a trial function to solve ordinary and partial differential equations. Usually the collocation method is used to solve the problem approximately at a set of sample points by minimizing the mean squared error with respect to the network parameters. Boundary conditions and other constraints can be taken into account by choosing proper trial functions or by adding penalty terms to the mean squared error. This approach has the advantage that derivatives of the trial function are analytically available and can be calculated by the network automatically. In accompanying computer experiments the reader can watch a simple ANN while solving differential equations.

13.1 Classification of Differential Equations

An ordinary differential equation (ODE) is a differential equation for a function of one single variable, like Newton's law for the motion of a body under the influence of a force field

$$m\frac{d^2}{dt^2}\mathbf{x}(t) = \mathbf{F}(\mathbf{x}, t), \tag{13.1}$$

a typical initial value problem where the solution in the domain $t_0 \leq t \leq T$ is determined by position and velocity at the initial time

$$\mathbf{x}(t = t_0) = \mathbf{x}_0 \quad \frac{d}{dt}\mathbf{x}(t = t_0) = \mathbf{v}_0. \tag{13.2}$$

Such equations of motion are discussed in Chap. 14. They also appear if the spatial derivatives of a partial differential equation have been discretized. Usually this kind of equation is solved by numerical integration over finite time steps $\Delta t = t_{n+1} - t_n$. Boundary value problems, on the other hand, require certain boundary conditions[1] to be fulfilled, for instance the linearized Poisson-Boltzmann equation in one dimension (Chap. 8 in Vol. 2)

$$\frac{d^2}{dx^2}\Phi - \kappa^2\Phi = -\frac{1}{\varepsilon}\rho(x) \tag{13.3}$$

where the value of the potential is prescribed on the boundary of the domain $x_0 \leq x \leq x_1$

[1] Dirichlet b.c concern the function values, Neumann b.c. the derivative, Robin b.c. a linear combination of both, Cauchy b.c the function value and the normal derivative and mixed b.c. have different character on different parts of the boundary.

$$\Phi(x_0) = \Phi_0 \quad \Phi(x_1) = \Phi_1. \tag{13.4}$$

Partial differential equations (PDE) finally involve partial derivatives with respect to at least two different variables, in many cases time and spatial coordinates.

Linear Second Order PDE

A very important class are second order linear partial differential equations of the general form

$$\left[\sum_{i=1}^{N}\sum_{j=1}^{N} a_{ij}\frac{\partial^2}{\partial x_i \partial x_j} + \sum_{i=1}^{N} b_i \frac{\partial}{\partial x_i} + c\right] f(x_1 \dots x_N) + d = 0 \tag{13.5}$$

where the coefficients a_{ij}, b_i, c, d are functions of the variables $x_1 \dots x_N$ but do not depend on the function f itself. The equation is classified according to the eigenvalues of the coefficient matrix a_{ij} as [226]

Elliptical

If all eigenvalues are positive or all eigenvalues are negative, like for the Poisson equation (Chap. 5 in Vol. 2)

$$\left(\frac{\partial^2}{\partial x^2} + \frac{\partial^2}{\partial y^2} + \frac{\partial^2}{\partial z^2}\right)\Phi(x, y, z) = -\frac{1}{\varepsilon}\varrho(x, y, z). \tag{13.6}$$

Hyperbolic

If one eigenvalue is negative and all the other eigenvalues are positive or vice versa, for example the wave equation in one spatial dimension (Chap. 7 in Vol. 2)

$$\frac{\partial^2}{\partial t^2} f - c^2 \frac{\partial^2}{\partial x^2} f = 0. \tag{13.7}$$

Parabolic

If at least one eigenvalue is zero, like for the diffusion equation (Chap. 8 in Vol. 2)

$$\frac{\partial}{\partial t} f(x, y, z, t) - D\left(\frac{\partial^2}{\partial x^2} + \frac{\partial^2}{\partial y^2} + \frac{\partial^2}{\partial z^2}\right) f(x, y, z, t) = S(x, y, z, t). \tag{13.8}$$

Ultra-hyperbolic

If there is no zero eigenvalue and more than one positive as well as more than one negative eigenvalue. Obviously the dimension then must be 4 at least.

Conservation Laws

One of the simplest first order partial differential equations is the one-dimensional advection equation

$$\frac{\partial}{\partial t} f(x,t) + u \frac{\partial}{\partial x} f(x,t) = 0 \tag{13.9}$$

which describes transport of a conserved quantity with density f (for instance mass, number of particles, charge etc.) in a medium streaming with velocity u. This is a special case of the class of conservation laws (also called continuity equations)

$$\frac{\partial}{\partial t} f(\mathbf{x},t) + \mathrm{div}\mathbf{J}(\mathbf{x},t) = g(\mathbf{x},t) \tag{13.10}$$

which are very common in physics. Here $\mathbf{J}$ describes the corresponding flux and g is an additional source (or sink) term. For instance the advection-diffusion equation (also known as convection equation) has this form which describes quite general transport processes:

$$\frac{\partial}{\partial t} C = \mathrm{div}\ (D\,\mathrm{grad}\ C - \mathbf{u}C) + S(\mathbf{x},t) = -\ \mathrm{div}\ \mathbf{J} + S(\mathbf{x},t) \tag{13.11}$$

where one contribution to the flux

$$\mathbf{J} = -D\ \mathrm{grad}\ C + \mathbf{u}C \tag{13.12}$$

is proportional to the gradient of the concentration C (Fick's first law) and the second part depends on the velocity field $\mathbf{u}$ of a streaming medium. The source term S represents the effect of chemical reactions. Equation (13.11) is also similar to the drift-diffusion equation in semiconductor physics and closely related to the Navier Stokes equations which are based on the Cauchy momentum equation [227]

$$\varrho \frac{d\mathbf{u}}{dt} = \varrho \left(\frac{\partial \mathbf{u}}{\partial t} + \mathbf{u}\ \mathrm{grad}\ \mathbf{u} \right) = \mathrm{div}\sigma + \mathbf{f} \tag{13.13}$$

where σ denotes the stress tensor. Equation (13.10) is the strong or differential form of the conservation law. The requirements on the smoothness of the solution are reduced by using the integral form which is obtained with the help of Gauss' theorem

$$\int_V \left(\frac{\partial}{\partial t} f(\mathbf{x},t) - g(\mathbf{x},t) \right) dV + \oint_{\partial V} \mathbf{J}(\mathbf{x},t) d\mathbf{A} = 0. \tag{13.14}$$

An alternative integral form results from Galerkin's [228] method of weighted residuals which introduces a weight function $w(\mathbf{x})$ and considers the equation

$$\int_V \left(\frac{\partial}{\partial t} f(\mathbf{x}, t) + \mathrm{div}\mathbf{J}(\mathbf{x}, t) - g(\mathbf{x}, t) \right) w(\mathbf{x})\, dV = 0 \tag{13.15}$$

or after applying Gauss' theorem

$$\int_V \left\{ \left(\frac{\partial}{\partial t} f(\mathbf{x}, t) - g(\mathbf{x}, t) \right) w(\mathbf{x}) - \mathbf{J}(\mathbf{x}, t)\ \mathrm{grad} w(\mathbf{x}) \right\} dV$$

$$+ \oint_{\partial V} w(\mathbf{x})\mathbf{J}(\mathbf{x}, t) d\mathbf{A} = 0. \tag{13.16}$$

The so called weak form of the conservation law states that this equation holds for arbitrary weight functions w.

13.2 Finite Differences

The simplest method to discretize a differential equation is to introduce a grid of equidistant points and to discretize the differential operators by finite differences (FDM) as described in Chap. 3. For instance, in one dimension the first and second derivatives can be discretized by

$$x \to x_m = m\Delta x \quad m = 1 \ldots M \tag{13.17}$$

$$f(x) \to f_m = f(x_m) \quad m = 1 \ldots M \tag{13.18}$$

$$\frac{\partial f}{\partial x} \to (\frac{\partial}{\partial x} f)_m = \frac{f_{m+1} - f_m}{\Delta x} \text{ or } (\frac{\partial}{\partial x} f)_m = \frac{f_{m+1} - f_{m-1}}{2\Delta x} \tag{13.19}$$

$$\frac{\partial^2 f}{\partial x^2} \to (\frac{\partial^2}{\partial x^2} f)_m = \frac{f_{m+1} + f_{m-1} - 2 f_m}{\Delta x^2} \quad . \tag{13.20}$$

These expressions are not well defined at the boundaries of the grid $m = 1, M$ unless the boundary conditions are taken into account. For instance, in case of a Dirichlet problem f_0 and f_{M+1} are given boundary values and

$$(\frac{\partial}{\partial x} f)_1 = \frac{f_2 - f_0}{2\Delta x} \quad (\frac{\partial^2}{\partial x^2} f)_1 = \frac{f_2 - 2 f_1 + f_0}{\Delta x^2} \tag{13.21}$$

$$(\frac{\partial}{\partial x} f)_M = \frac{f_{M+1} - f_M}{\Delta x} \text{ or } \frac{f_{M+1} - f_{M-1}}{2\Delta x} \quad (\frac{\partial^2}{\partial x^2} f)_M = \frac{f_{M-1} - 2 f_M + f_{M+1}}{\Delta x^2}. \tag{13.22}$$

Other kinds of boundary conditions can be treated in a similar way.

13.2.1 Finite Differences in Time

Time derivatives can be treated similarly using an independent time grid

$$t \to t_n = n\Delta t \quad n = 1 \ldots N \tag{13.23}$$

$$f(t,x) \to f_m^n = f(t_n, x_m) \tag{13.24}$$

and finite differences like the first order forward difference quotient

$$\frac{\partial f}{\partial t} \to \frac{f_m^{n+1} - f_m^n}{\Delta t} \tag{13.25}$$

or the symmetric difference quotient

$$\frac{\partial f}{\partial t} \to \frac{f_m^{n+1} - f_m^{n-1}}{2\Delta t} \tag{13.26}$$

to obtain a system of equations for the function values at the grid points f_m^n. For instance for the diffusion equation in one spatial dimension

$$\frac{\partial f(x,t)}{\partial t} = D\frac{\partial^2}{\partial x^2} f(x,t) + S(x,t). \tag{13.27}$$

The simplest discretization is the FTCS (forward in time, centered in space) scheme

$$(f_m^{n+1} - f_m^n) = D\frac{\Delta t}{\Delta x^2}(f_{m+1}^n + f_{m-1}^n - 2f_m^n) + S_m^n \Delta t \tag{13.28}$$

which can be written in matrix notation as

$$\mathbf{f}_{n+1} - \mathbf{f}_n = D\frac{\Delta t}{\Delta x^2} M\mathbf{f}_n + \mathbf{S}_n \Delta t \tag{13.29}$$

with

$$\mathbf{f}_n = \begin{pmatrix} f_1^n \\ f_2^n \\ f_3^n \\ \vdots \\ f_M^n \end{pmatrix} \quad \text{and} \quad M = \begin{pmatrix} -2 & 1 & & & \\ 1 & -2 & 1 & & \\ & 1 & -2 & 1 & \\ & & \ddots & \ddots & \ddots \\ & & & 1 & -2 \end{pmatrix}. \tag{13.30}$$

13.2.2 Stability Analysis

Fully discretized linear differential equations provide an iterative algorithm of the type[2]

$$\mathbf{f}_{n+1} = A\mathbf{f}_n + \mathbf{S}_n \Delta t \tag{13.31}$$

which propagates numerical errors according to

$$\mathbf{f}_{n+1} + \epsilon_{n+1} = A(\mathbf{f}_n + \epsilon_n) + \mathbf{S}_n \Delta t \tag{13.32}$$

$$\epsilon_{j+1} = A\epsilon_j. \tag{13.33}$$

Errors are amplified exponentially if the absolute value of at least one eigenvalue of A is larger than one. The algorithm is stable if all eigenvalues of A are smaller than one in absolute value (Sect. 1.4). If the eigenvalue problem is difficult to solve, the von Neumann analysis is helpful which decomposes the errors into a Fourier series and considers the Fourier components individually by setting

$$\mathbf{f}_n = g^n(k) \begin{pmatrix} e^{ik} \\ \vdots \\ e^{ikM} \end{pmatrix} \tag{13.34}$$

and calculating the amplification factor

$$\left| \frac{f_m^{n+1}}{f_m^n} \right| = |g(k)|. \tag{13.35}$$

The algorithm is stable if $|g(k)| \le 1$ for all k.

Example: For the discretized diffusion equation (13.28) we find

$$g^{n+1}(k) = g^n(k) + 2D\frac{\Delta t}{\Delta x^2} g^n(k)\,(\cos k - 1) \tag{13.36}$$

$$g(k) = 1 + 2D\frac{\Delta t}{\Delta x^2}(\cos k - 1) = 1 - 4D\frac{\Delta t}{\Delta x^2}\sin^2\left(\frac{k}{2}\right) \tag{13.37}$$

$$1 - 4D\frac{\Delta t}{\Delta x^2} \le g(k) \le 1 \tag{13.38}$$

hence stability requires

[2] Differential equations which are higher order in time can be always brought to first order by introducing the time derivatives as additional variables.

$$D\frac{\Delta t}{\Delta x^2} \le \frac{1}{2}. \tag{13.39}$$

13.2.3 Method of Lines

Alternatively time can be considered as a continuous variable. The discrete values of the function then are functions of time (so called lines)

$$f_m(t) \tag{13.40}$$

and a set of ordinary differential equations has to be solved. For instance for diffusion in one dimension (13.27) the equations

$$\frac{\mathrm{d}f_m}{\mathrm{d}t} = \frac{D}{h^2}\left(f_{m+1} + f_{m-1} - 2f_m\right) + S_m(t) \tag{13.41}$$

which can be written in matrix notation as

$$\frac{\mathrm{d}}{\mathrm{d}t}\begin{pmatrix} f_1 \\ f_1 \\ f_2 \\ \vdots \\ f_M \end{pmatrix} = \frac{D}{\Delta x^2}\begin{pmatrix} -2 & 1 & & & \\ 1 & -2 & 1 & & \\ & 1 & -2 & 1 & \\ & & \ddots & \ddots & \ddots \\ & & & 1 & -2 \end{pmatrix}\begin{pmatrix} f_1 \\ f_2 \\ f_3 \\ \vdots \\ f_M \end{pmatrix} + \begin{pmatrix} S_1 + \frac{D}{h^2}f_0 \\ S_2 \\ S_3 \\ \vdots \\ S_M + \frac{D}{h^2}f_{M+1} \end{pmatrix} \tag{13.42}$$

or briefly

$$\frac{\mathrm{d}}{\mathrm{d}t}\mathbf{f}(t) = \mathbf{A}\mathbf{f}(t) + \mathbf{S}(t). \tag{13.43}$$

Several methods to integrate such a semi-discretized equation will be discussed in Chap. 14. If eigenvectors and eigenvalues of A are easy available, an eigenvector expansion can be used.

13.2.4 Eigenvector Expansion

A homogeneous system

$$\frac{\mathrm{d}}{\mathrm{d}t}\mathbf{f}(t) = \mathbf{A}\mathbf{f}(t) \tag{13.44}$$

where the matrix A is obtained from discretizing the spatial derivatives, can be solved by an eigenvector expansion. From the eigenvalue problem

$$\mathbf{A}\mathbf{f} = \lambda\mathbf{f} \tag{13.45}$$

we obtain the eigenvalues λ and eigenvectors $\mathbf{f}_\lambda$ which provide the particular solutions

$$\mathbf{f}(t) = \mathrm{e}^{\lambda t}\, \mathbf{f}_\lambda \tag{13.46}$$

$$\frac{\mathrm{d}}{\mathrm{d}t}(\mathrm{e}^{\lambda t}\, \mathbf{f}_\lambda) = \lambda(\mathrm{e}^{\lambda t}\, \mathbf{f}_\lambda) = A(\mathrm{e}^{\lambda t}\, \mathbf{f}_\lambda). \tag{13.47}$$

These can be used to expand the general solution

$$\mathbf{f}(t) = \sum_\lambda C_\lambda \mathrm{e}^{\lambda t}\mathbf{f}_\lambda. \tag{13.48}$$

The coefficients C_λ follow from the initial values by solving the linear equations

$$\mathbf{f}(t=0) = \sum_\lambda C_\lambda \mathbf{f}_\lambda. \tag{13.49}$$

If the differential equation is second order in time

$$\frac{\mathrm{d}^2}{\mathrm{d}t^2}\mathbf{f}(t) = A\mathbf{f}(t) \tag{13.50}$$

the particular solutions are

$$\mathbf{f}(t) = \mathrm{e}^{\pm t\sqrt{\lambda}}\, \mathbf{f}_\lambda \tag{13.51}$$

$$\frac{\mathrm{d}^2}{\mathrm{d}t^2}(\mathrm{e}^{\pm t\sqrt{\lambda}}\, \mathbf{f}_\lambda) = \lambda(\mathrm{e}^{\pm t\sqrt{\lambda}}\, \mathbf{f}_\lambda) = A(\mathrm{e}^{\pm t\sqrt{\lambda}}\, \mathbf{f}_\lambda) \tag{13.52}$$

and the eigenvector expansion is

$$\mathbf{f}(t) = \sum_\lambda \left(C_{\lambda+}\mathrm{e}^{t\sqrt{\lambda}} + C_{\lambda-}\mathrm{e}^{-t\sqrt{\lambda}} \right) \mathbf{f}_\lambda. \tag{13.53}$$

The coefficients $C_{\lambda\pm}$ follow from the initial amplitudes and velocities

$$\begin{aligned} \mathbf{f}(t=0) &= \sum_\lambda (C_{\lambda+} + C_{\lambda-})\mathbf{f}_\lambda \\ \frac{\mathrm{d}}{\mathrm{d}t}\mathbf{f}(t=0) &= \sum_\lambda \sqrt{\lambda}(C_{\lambda+} - C_{\lambda-})\mathbf{f}_\lambda. \end{aligned} \tag{13.54}$$

For a first order inhomogeneous system

$$\frac{\mathrm{d}}{\mathrm{d}t}\mathbf{f}(t) = A\mathbf{f}(t) + \mathbf{S}(t) \tag{13.55}$$

the expansion coefficients have to be time dependent

$$\mathbf{f}(t) = \sum_{\lambda} C_{\lambda}(t) \mathrm{e}^{\lambda t} \mathbf{f}_{\lambda} \tag{13.56}$$

and satisfy

$$\frac{\mathrm{d}}{\mathrm{d}t}\mathbf{f}(t) - A\mathbf{f}(t) = \sum_{\lambda} \frac{\mathrm{d}C_{\lambda}}{\mathrm{d}t} \mathrm{e}^{\lambda t} \mathbf{f}_{\lambda} = \mathbf{S}(t). \tag{13.57}$$

After taking the scalar product with $\mathbf{f}_{\mu}$[3]

$$\frac{\mathrm{d}C_{\mu}}{\mathrm{d}t} = \mathrm{e}^{-\mu t} \left(\mathbf{f}_{\mu} \mathbf{S}(t) \right) \tag{13.58}$$

can be solved by a simple time integration. For a second order system

$$\frac{\mathrm{d}^2}{\mathrm{d}t^2}\mathbf{f}(t) = A\mathbf{f}(t) + \mathbf{S}(t) \tag{13.59}$$

we introduce the first time derivative as a new variable

$$\mathbf{g} = \frac{\mathrm{d}}{\mathrm{d}t}\mathbf{f} \tag{13.60}$$

to obtain a first order system of double dimension

$$\frac{\mathrm{d}}{\mathrm{d}t}\begin{pmatrix} \mathbf{f} \\ \mathbf{g} \end{pmatrix} = \begin{pmatrix} 0 & 1 \\ A & 0 \end{pmatrix} \begin{pmatrix} \mathbf{f} \\ \mathbf{g} \end{pmatrix} + \begin{pmatrix} \mathbf{S} \\ 0 \end{pmatrix} \tag{13.61}$$

where eigenvectors and eigenvalues can be found from those of A (13.45)

$$\begin{pmatrix} 0 & 1 \\ A & 0 \end{pmatrix} \begin{pmatrix} \mathbf{f}_{\lambda} \\ \pm\sqrt{\lambda}\mathbf{f}_{\lambda} \end{pmatrix} = \begin{pmatrix} \pm\sqrt{\lambda}\mathbf{f}_{\lambda} \\ \lambda\mathbf{f}_{\lambda} \end{pmatrix} = \pm\sqrt{\lambda} \begin{pmatrix} \mathbf{f}_{\lambda} \\ \pm\sqrt{\lambda}\mathbf{f}_{\lambda} \end{pmatrix} \tag{13.62}$$

$$\left(\pm\sqrt{\lambda}\mathbf{f}_{\lambda}^T \; \mathbf{f}_{\lambda}^T \right) \begin{pmatrix} 0 & 1 \\ A & 0 \end{pmatrix} = \left(\lambda \mathbf{f}_{\lambda}^T \; \pm\sqrt{\lambda}\mathbf{f}_{\lambda}^T \right) = \pm\sqrt{\lambda} \left(\pm\sqrt{\lambda}\mathbf{f}_{\lambda}^T \; \mathbf{f}_{\lambda}^T \right). \tag{13.63}$$

Insertion of

$$\sum_{\lambda} C_{\lambda+} \mathrm{e}^{\sqrt{\lambda}t} \begin{pmatrix} \mathbf{f}_{\lambda} \\ \sqrt{\lambda}\mathbf{f}_{\lambda} \end{pmatrix} + C_{\lambda-} \mathrm{e}^{-\sqrt{\lambda}t} \begin{pmatrix} \mathbf{f}_{\lambda} \\ -\sqrt{\lambda}\mathbf{f}_{\lambda} \end{pmatrix}$$

gives

[3] If A is not Hermitian we have to distinguish left- and right-eigenvectors.

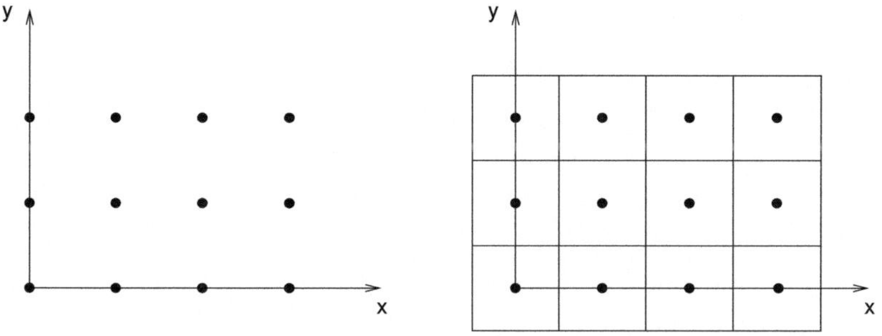

Fig. 13.1 Finite volume method. The domain V is divided into small control volumes V_r, in the simplest case cubes around the grid points $\mathbf{r}_{ijk}$

$$\sum_\lambda \frac{dC_{\lambda+}}{dt} e^{\sqrt{\lambda}t} \begin{pmatrix} \mathbf{f}_\lambda \\ \sqrt{\lambda}\mathbf{f}_\lambda \end{pmatrix} + \frac{dC_{\lambda-}}{dt} e^{\sqrt{\lambda}t} \begin{pmatrix} \mathbf{f}_\lambda \\ -\sqrt{\lambda}\mathbf{f}_\lambda \end{pmatrix} = \begin{pmatrix} \mathbf{S}(t) \\ 0 \end{pmatrix} \tag{13.64}$$

and taking the scalar product with one of the left-eigenvectors we end up with

$$\frac{dC_{\lambda+}}{dt} = \frac{1}{2} (\mathbf{f}_\lambda \mathbf{S}(t))\, e^{-\sqrt{\lambda}t} \tag{13.65}$$

$$\frac{dC_{\lambda-}}{dt} = -\frac{1}{2} (\mathbf{f}_\lambda \mathbf{S}(t))\, e^{\sqrt{\lambda}t}. \tag{13.66}$$

13.3 Finite Volumes

Whereas the finite differences method uses function values

$$f_{i,j,k} = f(x_i, y_j, z_k) \tag{13.67}$$

at the grid points

$$\mathbf{r}_{ijk} = (x_i, y_j, z_k), \tag{13.68}$$

the finite volume method (FVM) [229] averages function values and derivatives over small control volumes V_r which are disjoint and span the domain V (Fig. 13.1)

$$V = \bigcup_r V_r \qquad V_r \bigcap V_{r'} = \emptyset \qquad \forall r \neq r'. \tag{13.69}$$

The averages are

$$\overline{f_r} = \frac{1}{V_r} \int_{V_r} dV\, f(\mathbf{r}) \tag{13.70}$$

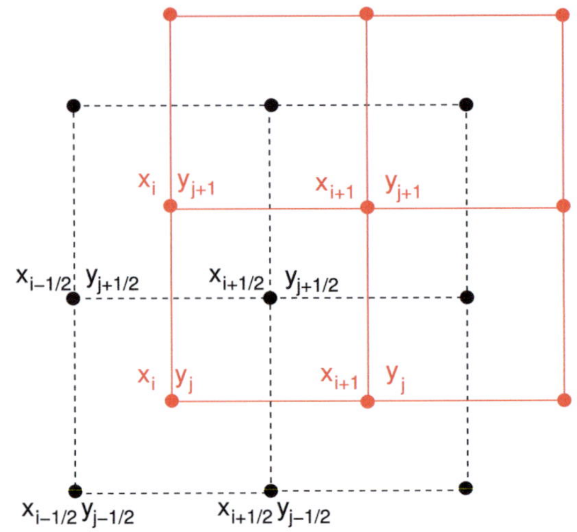

Fig. 13.2 Dual grid. The dual grid (black) is centered around the vertices of the original grid (red)

or in the simple case of cubic control volumes of equal size h^3

$$\overline{f}_{ijk} = \frac{1}{h^3} \int_{x_i-h/2}^{x_i+h/2} dx \int_{y_j-h/2}^{y_j+h/2} dy \int_{z_k-h/2}^{z_k+h/2} dz\, f(x, y, z). \tag{13.71}$$

Such average values have to be related to discrete function values by numerical integration (Chap. 4). The midpoint rule (4.17), for instance replaces the average by the central value

$$\overline{f}_{ijk} = f(x_i, y_j, z_k) + O(h^2) \tag{13.72}$$

whereas the trapezoidal rule (4.13) implies the average over the eight corners of the cube

$$\overline{f}_{ijk} = \frac{1}{8} \sum_{m,n,p=\pm 1} f(x_{i+m/2}, y_{j+n/2}, z_{k+p/2}) + O(h^2). \tag{13.73}$$

In (13.73) the function values refer to a dual grid (Fig. 13.2) [229] centered around the vertices of the original grid (13.68).

$$\mathbf{r}_{i+\frac{1}{2},j+\frac{1}{2},k+\frac{1}{2}} = (x_i + \frac{h}{2}, y_j + \frac{h}{2}, z_k + \frac{h}{2}). \tag{13.74}$$

The average gradient can be rewritten using the generalized Stokes' theorem as

$$\overline{\text{grad}\, f_{ijk}} = \frac{1}{V} \int_{V_{ijk}} dV \,\text{grad}\, f(\mathbf{r}) = \oint_{\partial V_{ijk}} f(\mathbf{r})\, d\mathbf{A}. \tag{13.75}$$

For a cubic grid we have to integrate over the six faces of the control volume

$$\overline{\text{grad } f_{ijk}} = \frac{1}{h^3}\begin{pmatrix} \int_{z_k-h/2}^{z_k+h/2} dz \int_{y_j-h/2}^{y_j+h/2} dy \left(f(x_i+\frac{h}{2},y,z) - f(x_i-\frac{h}{2},y,z)\right) \\ \int_{z_k-h/2}^{z_k+h/2} dz \int_{x_i-h/2}^{x_i+h/2} dx \left(f(x,y_j+\frac{h}{2},z) - f(x,y_j-\frac{h}{2},z)\right) \\ \int_{x_i-h/2}^{x_i+h/2} dx \int_{y_j-h/2}^{y_j+h/2} dy \left(f(x,y,z_k+\frac{h}{2}) - f(x,y,z_k-\frac{h}{2})\right) \end{pmatrix}. \tag{13.76}$$

The integrals have to be evaluated numerically. Applying as the simplest approximation the midpoint rule (4.17)

$$\int_{x_i-h/2}^{x_i+h/2} dx \int_{y_j-h/2}^{y_j+h/2} dy\, f(x,y) = h^2\left(f(x_i,y_j) + O(h^2)\right) \tag{13.77}$$

this becomes

$$\overline{\text{grad } f_{ijk}} = \frac{1}{h}\begin{pmatrix} f(x_i+\frac{h}{2},y_j,z_k) - f(x_i-\frac{h}{2},y_j,z_k) \\ f(x_i,y_j+\frac{h}{2},z_k) - f(x_i,y_j-\frac{h}{2},z_k) \\ f(x_i,y_j,z_k+\frac{h}{2}) - f(x_i,y_j,z_k-\frac{h}{2}) \end{pmatrix} \tag{13.78}$$

which involves symmetric difference quotients. However, the function values in (13.78) refer neither to the original nor to the dual grid. Therefore we interpolate (Fig. 13.3)

$$f(x_i \pm \frac{h}{2}, y_j, z_k) \approx \frac{1}{2}\left(f(x_{i+1},y_j,z_k) + f(x_{i\pm1},y_j,z_k)\right) \tag{13.79}$$

$$\frac{1}{2}\left(f(x_i+\frac{h}{2},y_j,z_k) - f(x_i-\frac{h}{2},y_j,z_k)\right)$$

$$\approx \frac{1}{2h}\left(f(x_{i+1},y_j,z_k) - f(x_{i-1},y_j,z_k)\right) \tag{13.80}$$

or

$$f(x_i \pm \frac{h}{2}, y_j, z_k) \approx \frac{1}{4}\sum_{m,n=\pm1} f(x_i \pm \frac{h}{2}, y_j + m\frac{h}{2}, z_k + n\frac{h}{2}). \tag{13.81}$$

The finite volume method is capable of treating discontinuities and is very flexible concerning the size and shape of the control volumes.

13.3.1 Discretization of Fluxes

Integration of (13.10) over a control volume and application of Gauss' theorem gives the integral form of the conservation law

$$\frac{1}{V}\oint \mathbf{J}d\mathbf{A} + \frac{\partial}{\partial t}\frac{1}{V}\int f\,dV = \frac{1}{V}\int g\,dV \tag{13.82}$$

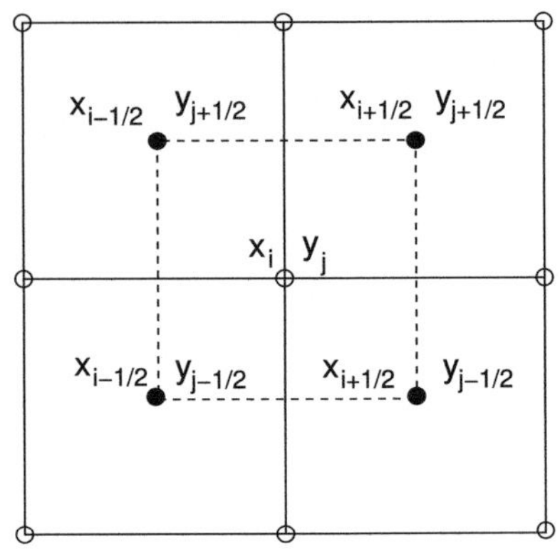

Fig. 13.3 Interpolation between grid points. Interpolation is necessary to relate the averaged gradient (13.78) to the original or dual grid

which involves the flux **J** of some property like particle concentration, mass, energy or momentum density or the flux of an electromagnetic field. The total flux through a control volume is given by the surface integral

$$\Phi = \oint_{\partial V} \mathbf{J} d\mathbf{A} \tag{13.83}$$

which in the special case of a cubic volume element of size h^3 becomes the sum over the six faces of the cube (Fig. 13.4)

$$\begin{aligned}
\Phi &= \sum_{r=1}^{6} \int_{A_r} \mathbf{J} d\mathbf{A} \\
&= \int_{x_i-h/2}^{x_i+h/2} dx \int_{y_j-h/2}^{y_j+h/2} dy \left(J_z(x, y, z_k + \frac{h}{2}) - J_z(x, y, z_k - \frac{h}{2}) \right) \\
&+ \int_{x_i-h/2}^{x_i+h/2} dx \int_{z_k-h/2}^{z_k+h/2} dz \left(J_y(x, y_j + \frac{h}{2}, z) - J_y(x, y_j - \frac{h}{2}, z) \right) \\
&+ \int_{z_k-h/2}^{z_k+h/2} dz \int_{y_j-h/2}^{y_j+h/2} dy \left(J_x(x_i + \frac{h}{2}, y, z) - J_x(x_i - \frac{h}{2}, y, z) \right).
\end{aligned} \tag{13.84}$$

The surface integral can be evaluated numerically (Chap. 4). Using the midpoint approximation (13.77) we obtain

$$\begin{aligned}
\frac{1}{V}\Phi(x_i, y_j, z_k) &= \frac{1}{h} \Big(J_y(x_i, y_j, z_{k+\frac{1}{2}}) - J_y(x_i, y_j, z_{k-\frac{1}{2}}) \\
&+ J_y(x_i y_{j+\frac{1}{2}}, z_k) - J_y(x_i, y_{j-\frac{1}{2}}, z_k) + J_x(x_{i+\frac{1}{2}}, y_j, z_k) - J_x(x_{i-\frac{1}{2}}, y_j, z_k) \Big).
\end{aligned} \tag{13.85}$$

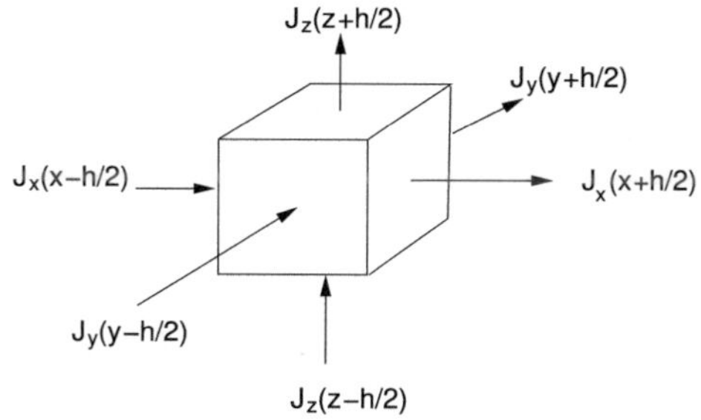

Fig. 13.4 Flux through a control volume

The trapezoidal rule (4.13) introduces an average over the four corners (Fig. 13.3)

$$\int_{x_i-h/2}^{x_i+h/2} dx \int_{y_j-h/2}^{y_j+h/2} dy\, f(x,y)$$

$$= h^2 \left(\frac{1}{4} \sum_{m,n=\pm 1} f(x_{i+m/2}, y_{j+n/2}) + O(h^2) \right). \tag{13.86}$$

which replaces the flux values in (13.85) by the averages

$$J_x(x_{i\pm\frac{1}{2}}, y_j, z_k) = \frac{1}{4} \sum_{m,n=\pm 1} J_z(x_{i\pm\frac{1}{2}}, y_{j+m/2}, z_{k+n/2}) \tag{13.87}$$

$$J_y(x_i, y_{j\pm\frac{1}{2}}, z_k) = \frac{1}{4} \sum_{m,n=\pm 1} J_z(x_{i+m/2}, y_{j\pm\frac{1}{2}}, z_{k+n/2}) \tag{13.88}$$

$$J_z(x_i, y_j, z_{k\pm\frac{1}{2}}) = \frac{1}{4} \sum_{m,n=\pm 1} J_z(x_{i+m/2}, y_{j+n/2}, z_{k\pm\frac{1}{2}}). \tag{13.89}$$

One advantage of the finite volume method is that the flux is strictly conserved.

13.4 Weighted Residual Based Methods

A general method to discretize partial differential equations is to approximate the solution within a finite-dimensional space of trial functions.[4] The partial differential equation is turned into a finite system of equations or a finite system of ordinary differential equations if time is treated as a continuous variable. This is the basis of spectral methods which make use of polynomials or Fourier series but also of the

[4] Also called expansion functions.

very successful finite element methods. Even finite difference methods and finite volume methods can be formulated as weighted residual based methods.

Consider a differential equation[5] on the domain V which is written symbolically with the differential operator $\mathcal{T}$

$$\mathcal{T}[u(\mathbf{r})] = f(\mathbf{r}) \quad \mathbf{r} \in V \tag{13.90}$$

and corresponding boundary conditions which are expressed with a boundary operator $\mathcal{B}$[6]

$$\mathcal{B}[u(\mathbf{r})] = g(\mathbf{r}) \quad \mathbf{r} \in \partial V. \tag{13.91}$$

The basic principle to obtain an approximate solution $\tilde{u}(\mathbf{r})$ is to choose a linear combination of expansion functions $N_i(\mathbf{r}) \quad i = 1 \dots r$ as a trial function which fulfills the boundary conditions[7]

$$\tilde{u} = \sum_{i=1}^{r} u_i N_i(\mathbf{r}) \tag{13.92}$$

$$\mathcal{B}[\tilde{u}(\mathbf{r})] = g(\mathbf{r}). \tag{13.93}$$

In general (13.92) is not an exact solution and the residual

$$R(\mathbf{r}) = \mathcal{T}\left[\tilde{u}\right](\mathbf{r}) - f(\mathbf{r}) \tag{13.94}$$

will not be zero throughout the whole domain V. The function $\tilde{u}$ has to be determined such that the residual becomes "small" in a certain sense. To that end weight functions[8] $w_j \quad j = 1 \dots r$ are chosen to define the weighted residuals

$$R_j(u_1 \dots u_r) = \int dV\, w_j(\mathbf{r}) \left(\mathcal{T}\left[\tilde{u}\right](\mathbf{r}) - f(\mathbf{r})\right). \tag{13.95}$$

The optimal parameters u_i are then obtained from the solution of the equations

$$R_j(u_1 \dots u_r) = 0 \quad j = 1 \dots r. \tag{13.96}$$

In the special case of a linear differential operator these equations are linear

---

[5] Generalization to systems of equations is straightforward.

[6] One or more linear differential operators, usually a combination of the function and its first derivatives.

[7] This requirement can be replaced by additional equations for the u_i, for instance with the tau method [230].

[8] Also called test functions.

$$\sum_{i=1}^{r} u_i \int dV\, w_j(\mathbf{r}) \mathcal{T}\left[N_i(\mathbf{r})\right] - \int dV\, w_j(\mathbf{r}) f(\mathbf{r}) = 0. \tag{13.97}$$

Several strategies are available to choose suitable weight functions.

13.4.1 Point Collocation Method

The collocation method uses the weight functions $w_j(\mathbf{r}) = \delta(\mathbf{r} - \mathbf{r}_j)$, with certain collocation points $\mathbf{r}_j \in V$. The approximation $\tilde{u}$ obeys the differential equation at the collocation points

$$0 = R_j = \mathcal{T}[\tilde{u}](\mathbf{r}_j) - f(\mathbf{r}_j) \tag{13.98}$$

and for a linear differential operator

$$0 = \sum_{i=1}^{r} u_i \mathcal{T}[N_i](\mathbf{r}_j) - f(\mathbf{r}_j). \tag{13.99}$$

The point collocation method is simple to use, especially for nonlinear problems. Instead of using trial functions satisfying the boundary conditions, extra collocation points on the boundary can be added (mixed collocation method).

13.4.2 Sub-domain Method

This approach uses weight functions which are the characteristic functions of a set of control volumes V_i which are disjoint and span the whole domain similar as for the finite volume method

$$V = \bigcup_j V_j \qquad V_j \bigcap V_{j'} = \emptyset \qquad \forall j \neq j' \tag{13.100}$$

$$w_j(\mathbf{r}) = \begin{cases} 1 & \mathbf{r} \in V_j \\ 0 & \text{else} \end{cases}. \tag{13.101}$$

The residuals then are integrals over the control volumes and

$$0 = R_j = \int_{V_j} dV \left(\mathcal{T}\left[\tilde{u}\right](\mathbf{r}) - f(\mathbf{r}) \right) \tag{13.102}$$

respectively

$$0 = \sum_i u_i \int_{V_j} dV\, \mathcal{T}[N_i](\mathbf{r}) - \int_{V_j} dV\, f(\mathbf{r}). \tag{13.103}$$

13.4.3 Least Squares Method

Least squares methods have become popular for first order systems of differential equations in computational fluid dynamics and computational electrodynamics [231, 232].

The L2-norm of the residual (13.94) is given by the integral

$$S = \int_V dV\, R(\mathbf{r})^2. \tag{13.104}$$

It is minimized by solving the equations

$$0 = \frac{\partial S}{\partial u_j} = 2 \int_V dV\, \frac{\partial R}{\partial u_j} R(\mathbf{r}) \tag{13.105}$$

which is equivalent to choosing the weight functions

$$w_j(\mathbf{r}) = \frac{\partial R}{\partial u_j} R(\mathbf{r}) = \frac{\partial}{\partial u_j} \mathcal{T}\left[\sum_i u_i N_i(\mathbf{r})\right] \tag{13.106}$$

or for a linear differential operator simply

$$w_j(\mathbf{r}) = \mathcal{T}\left[N_j(\mathbf{r})\right]. \tag{13.107}$$

Advantages of the least squares method are that boundary conditions can be incorporated into the residual and that S provides a measure for the quality of the solution which can be used for adaptive grid size control. On the other hand S involves a differential operator of higher order and therefore much smoother trial functions are necessary.

13.4.4 Galerkin Method

Galerkin's widely used method [228, 233] chooses the basis functions as weight functions

$$w_j(\mathbf{r}) = N_j(\mathbf{r}) \tag{13.108}$$

and solves the following system of equations

$$\int dV\, N_j(\mathbf{r})\mathcal{T}\left[\sum_i u_i N_i(\mathbf{r})\right] - \int dV\, N_j(\mathbf{r}) f(\mathbf{r}) = 0 \tag{13.109}$$

or in the simpler linear case

$$\sum u_i \int_V dV\, N_j(\mathbf{r})\mathcal{T}[N_i(\mathbf{r})] = \int_V dV\, N_j(\mathbf{r}) f(\mathbf{r}). \tag{13.110}$$

13.5 Spectral and Pseudo-spectral Methods

Spectral methods use basis functions which are nonzero over the whole domain, the trial functions being mostly polynomials or Fourier sums [234]. They can be used to solve ordinary as well as partial differential equations. The combination of a spectral method with the point collocation method is also known as pseudo-spectral method.

13.5.1 Fourier Pseudo-spectral Methods

Linear differential operators become diagonal in Fourier space. Combination of Fourier series expansion and point collocation leads to equations involving a discrete Fourier transformation, which can be performed very efficiently with the Fast Fourier Transform methods.

For simplicity we consider only the one-dimensional case. We choose equidistant collocation points

$$x_m = m\Delta x \quad m = 0, 1 \ldots M-1 \tag{13.111}$$

and expansion functions

$$N_j(x) = \mathrm{e}^{\mathrm{i}k_j x} \quad k_j = \frac{2\pi}{M\Delta x} j \quad j = 0, 1 \ldots M-1. \tag{13.112}$$

For a linear differential operator

$$\mathcal{L}[\mathrm{e}^{\mathrm{i}k_j x}] = l(k_j)\mathrm{e}^{\mathrm{i}k_j x} \tag{13.113}$$

and the condition on the residual becomes

$$0 = R_m = \sum_{j=0}^{M-1} u_j\, l(k_j)\mathrm{e}^{\mathrm{i}k_j x_m} - f(x_m) \tag{13.114}$$

or

$$f(x_m) = \sum_{j=0}^{M-1} u_j\, l(k_j) e^{i2\pi mj/M} \tag{13.115}$$

which is nothing but a discrete Fourier back transformation (Sect. 7.2 and Eq. 7.21) which can be inverted to give

$$u_j\, l(k_j) = \frac{1}{N} \sum_{m=0}^{M-1} f(x_m)\, e^{-i2\pi mj/M}. \tag{13.116}$$

Instead of exponential expansion functions, sine and cosine functions can be used to satisfy certain boundary conditions, for instance to solve the Poisson equation within a cube (Sect. 5.1.2 in Vol. 2) .

13.5.2 Example: Polynomial Approximation

Let us consider the initial value problem (Fig. 13.5)

$$\frac{d}{dx} u(x) - u(x) = 0 \quad u(0) = 1 \quad \text{for } 0 \le x \le 1 \tag{13.117}$$

with the well known solution

$$u(x) = e^x. \tag{13.118}$$

We choose a polynomial trial function with the proper initial value

$$\tilde{u}(x) = 1 + u_1 x + u_2 x^2. \tag{13.119}$$

The residual is

$$R(x) = u_1 + 2u_2 x - \left(1 + u_1 x + u_2 x^2\right) = (u_1 - 1) + (2u_2 - u_1)x - u_2 x^2. \tag{13.120}$$

Point Collocation Method

For our example we need two collocation points to obtain two equations for the two unknowns $u_{1,2}$. We choose $x_1 = 0$, $x_2 = \frac{1}{2}$. Then we have to solve the equations

$$R(x_1) = u_1 - 1 = 0 \tag{13.121}$$

$$R(x_2) = \frac{1}{2} u_1 + \frac{3}{4} u_2 - 1 = 0 \tag{13.122}$$

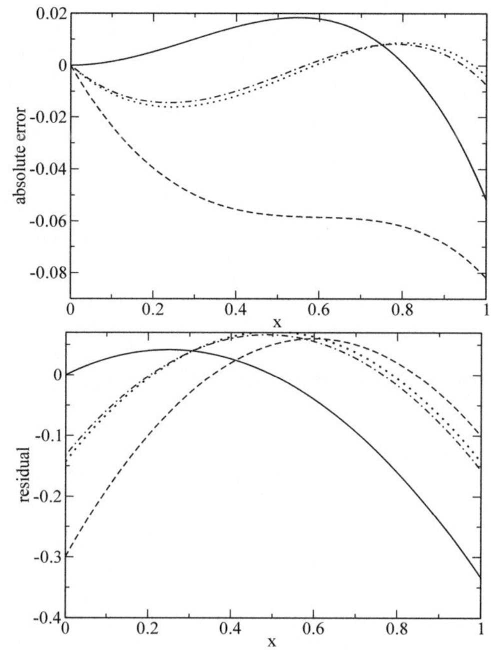

Fig. 13.5 Approximate solution of a simple differential equation. The initial value problem $\frac{\mathrm{d}}{\mathrm{d}x}u(x) - u(x) = 0 \quad u(0) = 1 \quad \text{for } 0 \le x \le 1$ is approximately solved with a polynomial trial function $\tilde{u}(x) = 1 + u_1 x + u_2 x^2$. The parameters $u_{1,2}$ are optimized with the method of weighted residuals using point collocation (full curve), sub-domain collocation (dotted curve), Galerkin's method (dashed curve) and least squares (dash-dotted curve). The absolute error $\tilde{u}(x) - \mathrm{e}^x$ (top) and the residual $R(x) = \frac{\mathrm{d}}{\mathrm{d}x}\tilde{u}(x) - \tilde{u}(x) = (u_1 - 1) + (2u_2 - u_1)x - u_2 x^2$ both are smallest for the least squares and sub-domain collocation methods

which gives

$$u_1 = 1 \quad u_2 = \frac{2}{3} \tag{13.123}$$

$$u_c = 1 + x + \frac{2}{3}x^2. \tag{13.124}$$

Sub-domain Method

We need two sub-domains to obtain two equations for the two unknowns $u_{1,2}$. We choose $V_1 = \{x, 0 < x < \frac{1}{2}\}$, $V_2 = \{x, \frac{1}{2} < x < 1\}$. Integration gives

$$R_1 = \frac{3}{8}u_1 + \frac{5}{24}u_2 - \frac{1}{2} = 0 \tag{13.125}$$

$$R_2 = \frac{1}{8}u_1 + \frac{11}{24}u_2 - \frac{1}{2} = 0 \tag{13.126}$$

$$u_1 = u_2 = \frac{6}{7} \tag{13.127}$$

$$u_{sdc} = 1 + \frac{6}{7}x + \frac{6}{7}x^2. \tag{13.128}$$

Galerkin Method

Galerkin's method uses the weight functions $w_1(x) = x$, $w_2(x) = x^2$. The equations

$$\int_0^1 dx\, w_1(x)R(x) = \frac{1}{6}u_1 + \frac{5}{12}u_2 - \frac{1}{2} = 0 \tag{13.129}$$

$$\int_0^1 dx\, w_2(x)R(x) = \frac{1}{12}u_1 + \frac{3}{10}u_2 - \frac{1}{3} = 0 \tag{13.130}$$

have the solution

$$u_1 = \frac{8}{11} \quad u_2 = \frac{10}{11} \tag{13.131}$$

$$u_G = 1 + \frac{8}{11}x + \frac{10}{11}x^2. \tag{13.132}$$

Least Squares Method

The integral of the squared residual

$$S = \int_0^1 dx\, R(x)^2 = 1 - u_1 - \frac{4}{3}u_2 + \frac{1}{3}u_1^2 + \frac{1}{2}u_1u_2 + \frac{8}{15}u_2^2 \tag{13.133}$$

is minimized by solving

$$\frac{\partial S}{\partial u_1} = \frac{2}{3}u_1 + \frac{1}{2}u_2 - 1 = 0 \tag{13.134}$$

$$\frac{\partial S}{\partial u_2} = \frac{1}{2}u_1 + \frac{16}{15}u_2 - \frac{4}{3} = 0 \tag{13.135}$$

which gives

$$u_1 = \frac{72}{83} \quad u_2 = \frac{70}{83} \tag{13.136}$$

$$u_{LS} = 1 + \frac{72}{83}x + \frac{70}{83}x^2. \tag{13.137}$$

13.6 Finite Elements

The method of finite elements (FEM) is a very flexible method to discretize partial differential equations [235, 236]. It is rather dominant in a variety of engineering sciences. Usually the expansion functions N_i are chosen to have compact support. The integration volume is divided into disjoint sub-volumes

$$V = \bigcup_{i=1}^{r} V_i \qquad V_i \bigcap V_{i'} = \emptyset \qquad \forall i \neq i'. \tag{13.138}$$

The $N_i(\mathbf{x})$ are piecewise continuous polynomials which are nonzero only inside V_i and a few neighbor cells.

13.6.1 One-Dimensional Elements

In one dimension the domain is an interval $V = \{x; a \le x \le b\}$ and the sub-volumes are small sub-intervals $V_i = \{x; x_i \le x \le x_{i+1}\}$. The one-dimensional mesh is the set of nodes $\{a = x_0, x_1 \ldots x_r = b\}$. Piecewise linear basis functions (Fig. 13.6) are in the one-dimensional case given by

$$N_i(x) = \begin{cases} \frac{x_{i+1}-x}{x_{i+1}-x_i} & \text{for } x_i < x < x_{i+1} \\ \frac{x-x_{i-1}}{x_i-x_{i-1}} & \text{for } x_{i-1} < x < x_i \\ 0 & \text{else} \end{cases} \tag{13.139}$$

and the derivatives are (except at the nodes x_i)

$$N_i'(x) = \begin{cases} -\frac{1}{x_{i+1}-x_i} & \text{for } x_i < x < x_{i+1} \\ \frac{1}{x_i-x_{i-1}} & \text{for } x_{i-1} < x < x_i \\ 0 & \text{else} \end{cases} . \tag{13.140}$$

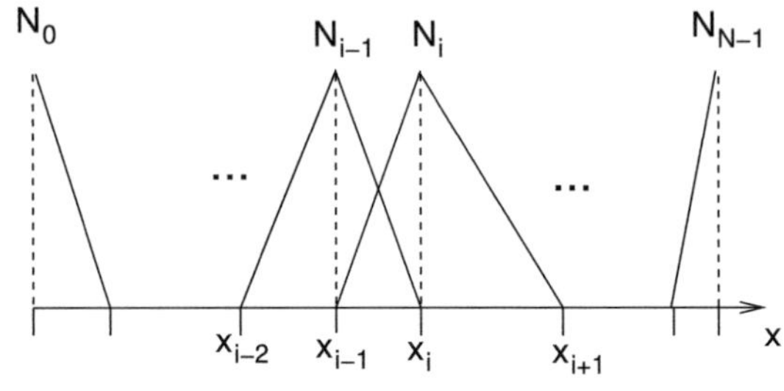

Fig. 13.6 Finite elements in one dimension. The basis functions N_i are piecewise continuous polynomials and have compact support. In the simplest case they are composed of two linear functions over the sub-intervals $x_{i-1} \leq x \leq x_i$ and $x_i \leq x \leq x_{i+1}$

13.6.2 Two- and Three-Dimensional Elements

In two dimensions the mesh is defined by a finite number of points $(x_i, y_i) \in V$ (the nodes of the mesh). There is considerable freedom in the choice of these points and they need not be equally spaced.

Triangulation

The nodes can be regarded as forming the vertices of a triangulation[9] of the domain V (Fig. 13.7).

The piecewise linear basis function in one dimension (13.139) can be generalized to the two-dimensional case by constructing functions $N_i(x, y)$ which are zero at all nodes except (x_i, y_i)

$$N_i(x_j, y_j) = \delta_{i,j}. \tag{13.141}$$

These functions are linear over each triangle which contains the vertex i and can be combined as the sum of small pyramids (Fig. 13.8). Let one of the triangles be denoted by its three vertices as T_{ijk}.[10] The corresponding linear function then is

$$n_{ijk}(x, y) = \alpha + \beta_x(x - x_i) + \beta_y(y - y_i) \tag{13.142}$$

where the coefficients follow from the conditions

$$n_{ijk}(x_i, y_i) = 1 \quad n_{ijk}(x_j, y_j) = n_{ijk}(x_k, y_k) = 0 \tag{13.143}$$

as

$$\alpha = 1 \quad \beta_x = \frac{y_j - y_k}{2A_{ijk}} \quad \beta_y = \frac{x_k - x_j}{2A_{ijk}} \tag{13.144}$$

with

---

[9] The triangulation is not determined uniquely by the nodes.

[10] The order of the indices does matter.

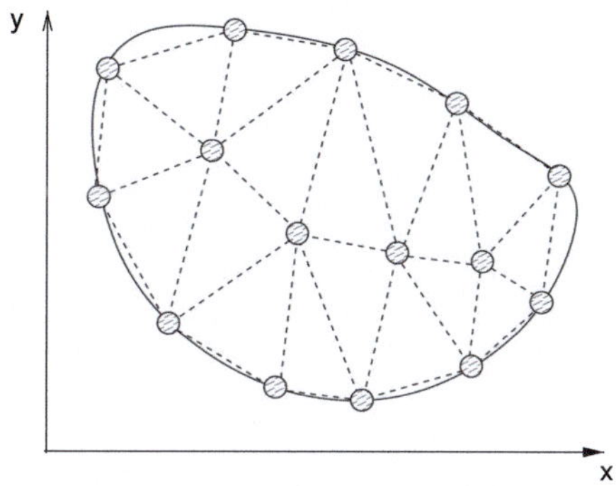

Fig. 13.7 Triangulation of a two-dimensional domain. A two-dimensional mesh is defined by a set of node points which can be regarded to form the vertices of a triangulation

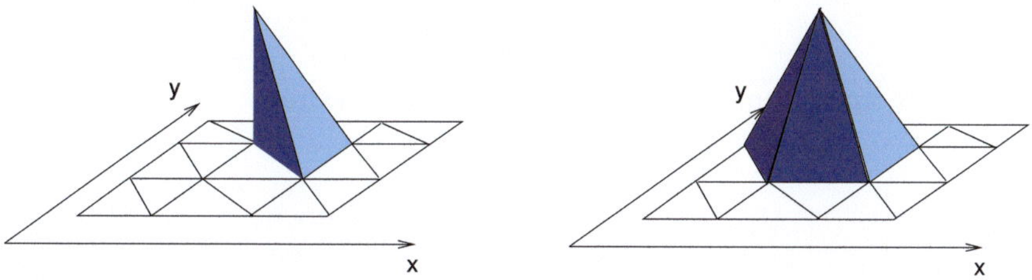

Fig. 13.8 Finite elements in two dimensions. **Left:** Small pyramids built upon one of the triangles that contains (x_i, y_i); **Right:** The simplest finite element in two dimensions is a piecewise linear function $N_i(x, y)$ which is non-vanishing only at one node (x_i, y_i)

$$A_{ijk} = \frac{1}{2}\det\begin{vmatrix} x_j - x_i & x_k - x_i \\ y_j - y_i & y_k - y_i \end{vmatrix} \tag{13.145}$$

which, apart from sign, is the area of the triangle T_{ijk}. The basis function N_i now is given by

$$N_i(x, y) = \begin{cases} n_{ijk}(x, y) & (x, y) \in T_{ijk} \\ 0 & \text{else} \end{cases}$$

In three dimensions we consider tetrahedrons (Fig. 13.9) instead of triangles. The corresponding linear function of three arguments has the form

$$n_{i,j,k,l}(x, y, z) = \alpha + \beta_x(x - x_i) + \beta_y(y - y_i) + \beta_z(z - z_i) \tag{13.146}$$

and from the conditions $n_{i,j,k,l}(x_i, y_i, z_i) = 1$ and $n_{i,j,k,l} = 0$ on all other nodes we find (an algebra program is helpful at that point)

$$\alpha = 1$$

$$\beta_x = \frac{1}{6V_{ijkl}} \det\begin{vmatrix} y_k - y_j & y_l - y_j \\ z_k - z_j & z_l - z_j \end{vmatrix}$$

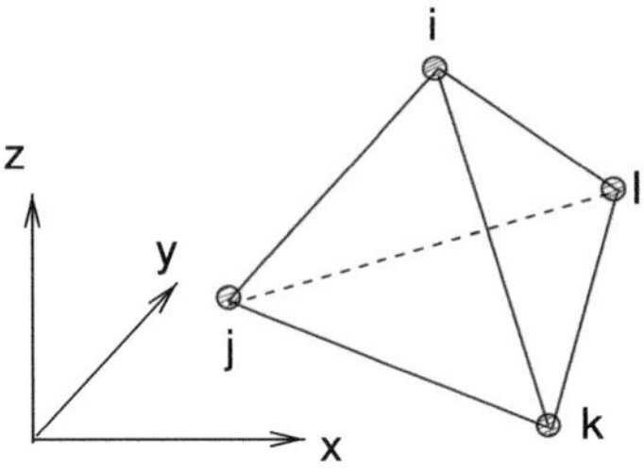

Fig. 13.9 Tetrahedron. The tetrahedron is the three-dimensional case of an Euclidean simplex, i.e. the simplest polytop

$$\beta_y = \frac{1}{6V_{ijkl}} \det \begin{vmatrix} z_k - z_j & z_l - z_j \\ x_k - x_j & x_l - x_j \end{vmatrix}$$

$$\beta_z = \frac{1}{6V_{ijkl}} \det \begin{vmatrix} x_k - x_j & x_l - x_j \\ y_k - y_j & y_l - y_j \end{vmatrix} \tag{13.147}$$

where V_{ijkl} is, apart from sign, the volume of the tetrahedron

$$V_{ijkl} = \frac{1}{6} \det \begin{vmatrix} x_j - x_i & x_k - x_i & x_l - x_i \\ y_j - y_i & y_k - y_i & y_l - y_i \\ z_j - z_i & z_k - z_i & z_l - z_i \end{vmatrix}. \tag{13.148}$$

Rectangular Elements

For a rectangular grid rectangular elements offer a practical alternative to triangles. Since equations for four nodes have to be fulfilled, the basic element needs four parameters, which is the case for a bilinear expression. Let us denote one of the rectangles which contains the vertex i as $R_{i,j,k,l}$. The other three edges are

$$(x_j, y_j) = (x_i + b_x, y_i) \quad (x_k, y_k) = (x_i, y_i + b_y) \quad (x_l, y_l) = (x_i + b_x, y_i + b_y) \tag{13.149}$$

where $b_x = \pm h_x, b_y = \pm h_y$ corresponding to the four rectangles with the common vertex i (Fig. 13.10).

The bilinear function (Fig. 13.11) corresponding to R_{ijkl} is

$$n_{i,j,k,l}(x, y) = \alpha + \beta(x - x_i) + \gamma(y - y_i) + \eta(x - x_i)(y - y_i). \tag{13.150}$$

It has to fulfill

$$n_{i,j,k,l}(x_i, y_i) = 1 \quad n_{i,j,k,l}(x_j, y_j) = n_{i,j,k,l}(x_k, y_k) = n_{i,j,k,l}(x_l, y_l) = 0 \tag{13.151}$$

from which we find

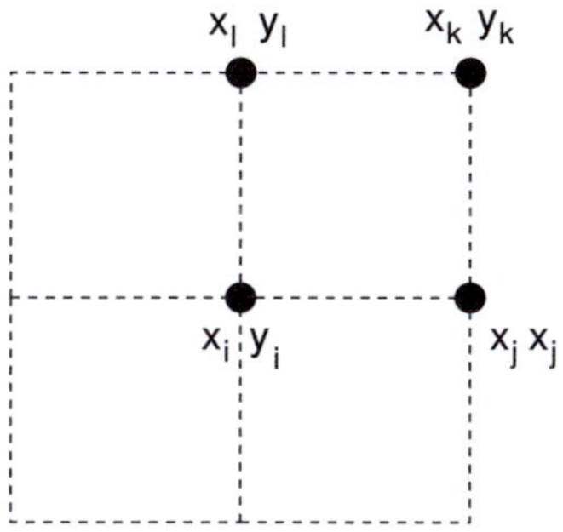

Fig. 13.10 Rectangular elements around one vertex. The basis function N_i is a bilinear function on each of the four rectangles containing the vertex (x_i, y_i)

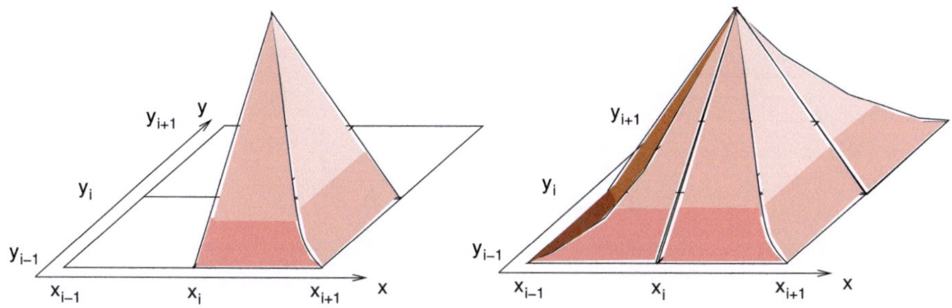

Fig. 13.11 Bilinear elements on a rectangular grid. **Left:** Bilinear function; **Right:** Basis function $N_i(x, y)$ is a piecewise bilinear function vanishing at all nodes except (x_i, y_i)

$$\alpha = 1 \quad \beta = -\frac{1}{b_x} \quad \gamma = -\frac{1}{b_y} \quad \eta = \frac{1}{b_x b_y} \tag{13.152}$$

$$n_{i.j.k.l}(x, y) = 1 - \frac{x - x_i}{b_x} - \frac{y - y_i}{b_y} + \frac{(x - x_i)}{b_x}\frac{(y - y_i)}{b_y}. \tag{13.153}$$

The basis function centered at node i then is

$$N_i(x, y) = \begin{cases} n_{i,j,k,l}(x, y) & (x, y) \in R_{i,j,k,l} \\ 0 & \text{else} \end{cases}. \tag{13.154}$$

Generalization to a three-dimensional grid is straightforward (Fig. 13.12). We denote one of the eight cuboids containing the node (x_i, y_i, z_i) as $C_{i,j_1\ldots j_7}$ with $(x_{j_1}, y_{j_1}, z_{j_1}) = (x_i + b_x, y_i, z_i) \ldots (x_{j_7}, y_{j_7}, z_{j_7}) = (x_i + b_x, y_i + b_y, z_i + b_z)$. The corresponding trilinear function is

$$n_{i,j_1\ldots j_7} = 1 - \frac{x - x_i}{b_x} - \frac{y - y_i}{b_y} - \frac{z - z_i}{b_z}$$

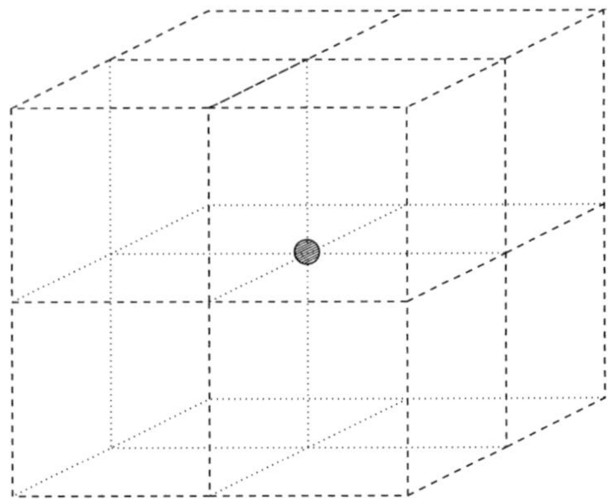

Fig. 13.12 Three-dimensional rectangular grid. The basis function N_i is trilinear on each of the eight cuboids containing the vertex i. It vanishes on all nodes except (x_i, y_i, z_i)

$$+\frac{(x-x_i)}{b_x}\frac{(y-y_i)}{b_y}+\frac{(x-x_i)}{b_x}\frac{(z-z_i)}{b_z}+\frac{(z-z_i)}{b_z}\frac{(y-y_i)}{b_y}$$

$$-\frac{(x-x_i)}{b_x}\frac{(y-y_i)}{b_y}\frac{(z-z_i)}{b_z}. \tag{13.155}$$

13.6.3 One-Dimensional Galerkin FEM

As an example we consider the one-dimensional linear differential equation (13.5)

$$\left(a\frac{\partial^2}{\partial x^2}+b\frac{\partial}{\partial x}+c\right)u(x)=f(x) \tag{13.156}$$

in the domain $0 \le x \le 1$ with boundary conditions

$$u(0)=u(1)=0. \tag{13.157}$$

We use the basis functions from (13.139) on a one-dimensional grid with

$$x_{i+1}-x_i=h_i \tag{13.158}$$

and apply the Galerkin method [237]. The boundary conditions require

$$u_0=u_{N-1}=0. \tag{13.159}$$

The weighted residual is

$$0 = R_j = \sum_i u_i \int_0^1 dx\, N_j(x) \left(a \frac{\partial^2}{\partial x^2} + b \frac{\partial}{\partial x} + c \right) N_i(x) - \int_0^1 dx\, N_j(x) f(x). \tag{13.160}$$

First we integrate

$$\int_0^1 N_j(x) N_i(x)\, dx = \int_{x_{i-1}}^{x_{i+1}} N_j(x) N_i(x) dx = \begin{cases} \frac{h_i + h_{i-1}}{3} & j = i \\ \frac{h_i}{6} & j = i+1 \\ \frac{h_{i-1}}{6} & j = i-1 \\ 0 & |i-j| > 1 \end{cases} \tag{13.161}$$

Integration of the first derivative gives

$$\int_0^1 dx\, N_j(x) N_i'(x) = \begin{cases} 0 & j = i \\ \frac{1}{2} & j = i-1 \\ -\frac{1}{2} & j = i+1 \\ 0 & \text{else} \end{cases}. \tag{13.162}$$

For the second derivative partial integration gives

$$\int_0^1 dx\, N_j(x) \frac{\partial^2}{\partial x^2} N_i(x)$$

$$= N_j(1) N_i'(1-\varepsilon) - N_j(0) N_i'(0+\varepsilon) - \int_0^1 dx\, N_j'(x) N_i'(x) \tag{13.163}$$

where the first two summands are zero due to the boundary conditions. Since N_i and N_i' are nonzero only for $x_{i-1} < x < x_{i+1}$ we find

$$\int_0^1 dx\, N_j(x) \frac{\partial^2}{\partial x^2} N_i(x) = - \int_{x_{i-1}}^{x_{i+1}} dx\, N_j'(x)\, N_i'(x) = \begin{cases} \frac{1}{h_{i-1}} & j = i-1 \\ -\frac{1}{h_i} - \frac{1}{h_{i-1}} & i = j \\ \frac{1}{h_i} & j = i+1 \\ 0 & \text{else} \end{cases}. \tag{13.164}$$

Integration of the last term in (13.160) gives

$$\int_0^1 dx\, N_j(x) f(x) = \int_{x_{i-1}}^{x_{i+1}} dx\, N_j(x) f(x)$$

$$= \int_{x_{j-1}}^{x_j} dx\, \frac{x - x_{j-1}}{x_j - x_{j-1}} f(x) + \int_{x_j}^{x_{j+1}} dx\, \frac{x_{j+1} - x}{x_{j+1} - x_j} f(x). \tag{13.165}$$

Applying the trapezoidal rule[11] for both integrals we find

$$\int_{x_{j-1}}^{x_{j+1}} dx\, N_j(x) f(x) \approx f(x_j) \frac{h_j + h_{j-1}}{2}. \tag{13.166}$$

The discretized equation finally reads

$$\begin{aligned} a & \left\{ \frac{1}{h_{j-1}} u_{j-1} - \left(\frac{1}{h_j} + \frac{1}{h_{j-1}} \right) u_j + \frac{1}{h_j} u_{j+1} \right\} \\ & + b \left\{ -\frac{1}{2} u_{j-1} + \frac{1}{2} u_{j+1} \right\} \\ & + c \left\{ \frac{h_{j-1}}{6} u_{j-1} + \frac{h_j + h_{j-1}}{3} u_j + \frac{h_j}{6} u_{j+1} \right\} \\ & = f(x_j) \frac{h_j + h_{j-1}}{2} \end{aligned} \tag{13.167}$$

which can be written in matrix notation as

$$A\mathbf{u} = B\mathbf{f} \tag{13.168}$$

where the matrix A is tridiagonal as a consequence of the compact support of the basis functions

$$A = a \begin{pmatrix} -\frac{1}{h_1} - \frac{1}{h_0}, & \frac{1}{h_1} & & & \\ & \ddots & & & \\ & \frac{1}{h_{j-1}}, & -\frac{1}{h_j} - \frac{1}{h_{j-1}}, & \frac{1}{h_j} & \\ & & & \ddots & \\ & & & \frac{1}{h_{N-3}}, & -\frac{1}{h_{N-2}} - \frac{1}{h_{N-3}} \end{pmatrix}$$

$$+ b \begin{pmatrix} 0, & \frac{1}{2} & & \\ & \ddots & & \\ & -\frac{1}{2}, & 0, & \frac{1}{2} \\ & & & \ddots \\ & & & -\frac{1}{2}, & 0 \end{pmatrix}$$

[11] Higher accuracy can be achieved, for instance, by Gaussian integration.

$$+c\begin{pmatrix} \frac{(h_1+h_0)}{3}, & \frac{h_1}{6} & & & & \\ & \ddots & & & & \\ & \frac{h_{j-1}}{6}, & \frac{(h_j+h_{j-1})}{3}, & \frac{h_j}{6} & & \\ & & & \ddots & & \\ & & & \frac{h_{N-3}}{6}, & \frac{(h_{N-2}+h_{N-3})}{3}, \end{pmatrix}$$

$$B = \begin{pmatrix} \frac{h_0+h_1}{2} & & & & \\ & \ddots & & & \\ & & \frac{h_{j-1}+h_j}{2} & & \\ & & & \ddots & \\ & & & & \frac{h_{N-2}+h_{N-3}}{2} \end{pmatrix}. \tag{13.169}$$

For equally spaced nodes $h_i = h_{i-1} = h$ and after division by h (13.169) reduces to a system of equations where the derivatives are replaced by finite differences (13.20)

$$\left\{a\frac{1}{h^2}M_2 + b\frac{1}{h}M_1 + cM_0\right\}\mathbf{u} = \mathbf{f} \tag{13.170}$$

with the so called consistent mass matrix

$$M_0 = \begin{pmatrix} \ddots & & & & \\ \ddots & \ddots & \ddots & & \\ & \frac{1}{6} & \frac{2}{3} & \frac{1}{6} & \\ & & \ddots & \ddots & \ddots \\ & & & & \ddots \end{pmatrix} \tag{13.171}$$

and the derivative matrices

$$M_1 = \begin{pmatrix} \ddots & & & & \\ \ddots & \ddots & \ddots & & \\ & -\frac{1}{2} & 0 & \frac{1}{2} & \\ & & \ddots & \ddots & \ddots \\ & & & & \ddots \end{pmatrix} \quad M_2 = \begin{pmatrix} \ddots & & & & \\ \ddots & \ddots & \ddots & & \\ & 1 & -2 & 1 & \\ & & \ddots & \ddots & \ddots \\ & & & & \ddots \end{pmatrix}. \tag{13.172}$$

The vector $\mathbf{u}$ is replaced by

$$M_0\mathbf{u} = \left[1 + \frac{1}{6}M_2\right]\mathbf{u}. \tag{13.173}$$

Within the framework of the finite differences method

$$u_j + \frac{1}{6}\left(u_{j-1} - 2u_j + u_{j+1}\right) = u_j + \frac{h^2}{6}\left(\frac{d^2u}{dx^2}\right)_j + O(h^4) \tag{13.174}$$

hence replacing it by u_j (this is called mass lumping) introduces an error of the order $O(h^2)$.

13.7 Boundary Element Method

The boundary element method (BEM) [238, 239] is a method for linear partial differential equations which can be brought into boundary integral form[12] like Laplace's equation (Chap. 5 in Vol. 2)[13]

$$-\triangle\Phi(\mathbf{r}) = 0 \tag{13.175}$$

for which the fundamental solution

$$\triangle G(\mathbf{r}, \mathbf{r}') = -\delta(\mathbf{r} - \mathbf{r}')$$

is given by

$$G(\mathbf{r} - \mathbf{r}') = \frac{1}{4\pi|\mathbf{r} - \mathbf{r}'|} \quad \text{in three dimensions} \tag{13.176}$$

$$G(\mathbf{r} - \mathbf{r}') = \frac{1}{2\pi}\ln\frac{1}{|\mathbf{r} - \mathbf{r}'|} \quad \text{in two dimensions.} \tag{13.177}$$

We apply Gauss's theorem to the expression [240]

$$\begin{aligned} &\operatorname{div}\left[G(\mathbf{r} - \mathbf{r}')\operatorname{grad}(\Phi(\mathbf{r})) - \Phi(\mathbf{r})\operatorname{grad}(G(\mathbf{r} - \mathbf{r}'))\right] \\ &\quad = -\Phi(\mathbf{r})\triangle(G(\mathbf{r} - \mathbf{r}')). \end{aligned} \tag{13.178}$$

Integration over a volume V gives

$$\begin{aligned} &\oint_{\partial V} dA\left(G(\mathbf{r} - \mathbf{r}')\frac{\partial}{\partial n}(\Phi(\mathbf{r})) - \Phi(\mathbf{r})\frac{\partial}{\partial n}(G(\mathbf{r} - \mathbf{r}'))\right) \\ &\quad = -\int_V dV\,\left[\Phi(\mathbf{r})\triangle(G(\mathbf{r} - \mathbf{r}'))\right] = \Phi(\mathbf{r}'). \end{aligned} \tag{13.179}$$

---

[12] This is only possible if the fundamental solution or Green's function is available.

[13] The minus sign is traditionally used.

This integral equation determines the potential self-consistently by its value and normal derivative on the surface of the cavity. It can be solved numerically by dividing the surface into a finite number of boundary elements. The resulting system of linear equations often has smaller dimension than corresponding finite element approaches. However, the coefficient matrix is in general full and not necessarily symmetric.

13.8 Neural Networks as Solvers for Differential Equations

Artificial neural networks (ANN) can be used to provide an analytic and differentiable expression as a trial function to solve ordinary and partial differential equations [241–247]. The parameters of the network have to be optimized to find an approximate solution $u(\mathbf{r})$ to the equation[14]

$$\mathcal{T}\,[u_{sol}(\mathbf{r})] = f(\mathbf{r}) \quad \mathbf{r} \in V \tag{13.180}$$

13.90 with boundary conditions 13.91

$$\mathcal{B}[u_{sol}(\mathbf{r})] = g(\mathbf{r}) \quad \mathbf{r} \in \partial V. \tag{13.181}$$

Usually the collocation method is used to solve the problem approximately at a set of sample points by minimizing the mean squared error of a trial function with respect to the network parameters (weights and biases)[15]

$$L = \frac{1}{N_{\text{samples}}} \sum_{\mathbf{r}_i} \frac{1}{2} \left(\mathcal{T}\,[u_T(\mathbf{r}_i)] - f(\mathbf{r}_i)\right)^2 . \tag{13.182}$$

The sample points $\mathbf{r}_i$ may be chosen on a regular grid or randomly (especially for problems of high dimension). Boundary conditions can be taken into account by choosing proper trial functions [241, 242]

$$u_T(\mathbf{r}) = \alpha(\mathbf{r}) + \beta(\mathbf{r}) u_{\text{NN}}(\mathbf{r}) \tag{13.183}$$

with

$$\mathcal{B}[\alpha(\mathbf{r})] = g(\mathbf{r}), \quad \mathcal{B}[\beta(\mathbf{r})] = 0 \quad \mathbf{r} \in \partial V \tag{13.184}$$

or by adding penalty terms at a set of boundary points $\mathbf{r}_{bi}$ to the mean squared error

$$L_{bc} = \frac{1}{N_{\text{points}}} \sum_{\mathbf{r}_{bi}} \frac{1}{2} \left(\mathcal{B}[u_T(\mathbf{r}_{bi})] - g(\mathbf{r}_{bi})\right)^2 . \tag{13.185}$$

[14] The simpler task of fitting a given function is discussed in 12.6.

[15] The prefactor 1/2 is only for convenience.

This approach has the advantage that the derivatives do not have to be discretized since derivatives of the trial function are analytically available and can be calculated by the network automatically (algorithmic or automatic differentiation, see Sect. 3.6)

13.8.1 Derivatives of the Trial Function

As a first example we consider a simple feed-forward network Sect. 12.6 with one hidden layer to solve a one-dimensional first order problem (Fig. 13.13). The output function $u(x)$ is a chain of three nested functions

$$u(x) = b_o + \sum_{n=1}^{\mathrm{N_{nodes}}} w_{on} h_n(x) \tag{13.186}$$

$$h_n(x) = \sigma(s_n) \tag{13.187}$$

$$s_n(x) = w_n x + b_n \tag{13.188}$$

with weight and bias parameters w_{on}, w_n, b_o, b_n and the activation function σ, which will be taken as the hyperbolic tangent in the following examples

$$\sigma(s_n) = \tanh(s_n). \tag{13.189}$$

Gradients with respect to the parameters but also derivative values can be obtained by variation of the trial function at the sample points x_i

$$\delta u^{(i)} = \sum \left[\delta w_{on} h_n^{(i)} + w_{on} \delta h_n^{(i)}\right] + \delta b_o \tag{13.190}$$

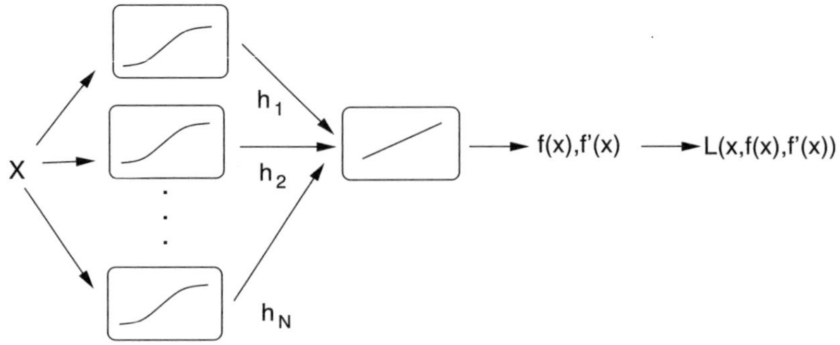

Fig. 13.13 Simple feed forward network for solution of a first order ODE

$$\delta h_n^{(i)} = \sigma'(s_n^{(i)})\delta s_n^{(i)} \tag{13.191}$$

$$\delta s_n^{(i)} = \delta w_n x^{(i)} + \delta b_n + w_n \delta x^{(i)}. \tag{13.192}$$

Varying only one parameter at a time and keeping all other fixed we obtain

$$u' = \frac{\partial u}{\partial x} = \sum w_{on}\delta h_n = \sum w_{on}\sigma'(s_n)\delta s_n = \sum_n w_{on}\sigma'(s_n)w_n \tag{13.193}$$

$$\frac{\partial u}{\partial b_o} = 1 \tag{13.194}$$

$$\frac{\partial u}{\partial w_{on}} = h_n \tag{13.195}$$

$$\frac{\partial u}{\partial b_n} = w_{on}\delta h_n = w_{on}\sigma'(s_n)\delta s_n = w_{on}\sigma'(s_n) \tag{13.196}$$

$$\frac{\partial u}{\partial w_n} = w_{on}\delta h_n = w_{on}\sigma'(s_n)\delta s_n = w_{on}\delta h_n = w_{on}\sigma'(s_n)x. \tag{13.197}$$

Gradients of the derivative can be similarly obtained by further variation

$$\delta u' = \sum(\delta w_{on} w_n + \delta w_n w_{on})\sigma'(s_n) + w_{on}w_n\sigma''(s_n)\left(\delta w_n x_i + \delta b_n + w_n \delta x_i\right) \tag{13.198}$$

and collecting the prefactors of the parameter variations

$$\frac{\partial u'}{\partial b_o} = 0 \tag{13.199}$$

$$\frac{\partial u'}{\partial w_{on}} = w_n\sigma'(s_n) \tag{13.200}$$

$$\frac{\partial u'}{\partial b_n} = w_{on}w_n\sigma''(s_n) \tag{13.201}$$

$$\frac{\partial u'}{\partial w_n} = w_{on}\sigma'(s_n) + w_{on}w_n\sigma''(s_n)x \tag{13.202}$$

$$u'' = \frac{\partial u'}{\partial x} = w_{on}w_n^2\sigma''(s_n). \tag{13.203}$$

For a more complicated model with two hidden layers (Fig. 13.14) function values and derivatives can be calculated by forward propagation according to the equations (see also Sect. 12.7 for a general feed forward network)

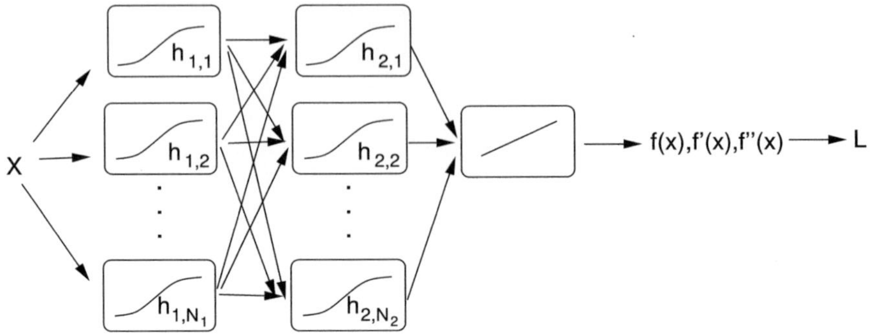

Fig. 13.14 Network with two hidden layers

$$\mathbf{s}_1 = \mathbf{b}_1 + \mathbf{w}_1 x \tag{13.204}$$

$$\mathbf{s}_1' = \mathbf{w}_1 \tag{13.205}$$

$$\mathbf{s}_1'' = 0 \tag{13.206}$$

$$\mathbf{h}_1 = \sigma(\mathbf{s}_1) \tag{13.207}$$

$$\mathbf{h}_1' = \sigma'(\mathbf{s}_1) \odot \mathbf{s}_1' = \sigma'(\mathbf{s}_1) \odot \mathbf{w}_1 \tag{13.208}$$

$$\mathbf{h}_1'' = \sigma''(\mathbf{s}_1) \odot \mathbf{s}_1' \odot \mathbf{s}_1' + \sigma'(\mathbf{s}_1) \odot \mathbf{s}_1'' = \sigma''(\mathbf{s}_1) \odot \mathbf{w}_1 \odot \mathbf{w}_1 \tag{13.209}$$

$$\mathbf{s}_2 = \mathbf{b}_2 + W_2 \mathbf{h}_1 \tag{13.210}$$

$$\mathbf{s}_2' = W_2 \mathbf{h}_1' \tag{13.211}$$

$$\mathbf{s}_2'' = W_2 \mathbf{h}_1'' \tag{13.212}$$

$$\mathbf{h}_2 = \sigma(\mathbf{s}_2) \tag{13.213}$$

$$\mathbf{h}_2' = \sigma'(\mathbf{s}_2) \odot \mathbf{s}_2' \tag{13.214}$$

$$\mathbf{h}_2'' = \sigma''(\mathbf{s}_2) \odot \mathbf{s}_2' \odot \mathbf{s}_2' + \sigma'(\mathbf{s}_2) \odot \mathbf{s}_2'' \tag{13.215}$$

$$u = b_o + \mathbf{w}_o \mathbf{h}_2 \tag{13.216}$$

$$u' = \mathbf{w}_o \mathbf{h}_2' \tag{13.217}$$

$$u'' = \mathbf{w}_o \mathbf{h}_2''. \tag{13.218}$$

Gradients are obtained with the help of the variations

$$\delta u = \delta b_o + \delta \mathbf{w}_o \mathbf{h}_2 + \mathbf{w}_o \sigma'(\mathbf{s}_2) \odot (\delta \mathbf{b}_2 + \delta W_2 \mathbf{h}_1 + W_2 \sigma'(s_1) \odot (\delta b_1 + \delta w_1 x)) \tag{13.219}$$

$$\delta u' = \delta \mathbf{w}_o \mathbf{h}_2' c$$

$$+\mathbf{w}_o \sigma'(\mathbf{s}_2) \odot [\delta W_2 \mathbf{h}_1' + W_2(\sigma'(s_1) \odot \delta w_1 + \sigma''(s_1) \odot (\delta b_1 + \delta w_1 x) \odot w_1)]$$

$$+\mathbf{w}_o \sigma''(\mathbf{s}_2) \odot [\delta \mathbf{b}_2 + \delta W_2 \mathbf{h}_1 + W_2 \sigma'(s_1) \odot (\delta b_1 + \delta w_1 x)] \odot \mathbf{s}_2' \tag{13.220}$$

$$\delta u'' = \delta \mathbf{w}_o \mathbf{h}_2''$$

$$+\mathbf{w}_o \sigma'''(\mathbf{s}_2) \odot \mathbf{s}_2' \odot \mathbf{s}_2' \odot (\delta \mathbf{b}_2 + \delta W_2 \mathbf{h}_1 + W_2 \sigma'(s_1) \odot (\delta b_1 + \delta w_1 x))$$

$$+\mathbf{w}_o 2\sigma''(\mathbf{s}_2) \odot \mathbf{s}_2' \odot (\delta W_2 \mathbf{h}_1' + W_2(\sigma'(s_1) \odot \delta w_1 + \sigma''(s_1) \odot (\delta b_1 + \delta w_1 x) \odot w_1))$$

$$+\mathbf{w_o} \sigma''(\mathbf{s}_2) \odot \mathbf{s}_2'' \odot (\delta \mathbf{b}_2 + \delta W_2 \mathbf{h}_1 + W_2 \sigma'(s_1) \odot (\delta b_1 + \delta w_1 x))$$

$$+\mathbf{w}_o \sigma'(\mathbf{s}_2) \odot (\delta W_2 \mathbf{h}_1'' + W_2(\sigma'''(s_1)(\delta b_1 + \delta w_1 x) + 2\sigma''(s_1) \odot w_1 \odot \delta w_1)) \tag{13.221}$$

13.8.2 First Order Initial Value Problems

In Exercise 13.1 we solve the initial value problem

$$\frac{du(x)}{dx} = f(x) \quad u(0) = 1. \tag{13.222}$$

Since the derivative of the trial function u' does not depend on b_o, we can adjust this parameter to fulfill the initial condition and minimize the loss function

$$L = \frac{1}{N_{\text{samples}}} \sum_{x^{(i)}} \frac{1}{2} (u'(x^{(i)}) - f(x^{(i)}))^2 \tag{13.223}$$

only with respect to w_{on}, w_n and b_n. Figures 13.15 and 13.16 show two examples.

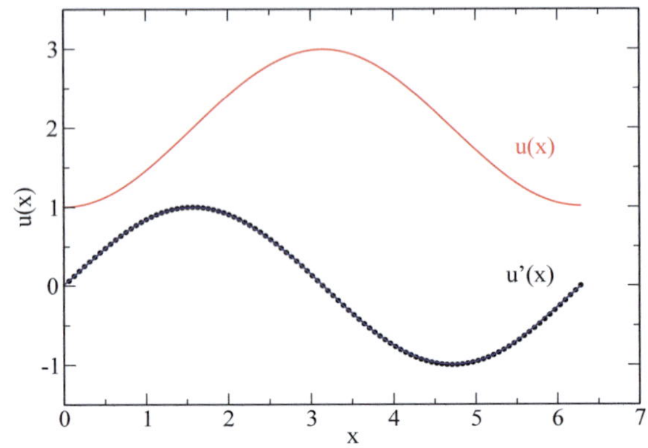

Fig. 13.15 Solving a simple first order ODE with one hidden layer. The equation $\frac{du(x)}{dx} = \sin(x)$ with initial value $u(0) = 1$ is solved by a network with 25 nodes in one hidden layer (13.13). The initial condition is established by adjusting the parameter b_o. Black circles show the 110 data points, the red curve shows the solution $u(x)$ and the blue curve its derivative $u'(x)$. A simple gradient descent with decreasing learning rate $\eta = 0.01 \cdots 0.0004$ reduces the loss function to $L = 10^{-6}$ after 1000 000 epochs. See Exercise 13.1

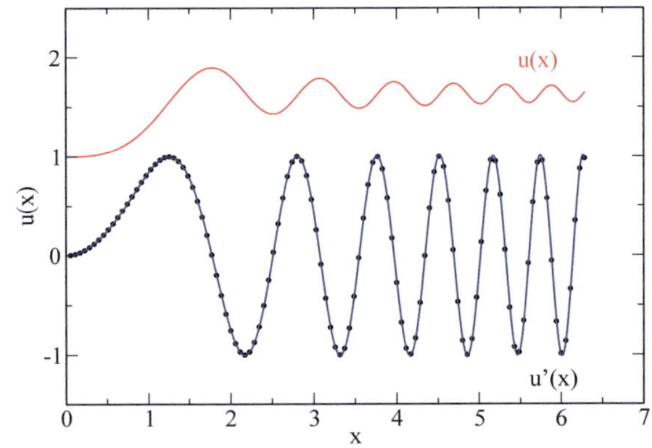

Fig. 13.16 Solving another simple first order ODE with one hidden layer. The equation $\frac{du(x)}{dx} = \sin(x^2)$ with initial value $u(0) = 1$ is solved by a network with 25 nodes in one hidden layer. Black dots show the 110 training points, the red curve the trial function and the blue curve its derivative. A simple gradient descent with decreasing learning rate $\eta = 0.01 \cdots 0.0004$ reduces the loss function to $L = 2 \times 10^{-4}$ after 10^7 epochs. See Exercise 13.1

Trial Functions with Proper Initial or Boundary Values

A more general first order ODE of the form

$$G(x, u(x), u'(x)) = 0 \tag{13.224}$$

with initial value

$$u(0) = u_0 \tag{13.225}$$

can be treated by using a trial function of the form [241, 242, 246]

$$u_t = \alpha(x) + \beta(x) f(x) \tag{13.226}$$

where α and β are chosen such that $\alpha(0) = u_0$ and $\beta(0) = 0$ which establishes the proper initial value and $f(x)$ is the output from the network. A simple choice is given by $\alpha = u_0$ and $\beta = x$. A similar method can be also used for partial differential equations with regular boundaries [246]. However, this parametrization is not bounded for large $|x|$ which can cause numerical problems. These could be avoided by choosing $\alpha = u_0$ and $\beta = (1 - e^{-x})$ [248].

Using the collocation method for a set of sample points $x^{(i)}$, $i = 1 \cdots N_{\text{samples}}$ and the mean square loss function we have to optimize

$$L = \frac{1}{N_{\text{samples}}} \sum_{x^{(i)}} L_i = \frac{1}{N_{\text{samples}}} \sum_{i=1}^{N_{\text{samples}}} \frac{1}{2} G(x^{(i)}, u_t(x^{(i)}), {u'}_t(x^{(i)}))^2. \tag{13.227}$$

As an example (Fig. 13.17, Exercise 13.1) we consider the equation

$$u' = u, \quad u(0) = 1 \tag{13.228}$$

with the exact solution $u = \exp(x)$. The loss function for one sample is

$$L_i = \frac{1}{2} \left[u_0 + (x^{(i)} - 1) f(x^{(i)}) - x^{(i)} f'(x^{(i)}) \right]^2 \tag{13.229}$$

where the gradients can be derived from

$$\delta L_i = \left[u_0 + (x^{(i)} - 1) f(x^{(i)}) - x^{(i)} f'(x^{(i)}) \right] \left[(x^{(i)} - 1) \delta f(x^{(i)}) - x^{(i)} \delta f'(x^{(i)}) \right]. \tag{13.230}$$

Penalty Term for the Constraints

A general method to apply additional constraints[16] to the trial function adds an extra penalty term to the loss function. For the example (13.222) the initial value $u(0) = 1$ can be taken into account by minimizing the total loss[17]

$$L = \frac{1}{N_{\text{samples}}} \sum_{x^{(i)}} \frac{1}{2} (u'(x^{(i)}) - f(x^{(i)}))^2 + \frac{1}{2} (u(0) - 1)^2 \tag{13.231}$$

[16] Besides initial or boundary values this could also involve conserved quantities like energy.

[17] A weight parameter could be introduced to determine the relative importance of the constraint.

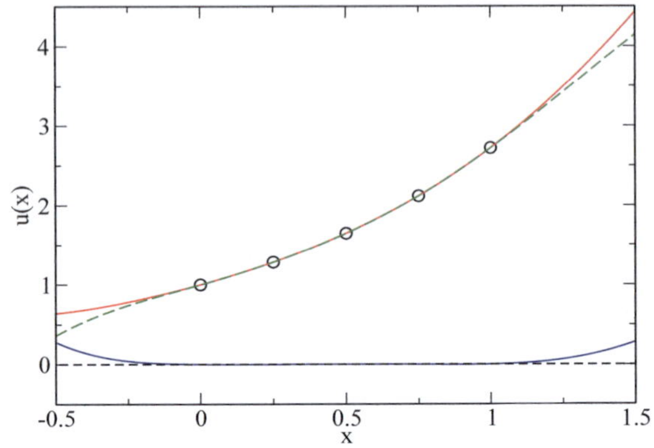

Fig. 13.17 Solving a simple ODE by a network with one hidden layer. The equation $u' = u$, $u(0) = 1$ is solved by a network with one hidden layer containing 50 nodes. The initial value is established by the special form of the trial function. A simple gradient descent with decreasing learning rate $\eta = 0.01 \cdots 0.00001$ reduces the loss function to $L = 10^{-7}$ after 10^7 epochs. Black circles show the 5 training data, the black dash-dotted curve the exact solution, the red curve the function $u(x)$, the dashed green curve the derivative $u'(x)$ and the blue curve the residual $u(x) - u'(x)$. Outside the range of training data deviations from the exact solution grow rapidly

with the variation

$$\delta L = \frac{1}{N_{\text{samples}}} \sum_{x^{(i)}} (u'(x^{(i)}) - f(x^{(i)}))\delta u'(x^{(i)}) + (u(0) - 1)\delta u(0). \qquad (13.232)$$

Figure 13.18 shows results for the equation $u'(x) = \sin(x), \quad u(0) = 1$ using a network with two hidden layers (compare with Fig. 13.15).

13.8.3 Second Order Initial Value Problems

The initial values

$$u_T(0) = u_0 \quad u_T'(0) = u_0' \qquad (13.233)$$

of a general linear second order equation

$$c_0 u(x) + c_1 u'(x) + c_2 u''(x) = 0 \qquad (13.234)$$

can be enforced by using a trial function of the form[18]

$$u_T(x) = x + x^2 u(x) \qquad (13.235)$$

[18] A trial function with constraints is used in Sect. 12.1.1 in Vol. 2.

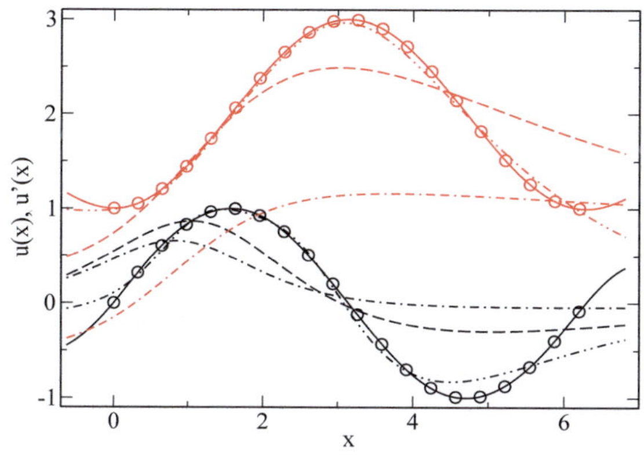

Fig. 13.18 Solving a simple ODE by a network with two hidden layers. The equation $u'(x) = \sin(x), \quad u(0) = 1$ is solved by a network with two hidden layers, each containing 10 nodes. The initial condition is established by a penalty term. A simple gradient descent with decreasing learning rate $\eta = 0.001 \cdots 0.0001$ decreases the loss function to $L = 5 \times 10^{-5}$ after 10^6 epochs. Black circles show the 20 training data, the black curves the resulting derivative $u'(x)$, red circles the exact solution $2 - \cos(x)$ and red curves the resulting function $u(x)$ after 100 (dash-dotted), 1000 (dashed), 10,000 (dash-double dotted) and 1000,000 (full) epochs

where $u(x)$ is optimized by the network. From the derivatives

$$u'_T(x) = 1 + 2xu(x) + x^2u'(x) \tag{13.236}$$

$$u''_T(x) = 2u(x) + 4xu'(x) + x^2u''(x) \tag{13.237}$$

we obtain the residual

$$r(x) = c_0u_T(x) + c_1u'_T(x) + c_2u''_T(x)$$

$$= c_0x + c_1 + u(x)\left[c_0x^2 + 2c_1x + 2c_2\right] + u'(x)\left[c_1x^2 + +4c_2x\right] + c_2x^2u''(x). \tag{13.238}$$

The network parameters are optimized to minimize the loss function

$$L = \frac{1}{N_{\text{samples}}} \sum_{x^{(i)}} |r(x^{(i)})|^2. \tag{13.239}$$

Figure 13.19 shows an example from Exercise 13.2.

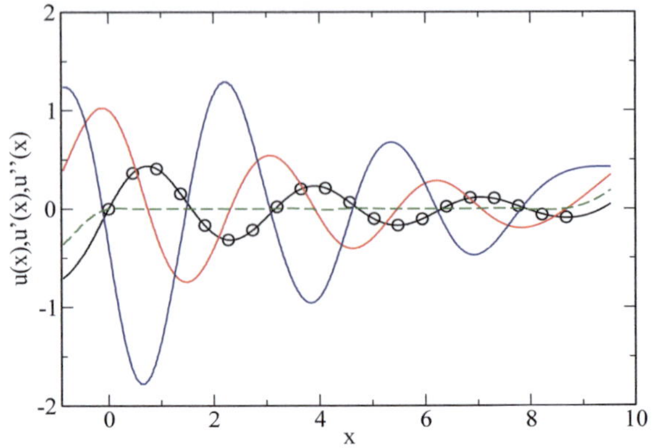

Fig. 13.19 Solving a simple ODE by a network with two hidden layers. Solution of the homogeneous equation $0.25u''(x) + 0.1u'(x) + u(x) = 0$ by a network with two hidden layers, each containing 10 nodes. AMAdam (Sect. 6.3.3) $\alpha = 10^{-4}$, $\beta = 0.9$, $L = 10^{-5}$ after 10^6 epochs. Black circles show the exact solution at the 20 sample points, the black curve the resulting function $u(x)$, the red curve the first derivative $u'(x)$ and the blue curve the second derivative $u''(x)$. The dashed green curve shows the residual $0.25u''(x) + 0.1u'(x) + u(x)$

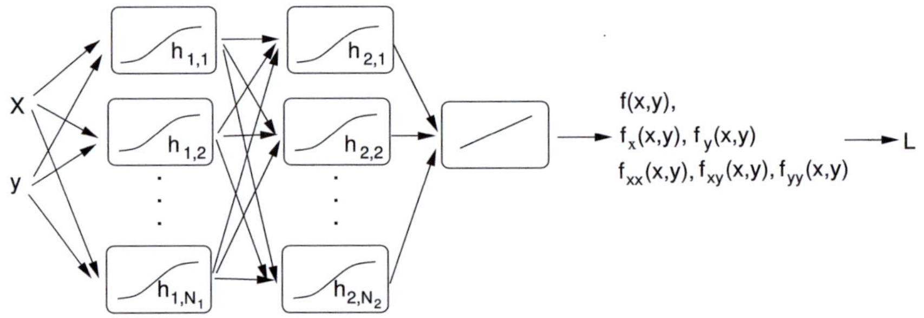

Fig. 13.20 Network for a partial differential equation. PDEs can be solved by networks with one input for each variable. The figure shows the example of a second order PDE of two variables

13.8.4 Systems of Partial Differential Equations

A partial differential equation can be solved by a network with several input variables (Fig. 13.20), systems of differential equations need several output variables and a sum of loss functions (Fig. 13.21) [249].

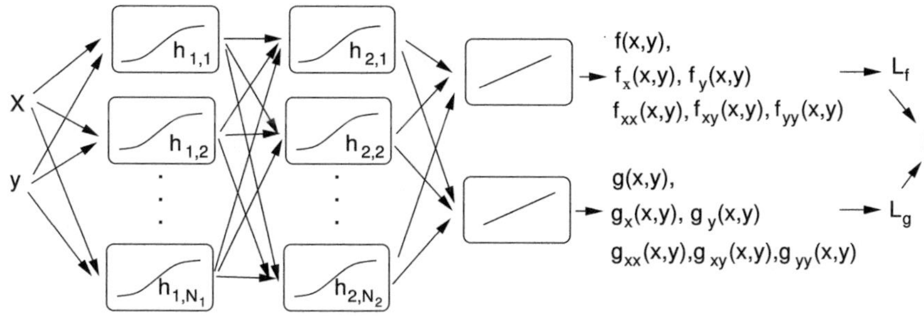

Fig. 13.21 Network for a system of partial differential equations. Systems of PDEs can be solved by networks with one output for each equation. The figure shows the example of two second order PDEs of two variables

Exercises

Exercise 13.1 First Order ODE

In this computer exercise a network with one hidden layer (Fig. 13.13) is used to solve the differential equations

$$\frac{dy}{dx} = \sin x \tag{13.240}$$

and

$$\frac{dy}{dx} = \sin(x^2) \tag{13.241}$$

with a simple gradient descent method (Figs. 13.15 and 13.16). The learning rate decreases according to

$$\eta_t = \eta_0 t^{-0.2} \tag{13.242}$$

where t counts the epochs and $\eta_0 = 0.01$. Convergence can be very slow, so take a coffee!

You can choose between the two functions and vary number of data points and units.

Exercise 13.2 Second Order ODE

A feed forward network with two hidden layers (Fig. 13.14) is used to solve the differential equation of a damped oscillator

$$y + Ay' + By'' = 0. \tag{13.243}$$

Initial conditions

$$y(0) = 0, \quad y'(0) = 1 \tag{13.244}$$

are established by the trial function

$$u_T = x + x^2 u(x). \tag{13.245}$$

The loss function

$$L = \frac{1}{N_{\text{samples}}} \sum_{x^{(i)}} |c_0 u_T(x^{(i)}) + c_1 u'_T(x^{(i)}) + c_2 u''_T(x^{(i)})|^2 \tag{13.246}$$

is minimized by the adaptive AMAdam method [105] (Fig. 13.19). You can change the prefactors of y, y', y'' and compare the approximation with the exact solution. This Exercise is similar to Exercise 26.1 where the boundary values are established by penalty terms.

Chapter 14
Equations of Motion

Simulation of a physical system means to calculate the time evolution of a model system in many cases. We consider a large class of models which can be described by a first order initial value problem

$$\frac{dY}{dt} = f(Y(t), t) \quad Y(t = 0) = Y_0 \tag{14.1}$$

where Y is the state vector (possibly of very high dimension) which contains all information about the system. Our goal is to calculate the time evolution of the state vector $Y(t)$ numerically. For obvious reasons this can be done only for a finite number of values of t and we have to introduce a grid of discrete times t_n which for simplicity are assumed to be equally spaced[1]:

$$t_{n+1} = t_n + \Delta t. \tag{14.2}$$

Advancing time by one step involves the calculation of the integral

$$Y(t_{n+1}) - Y(t_n) = \int_{t_n}^{t_{n+1}} f(Y(t'), t')dt' \tag{14.3}$$

which can be a formidable task since $f(Y(t), t)$ depends on time via the time dependence of all the elements of $Y(t)$. In this chapter we discuss several strategies for the time integration. The explicit Euler forward difference has low error order but is useful as a predictor step for implicit methods. A symmetric difference quotient is much more accurate. It can be used as the corrector step in combination with an explicit

[1] Control of the step width will be discussed later.

Supplementary Information The online version contains supplementary material available at https://doi.org/10.1007/978-3-032-07856-8_14.

P. O. J. Scherer, *Computational Physics I*, Graduate Texts in Physics,
https://doi.org/10.1007/978-3-032-07856-8_14

Euler predictor step and is often used for the time integration of partial differential equations. Methods with higher error order can be obtained from a Taylor series expansion, like the Nordsieck and Gear predictor-corrector methods which have been often applied in molecular dynamics calculations. Runge-Kutta methods are very important for ordinary differential equations. They are robust and allow an adaptive control of the step size. Very accurate results can be obtained for ordinary differential equations with extrapolation methods like the famous Gragg-Bulirsch-Stoer method. If the solution is smooth enough, multistep methods are applicable, which use information from several points. Most known are Adams-Bashforth-Moulton methods and Gear methods (also known as backward differentiation methods), which are especially useful for stiff problems. The class of Verlet methods has been developed for molecular dynamics calculations. They are symplectic and time reversible and conserve energy over long trajectories. While higher order ODE can be always brought to first order form by introducing the derivatives as new variables , there are direct integration methods which can be more efficient in certain cases. Numerov type methods are often used to integrate second order initial value problems where the first order derivative is absent. Alternatively, they can be solved directly and very efficiently with Runge-Kutta-Nyström methods. Kosteks method uses Hermite polynomial interpolation for a predictor-corrector method to solve second order ODEs.

14.1 The State Vector

The state of a classical N-particle system is given by the position in phase space, or equivalently by specifying position and velocity for all the N particles

$$Y = (\mathbf{r}_1, \mathbf{v}_1, \cdots \mathbf{r}_N, \mathbf{v}_N)\,. \tag{14.4}$$

The concept of a state vector is not restricted to a finite number of degrees of freedom. For instance a diffusive system can be described by the particle concentrations as a function of the coordinate, i.e. the elements of the state vector are now indexed by the continuous variable $\mathbf{x}$

$$Y = (c_1(\mathbf{x}), \cdots c_M(\mathbf{x}))\,. \tag{14.5}$$

Similarly, a quantum particle moving in an external potential can be described by the amplitude of the wave function

$$Y = (\Psi(\mathbf{x})). \tag{14.6}$$

Numerical treatment of continuous systems is not feasible since even the ultimate high end computer can only handle a finite number of data in finite time. Therefore discretization is necessary (Chap. 13), by introducing a spatial mesh (Sects. 13.2, 13.3, 13.6), which in the simplest case means a grid of equally spaced points

$$\mathbf{x}_{ijk} = (ih, jh, kh) \quad i = 1..i_{\max}, j = 1..j_{\max}, k = 1..k_{\max} \tag{14.7}$$

$$Y = \left(c_1(\mathbf{x}_{ijk}) \ldots c_M(\mathbf{x}_{ijk})\right) \tag{14.8}$$

$$Y = \left(\Psi(\mathbf{x}_{ijk})\right) \tag{14.9}$$

or by expanding the continuous function with respect to a finite set of basis functions (Sect. 13.5). The elements of the state vector then are the expansion coefficients

$$|\Psi >= \sum_{s=1}^{N} C_s|\Psi_s > \tag{14.10}$$

$$Y = (C_1, \ldots, C_N)\,. \tag{14.11}$$

If the density matrix formalism is used to take the average over a thermodynamic ensemble or to trace out the degrees of freedom of a heat bath, the state vector instead is composed of the elements of the density matrix

$$\rho = \sum_{s=1}^{N}\sum_{s'=1}^{N} \rho_{ss'}|\Psi_s >< \Psi_{s'}| = \sum_{s=1}^{N}\sum_{s'=1}^{N} \overline{C_{s'}^{*}C_s}|\Psi_s >< \Psi_{s'}| \tag{14.12}$$

$$Y = (\rho_{11} \cdots \rho_{1N}, \rho_{21} \cdots \rho_{2N}, \cdots, \rho_{N1} \cdots \rho_{NN})\,. \tag{14.13}$$

14.2 Time Evolution of the State Vector

We assume that all information about the system is included in the state vector. Then the simplest equation to describe the time evolution of the system gives the change of the state vector

$$\frac{\mathrm{d}Y}{\mathrm{d}t} = f(Y, t) \tag{14.14}$$

as a function of the state vector (or more generally a functional in the case of a continuous system). Explicit time dependence has to be considered for instance to describe the influence of an external time dependent field.

Some examples will show the universality of this equation of motion:

- N-particle system

The motion of N interacting particles is described by

$$\frac{\mathrm{d}Y}{\mathrm{d}t} = (\dot{\mathbf{r}}_1, \dot{\mathbf{v}}_1 \cdots) = (\mathbf{v}_1, \mathbf{a}_1 \cdots) \tag{14.15}$$

where the acceleration of a particle is given by the total force acting upon this particle and thus depends on all the coordinates and eventually time (velocity dependent forces could be also considered but are outside the scope of this book)

$$\mathbf{a}_i = \frac{\mathbf{F}_i(\mathbf{r}_1 \cdots \mathbf{r}_N, t)}{m_i}. \tag{14.16}$$

- Diffusion

Heat transport and other diffusive processes are described by the diffusion equation

$$\frac{\partial f}{\partial t} = D\Delta f + S(\mathbf{x}, t) \tag{14.17}$$

which in its simplest spatially discretized version for one-dimensional diffusion reads

$$\frac{\partial f(\mathbf{x}_i)}{\partial t} = \frac{D}{\Delta x^2}(f(\mathbf{x}_{i+1}) + f(\mathbf{x}_{i-1}) - 2f(\mathbf{x}_i)) + S(\mathbf{x}_i, t). \tag{14.18}$$

- Waves

Consider the simple one-dimensional wave equation

$$\frac{\partial^2 f}{\partial t^2} = c^2 \frac{\partial^2 f}{\partial x^2} \tag{14.19}$$

which by introducing the velocity $g(\mathbf{x}) = \frac{\partial}{\partial t} f(\mathbf{x})$ as an independent variable can be rewritten as

$$\frac{\partial}{\partial t}(f(\mathbf{x}), g(\mathbf{x})) = \left(g(\mathbf{x}), c^2 \frac{\partial^2}{\partial x^2} f(\mathbf{x})\right). \tag{14.20}$$

Discretization of space gives

$$\frac{\partial}{\partial t}(f(\mathbf{x}_i), g(\mathbf{x}_i)) = \left(g(\mathbf{x}_i), \frac{c^2}{\Delta x^2}(f(\mathbf{x}_{i+1}) + f(\mathbf{x}_{i-1}) - 2f(\mathbf{x}_i))\right). \tag{14.21}$$

- two-state quantum system

The Schrödinger equation for a two level system (for instance a spin-1/2 particle in a magnetic field) reads

$$\frac{\mathrm{d}}{\mathrm{d}t}\begin{pmatrix} C_1 \\ C_2 \end{pmatrix} = \begin{pmatrix} H_{11}(t) & H_{12}(t) \\ H_{21}(t) & H_{22}(t) \end{pmatrix}\begin{pmatrix} C_1 \\ C_2 \end{pmatrix}. \tag{14.22}$$

14.3 Explicit Forward Euler Method

The simplest method which is often discussed in elementary physics textbooks approximates the integrand by its value at the lower boundary (Fig. 14.1):

$$Y(t_{n+1}) - Y(t_n) \approx f(Y(t_n), t_n)\Delta t \tag{14.23}$$

The truncation error can be estimated from a Taylor series expansion

$$Y(t_{n+1}) - Y(t_n) = \Delta t \frac{dY}{dt}(t_n) + \frac{\Delta t^2}{2}\frac{d^2Y}{dt^2}(t_n) + \cdots$$
$$= \Delta t f(Y(t_n), t_n) + O(\Delta t^2). \tag{14.24}$$

The explicit Euler method has several serious drawbacks

- low error order

Suppose you want to integrate from the initial time t_0 to the final time $t_0 + T$. For a time step of Δt you have to perform $N = T/\Delta t$ steps. Assuming comparable error contributions from all steps the global error scales as $N\Delta t^2 = O(\Delta t)$. The error gets smaller as the time step is reduced but it may be necessary to use very small Δt to obtain meaningful results.

- loss of orthogonality and normalization

The simple Euler method can produce systematic errors which are very inconvenient if you want, for instance, to calculate the orbits of a planetary system. This can be most easily seen from a very simple example. Try to integrate the following equation of motion (see example Sect. 1.5):

$$\frac{dz}{dt} = i\omega z. \tag{14.25}$$

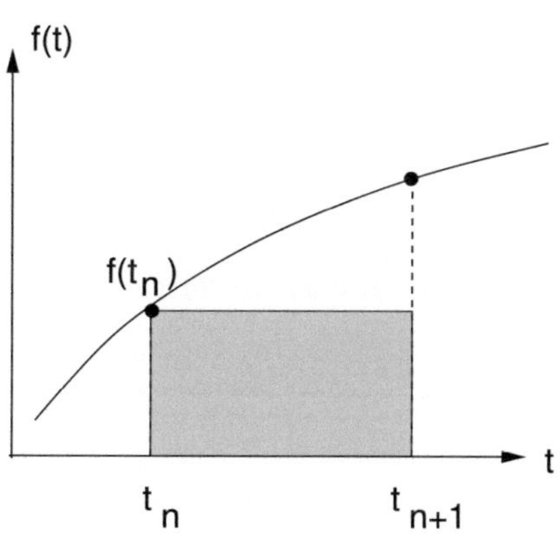

Fig. 14.1 Explicit Euler method

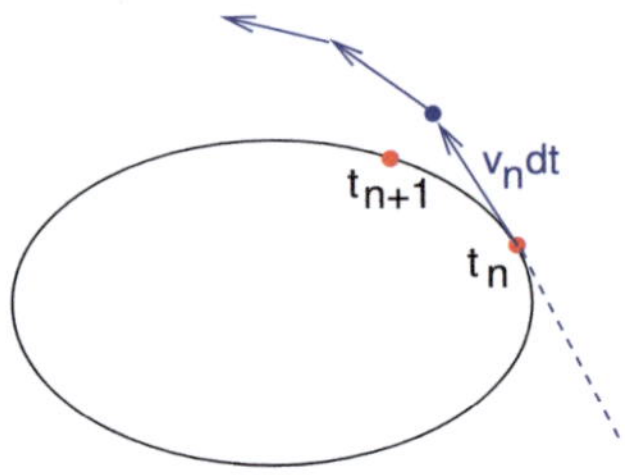

Fig. 14.2 Systematic errors of the Euler method

The exact solution is obviously given by a circular orbit in the complex plane:

$$z = z_0 \mathrm{e}^{\mathrm{i}\omega t} \tag{14.26}$$

$$|z| = |z_0| = \text{ const.} \tag{14.27}$$

Application of the Euler method gives

$$z(t_{n+1}) = z(t_n) + \mathrm{i}\omega \Delta t\, z(t_n) = (1 + \mathrm{i}\omega \Delta t) z(t_n) \tag{14.28}$$

and you find immediately

$$|z(t_n)| = \sqrt{1 + \omega^2 \Delta t^2}\, |z(t_{n-1})| = \left(1 + \omega^2 \Delta t^2\right)^{n/2} |z(t_0)| \tag{14.29}$$

which shows that the radius increases continually even for the smallest time step possible (Fig. 14.2).

The same kind of error appears if you solve the Schrödinger equation for a particle in an external potential or if you calculate the rotational motion of a rigid body. For the N-body system it leads to a violation of the conservation of phase space volume. This can introduce an additional sensitivity of the calculated results to the initial conditions. Consider a harmonic oscillator with the equation of motion

$$\frac{\mathrm{d}}{\mathrm{d}t} \begin{pmatrix} x(t) \\ v(t) \end{pmatrix} = \begin{pmatrix} v(t) \\ -\omega^2 x(t) \end{pmatrix}. \tag{14.30}$$

Application of the explicit Euler method gives

$$\begin{pmatrix} x(t + \Delta t) \\ v(t + \Delta t) \end{pmatrix} = \begin{pmatrix} x(t) \\ v(t) \end{pmatrix} + \begin{pmatrix} v(t) \\ -\omega^2 x(t) \end{pmatrix} \Delta t. \tag{14.31}$$

The change of the phase space volume (Fig. 14.3) is given by the Jacobi determinant

$$J = \left| \frac{\partial(x(t + \Delta t), v(t + \Delta t))}{\partial(x(t), v(t))} \right| = \begin{vmatrix} 1 & \Delta t \\ -\omega^2 \Delta t & 1 \end{vmatrix} = 1 + (\omega \Delta t)^2. \tag{14.32}$$

In this case the phase space volume increases continuously.

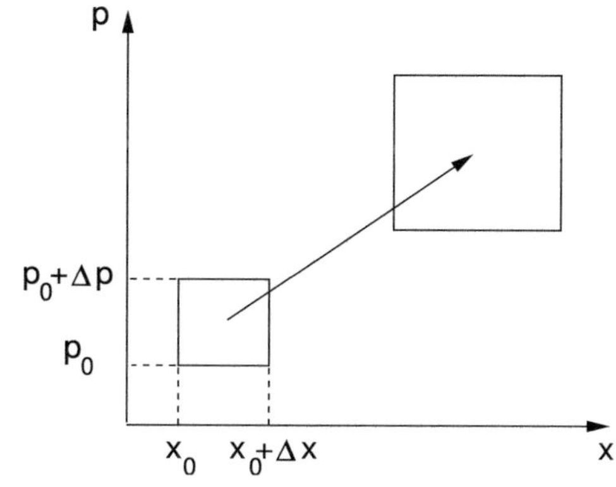

Fig. 14.3 Time evolution of the phase space volume

14.4 Implicit Backward Euler Method

Alternatively let us make a step backwards in time

$$Y(t_n) - Y(t_{n+1}) \approx -f(Y(t_{n+1}), t_{n+1})\Delta t \tag{14.33}$$

which can be written as (Fig. 14.4)

$$Y(t_{n+1}) \approx Y(t_n) + f(Y(t_{n+1}), t_{n+1})\Delta t. \tag{14.34}$$

Taylor series expansion gives

$$Y(t_n) = Y(t_{n+1}) - \frac{\mathrm{d}}{\mathrm{d}t}Y(t_{n+1})\Delta t + \frac{\mathrm{d}^2}{\mathrm{d}t^2}Y(t_{n+1})\frac{\Delta t^2}{2} + \cdots \tag{14.35}$$

which shows that the error order again is $O(\Delta t^2)$. The implicit method is sometimes used to avoid the inherent instability of the explicit method. For the examples in Sect. 14.3 it shows the opposite behavior. The radius of the circular orbit as well as the phase space volume decrease in time. The gradient at future time has to be estimated before an implicit step can be performed.

14.5 Improved Euler Methods

The quality of the approximation can be improved significantly by employing the midpoint rule (Fig. 14.5)

$$Y(t_{n+1}) - Y(t_n) \approx f(Y(t + \frac{\Delta t}{2}), t_n + \frac{\Delta t}{2})\Delta t. \tag{14.36}$$

Fig. 14.4 Implicit backward Euler method

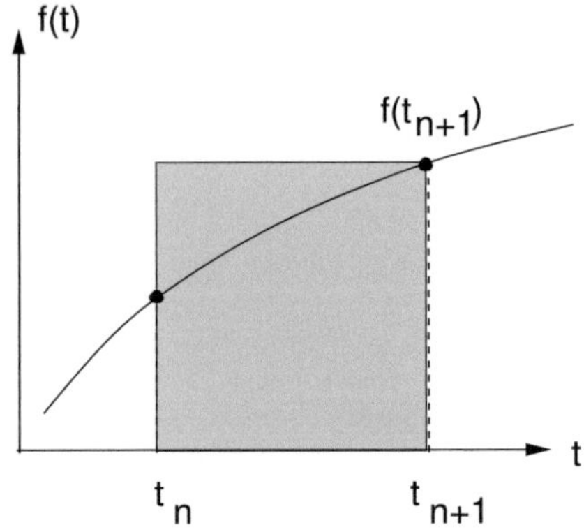

Fig. 14.5 Improved Euler method

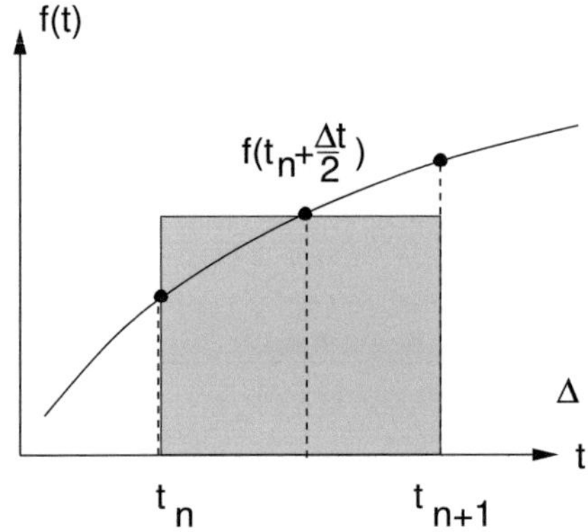

The error is smaller by one order of Δt:

$$
\begin{aligned}
&Y(t_n) + f(Y(t + \frac{\Delta t}{2}), t_n + \frac{\Delta t}{2})\Delta t \\
&= Y(t_n) + \left(\frac{\mathrm{d}Y}{\mathrm{d}t}(t_n) + \frac{\Delta t}{2}\frac{\mathrm{d}^2 Y}{\mathrm{d}t^2}(t_n) + \cdots \right) \Delta t \\
&= Y(t_n + \Delta t) + O(\Delta t^3) \qquad (14.37)
\end{aligned}
$$

The future value $Y(t + \frac{\Delta t}{2})$ can be obtained by two different approaches:

- explicit midpoint method (Predictor-corrector)

Since $f(Y(t + \frac{\Delta t}{2}), t_n + \frac{\Delta t}{2})$ is multiplied with Δt, it is sufficient to use an approximation with lower error order. Even the explicit Euler step is sufficient. Together the following algorithm (explicit midpoint method) results:

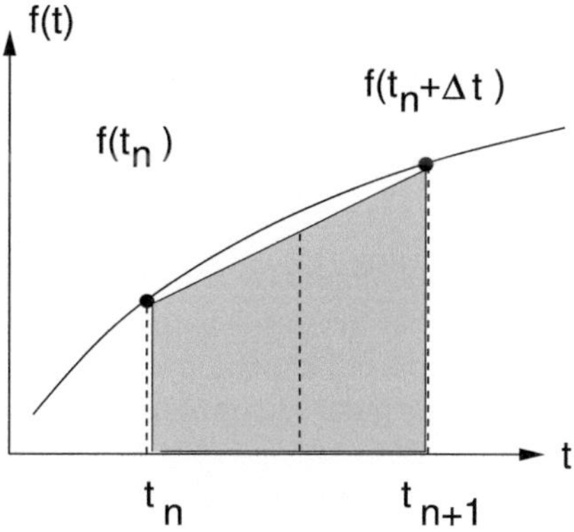

Fig. 14.6 Trapezoidal method

$$\begin{array}{lrl} \text{predictor step:} & Y^{(p)} &= Y(t_n) + \frac{\Delta t}{2} f(Y(t_n), t_n) \\ \text{corrector step:} & Y(t_n + \Delta t) &= Y(t_n) + \Delta t\, f(Y^{(p)}, t_n + \frac{\Delta t}{2}) \end{array} \tag{14.38}$$

- implicit midpoint method (averaging)

$$\begin{array}{lrl} \text{average:} & Y^{(p)} &= Y(t_n) + Y(t_n + \Delta t) \\ \text{integration:} & Y(t_n + \Delta t) &= Y(t_n) + \Delta t\, f(Y^{(p)}, t_n + \frac{\Delta t}{2}) \end{array}. \tag{14.39}$$

Application of the trapezoidal rule instead of the midpoint rule gives the Crank-Nicolson approximation

$$Y(t_n + \Delta t) = Y(t_n) + \frac{\Delta t}{2} \left(f(Y(t_n), t_n) + f(Y(t_n + \Delta t), t + \Delta t) \right) \tag{14.40}$$

where the midpoint value $f(Y(t_n + \Delta t/2), t_n + \Delta t/2)$ is replaced by the average of $f(Y(t_n), t_n)$ and $f(Y(t_n + \Delta t), t + \Delta t)$ (Fig. 14.6). This corresponds to a combination of explicit and implicit Euler step and is also an implicit method.

Expansion around $t_n + \Delta t/2$ gives

$$\frac{1}{2} \left(f(Y(t_n), t_n) + f(Y(t_n + \Delta t), t + \Delta t) \right)$$

$$= f(Y(t_n + \frac{\Delta t}{2}), t_n + \frac{\Delta t}{2}) + O(\Delta t^2). \tag{14.41}$$

In the special case of a linear function $f(Y(t), t) = F\, Y(t)$ (for instance rotational motion or diffusion) (14.40) can be solved formally by

$$Y(t_n + \Delta t) = \left(1 - \frac{\Delta t}{2} F\right)^{-1} \left(1 + \frac{\Delta t}{2} F\right) Y(t_n). \tag{14.42}$$

Numerically it is not necessary to perform the matrix inversion. Instead a linear system of equations is solved:

$$\left(1-\frac{\Delta t}{2}F\right)Y(t_n+\Delta t)=\left(1+\frac{\Delta t}{2}F\right)Y(t_n). \tag{14.43}$$

This method conserves the norm of the state vector if F has only imaginary eigenvalues (as for the one-dimensional Schrödinger equation, see Sect. 10.2.2 in Vol. 2).

The trapezoidal method can be made explicit with the help of a predictor step to estimate the state vector at $t_n + \Delta t$

$$Y^{(p)} = Y(t_n) + \Delta t\ f(Y(t_n), t_n) \tag{14.44}$$

$$Y(t_n+\Delta t) = Y(t_n) + \frac{\Delta t}{2}\left(f(Y(t_n), t_n) + f(Y^{(p)}, t+\Delta t)\right).$$

This is also known as Heun's method

14.6 Taylor Series Methods

Higher order methods can be obtained from a Taylor series expansion

$$Y(t_n+\Delta t) = Y(t_n) + \Delta t\ f(Y(t_n), t_n) + \frac{\Delta t^2}{2}\frac{\mathrm{d}f(Y(t_n), t_n)}{\mathrm{d}t} + \cdots. \tag{14.45}$$

The total time derivative can be expressed as

$$\frac{\mathrm{d}f}{\mathrm{d}t} = \frac{\partial f}{\partial Y}\frac{\mathrm{d}Y}{\mathrm{d}t} + \frac{\partial f}{\partial t} = f'f + \dot{f} \tag{14.46}$$

where the partial derivatives have been abbreviated in the usual way by $\frac{\partial f}{\partial t} = \dot{f}$ and $\frac{\partial f}{\partial Y} = f'$. Higher derivatives are given by

$$\frac{\mathrm{d}^2 f}{\mathrm{d}t^2} = f''f^2 + f'^2 f + 2\dot{f}'f + \ddot{f} \tag{14.47}$$

$$\frac{\mathrm{d}^3 f}{\mathrm{d}t^3} = \frac{\partial^3 f}{\partial t^3} + f'''f^3 + 3\dot{f}''f^2 + \ddot{f}f' + 3f''\dot{f}f$$
$$+ 3\ddot{f}' + 4f''f'f^2 + 5\dot{f}'f'f + f'^3 f + f'^2\dot{f}. \tag{14.48}$$

14.6.1 Nordsieck Predictor-Corrector Method

Nordsieck [250] determines an interpolating polynomial of degree m. As variables he uses the 0th to mth derivatives[2] evaluated at the current time t, for instance for $m = 5$ he uses the variables

$$Y(t) \tag{14.49}$$

$$g(t) = \frac{\mathrm{d}}{\mathrm{d}t}Y(t) \tag{14.50}$$

$$a(t) = \frac{\Delta t}{2}\frac{\mathrm{d}^2}{\mathrm{d}t^2}Y(t) \tag{14.51}$$

$$b(t) = \frac{\Delta t^2}{6}\frac{\mathrm{d}^3}{\mathrm{d}t^3}Y(t) \tag{14.52}$$

$$c(t) = \frac{\Delta t^3}{24}\frac{\mathrm{d}^4}{\mathrm{d}t^4}Y(t) \tag{14.53}$$

$$d(t) = \frac{\Delta t^4}{120}\frac{\mathrm{d}^5}{\mathrm{d}t^5}Y(t). \tag{14.54}$$

Taylor expansion gives approximate values at $t + \Delta t$

$$\begin{aligned} Y(t + \Delta t) &= Y(t) + \Delta t\,[g(t) + a(t) + b(t) + c(t) + d(t) + e(t)] \\ &= Y^p(t + \Delta t) + e(t)\Delta t \end{aligned} \tag{14.55}$$

$$g(t + \Delta t) = g(t) + 2a(t) + 3b(t) + 4c(t) + 5d(t) + 6e(t) = g^p(t + \Delta T) + 6e(t) \tag{14.56}$$

$$a(t + \Delta t) = a(t) + 3b(t) + 6c(t) + 10d(t) + 15e(t) = a^p(t + \Delta t) + 15e(t) \tag{14.57}$$

$$b(t + \Delta t) = b(t) + 4c(t) + 10d(t) + 20e(t) = b^p(t + \Delta t) + 20e(t) \tag{14.58}$$

$$c(t + \Delta t) = c(t) + 5d(t) + 15e(t) = c^p(t + \Delta t) + 15e(t) \tag{14.59}$$

$$d(t + \Delta t) = d(t) + 6e(t) = d^p(t + \Delta t) + 6e(t) \tag{14.60}$$

where the next term of the Taylor series $e(t) = \frac{\Delta t^5}{6!}\frac{\mathrm{d}^6}{\mathrm{d}t^6}Y(t)$ has been introduced as an approximation to the truncation error of the predicted values Y^p, g^p, etc. It can be estimated from the second equation

[2] In fact the derivatives of the interpolating polynomial which exist even if higher derivatives of f do not exist.

$$e = \frac{1}{6}\left[f(Y^p(t+\Delta t), t+\Delta t) - g^p(t+\Delta t)\right] = \frac{1}{6}\delta f. \tag{14.61}$$

This predictor-corrector method turns out to be rather unstable. However, stability can be achieved by slightly modifying the coefficients of the corrector step. Nordsieck suggested to use

$$Y(t+\Delta t) = Y^p(t+\Delta t) + \frac{95}{288}\delta f \tag{14.62}$$

$$a(t+\Delta t) = a^p(t+\Delta t) + \frac{25}{24}\delta f \tag{14.63}$$

$$b(t+\Delta t) = b^p(t+\Delta t) + \frac{35}{72}\delta f \tag{14.64}$$

$$c(t+\Delta t) = c^p(t+\Delta t) + \frac{5}{48}\delta f \tag{14.65}$$

$$d(t+\Delta t) = d^p(t+\Delta t) + \frac{1}{120}\delta f. \tag{14.66}$$

14.6.2 Gear Predictor-Corrector Methods

Gear [251] designed special methods for molecular dynamics simulations (Chap. 2 in Vol. 2) where Newton's law (14.15) has to be solved numerically. He uses again a truncated Taylor expansion for the predictor step

$$\mathbf{r}(t+\Delta t) = \mathbf{r}(t) + \mathbf{v}(t)\Delta t + \mathbf{a}(t)\frac{\Delta t^2}{2} + \dot{\mathbf{a}}(t)\frac{\Delta t^3}{6} + \ddot{\mathbf{a}}(t)\frac{\Delta t^4}{24} + \dots \tag{14.67}$$

$$\mathbf{v}(t+\Delta t) = \mathbf{v}(t) + \mathbf{a}(t)\Delta t + \dot{\mathbf{a}}(t)\frac{\Delta t^2}{2} + \ddot{\mathbf{a}}(t)\frac{\Delta t^3}{6} + \dots \tag{14.68}$$

$$\mathbf{a}(t+\Delta t) = \mathbf{a}(t) + \dot{\mathbf{a}}(t)\Delta t + \ddot{\mathbf{a}}(t)\frac{\Delta t^2}{2} + \dots \tag{14.69}$$

$$\dot{\mathbf{a}}(t+\Delta t) = \dot{\mathbf{a}}(t) + \ddot{\mathbf{a}}(t)\Delta t + \dots \tag{14.70}$$

$$\vdots$$

to calculate new coordinates etc. $\mathbf{r}^p_{n+1}, \mathbf{v}^p_{n+1}, \mathbf{a}^p_{n+1} \dots$ (Fig. 14.7). The difference between the predicted acceleration and that calculated using the predicted coordinates

$$\delta\mathbf{a}_{n+1} = \mathbf{a}(\mathbf{r}^P_{n+1}, t+\Delta t) - \mathbf{a}^p_{n+1} \tag{14.71}$$

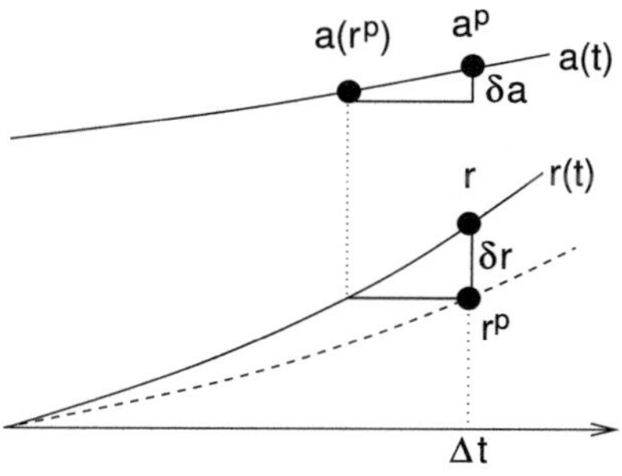

Fig. 14.7 Gear Predictor Corrector Method. The difference between predicted acceleration $\mathbf{a}^p$ and acceleration calculated for the predicted coordinates $\mathbf{a}(\mathbf{r}^p)$ is used as a measure of the error to estimate the correction $\delta\mathbf{r}$

is then used as a measure of the error to correct the predicted values according to

$$\mathbf{r}_{n+1} = \mathbf{r}_{n+1}^p + c_1 \delta\mathbf{a}_{n+1} \tag{14.72}$$

$$\mathbf{v}_{n+1} = \mathbf{v}_{n+1}^p + c_2 \delta\mathbf{a}_{n+1}. \tag{14.73}$$

$$\vdots$$

The coefficients c_i were determined to optimize stability and accuracy. For instance the fourth order Gear corrector reads

$$\mathbf{r}_{n+1} = \mathbf{r}_{n+1}^p + \frac{\Delta t^2}{12} \delta\mathbf{a}_{n+1} \tag{14.74}$$

$$\mathbf{v}_{n+1} = \mathbf{v}_{n+1}^p + \frac{5\Delta t}{12} \delta\mathbf{a}_{n+1} \tag{14.75}$$

$$\dot{\mathbf{a}}_{n+1} = \dot{\mathbf{a}}_n + \frac{1}{\Delta t} \delta\mathbf{a}_{n+1}. \tag{14.76}$$

Gear methods are generally not time reversible and show systematic energy drifts. A reversible symplectic predictor-corrector method has been presented recently by Martyna and Tuckerman [252].

14.7 Runge Kutta Methods

If higher derivatives are not so easily available, they can be approximated by numerical differences. f is evaluated at several trial points and the results are combined to reproduce the Taylor series as close as possible [253].

14.7.1 Second Order Runge Kutta Method

Let us begin with two function values. As common in the literature we will denote the function values as $K_1, K_2, \cdots$. From the gradient at time t_n

$$K_1 = f_n = f(Y(t_n), t_n) \tag{14.77}$$

we estimate the state vector at time $t_n + \Delta t$ as

$$Y(t_n + \Delta t) \approx \Delta t\, K_1. \tag{14.78}$$

The gradient at time $t_n + \Delta t$ is approximately

$$K_2 = f(Y(t_n) + \Delta t\, K_1, t_n + \Delta t) \tag{14.79}$$

which has the Taylor series expansion

$$K_2 = f_n + (\dot{f}_n + f_n' f_n)\Delta t + \cdots \tag{14.80}$$

and application of the trapezoidal rule (4.13) gives the 2nd order Runge Kutta method

$$Y_{n+1} = Y_n + \frac{\Delta t}{2}(K_1 + K_2) \tag{14.81}$$

which in fact coincides with the improved Euler or Heun method. Taylor series expansion shows how the combination of K_1 and K_2 leads to an expression of higher error order:

$$\begin{aligned} Y_{n+1} &= Y_n + \frac{\Delta t}{2}(f_n + f_n + (\dot{f}_n + f_n' f_n)\Delta t + \cdots) \\ &= Y_n + f_n \Delta t + \frac{\mathrm{d} f_n}{\mathrm{d}t}\frac{\Delta t^2}{2} + \cdots . \end{aligned} \tag{14.82}$$

14.7.2 Third Order Runge Kutta Method

The accuracy can be further improved by calculating one additional function value at mid-time. From (14.77) we estimate the gradient at mid-time by

$$\begin{aligned} K_2 &= f(Y(t) + \frac{\Delta t}{2}K_1, t + \frac{\Delta t}{2}) \\ &= f_n + (\dot{f}_n + f_n' f_n)\frac{\Delta t}{2} + (\ddot{f}_n + f_n'' f_n^2 + 2\dot{f}_n' f_n)\frac{\Delta t^2}{8} + \cdots . \end{aligned} \tag{14.83}$$

The gradient at time $t_n + \Delta t$ is then estimated as

$$K_3 = f(Y(t_n) + \Delta t(2K_2 - K_1), t_n + \Delta t)$$

$$= f_n + \dot{f}_n \Delta t + f_n'(2K_2 - K_1)\Delta t + \ddot{f}_n \frac{\Delta t^2}{2}$$

$$+ f_n'' \frac{(2K_2 - K_1)^2 \Delta t^2}{2} + 2\dot{f}_n' \frac{(2K_2 - K_1)\Delta t^2}{2} + \cdots . \tag{14.84}$$

Inserting the expansion (14.83) gives the leading terms

$$K_3 = f_n + (\dot{f}_n + f_n' f_n)\Delta t + (2f_n'^2 f_n + f_n'' f_n^2 + \ddot{f}_n + 2f_n'\dot{f}_n + 2\dot{f}_n^2)\frac{\Delta t^2}{2} + \cdots . \tag{14.85}$$

Applying Simpson's rule (4.14) we combine the three gradients to get the 3rd order Runge Kutta method

$$Y_{n+1} = Y(t_n) + \frac{\Delta t}{6}(K_1 + 4K_2 + K_3) \tag{14.86}$$

where the Taylor series

$$Y_{n+1} = Y(t_n) + \frac{\Delta t}{6}\left(6f_n + 3(\dot{f}_n + f_n f_n')\Delta t\right.$$

$$\left.+(f_n'^2 f_n + f_n'' f_n^2 + 2\dot{f}_n' f_n + f_n + \dot{f}_n \ddot{f}_n')\Delta t^2 + \cdots\right)$$

$$= Y(t_n + \Delta t) + O(\Delta t^4) \tag{14.87}$$

recovers the exact Taylor series (14.45) including terms of order $O(\Delta t^3)$.

14.7.3 Fourth Order Runge Kutta Method

The 4th order Runge Kutta method (RK4) is often used because of its robustness and accuracy. It uses two different approximations for the midpoint

$$K_1 = f(Y(t_n), t_n) \tag{14.88}$$

$$K_2 = f(Y(t_n) + \frac{K_1}{2}\Delta t, t_n + \frac{\Delta t}{2}) \tag{14.89}$$

$$K_3 = f(Y(t_n) + \frac{K_2}{2}\Delta t, t_n + \frac{\Delta t}{2}) \tag{14.90}$$

$$K_4 = f(Y(t_n) + K_3\Delta t, t_n + \Delta t) \tag{14.91}$$

and Simpson's rule (4.14) to obtain

$$Y_{n+1} = Y(t_n) + \frac{\Delta t}{6}(K_1 + 2K_2 + 2K_3 + K_4) = Y(t_n + \Delta t) + O(\Delta t^5). \quad (14.92)$$

Expansion of the Taylor series is cumbersome but with the help of an algebra program one can easily check that the error is of order Δt^5.

14.7.4 Quality Control and Adaptive Step Size Control

For practical applications it is necessary to have an estimate for the local error and to adjust the step size properly. With the Runge Kutta method this can be achieved by a step doubling procedure [254].

We calculate y_{n+1} first by one step Δt and then by two steps $\Delta t/2$. This needs 11 function evaluations as compared to 8 for the smaller step size only (Fig. 14.8). For the 4th order method we estimate the following errors:

$$\Delta\left(Y_{n+1}^{(\Delta t)}\right) = a\Delta t^5 \quad (14.93)$$

$$\Delta\left(Y_{n+1}^{(\Delta t/2)}\right) = 2a(\frac{\Delta t}{2})^5 = a\frac{\Delta t^5}{16}. \quad (14.94)$$

The local error can be estimated from

$$|Y_{n+1}^{(\Delta t)} - Y_{n+1}^{(\Delta t/2)}| = \frac{15}{16}|a|\Delta t^5 \quad (14.95)$$

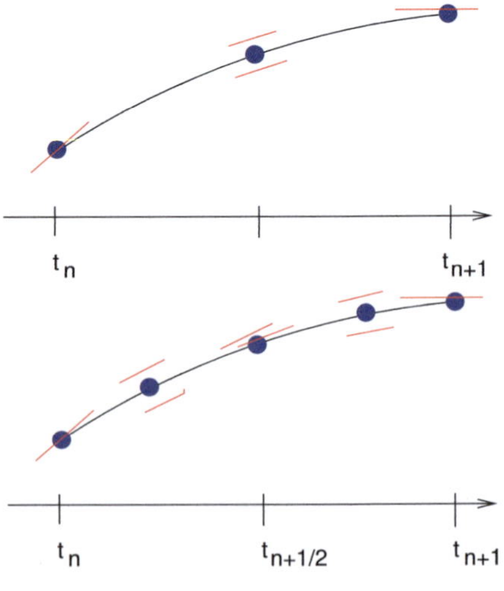

Fig. 14.8 Step doubling with the fourth order Runge Kutta method

Table 14.1 Butcher tableau for the coefficients of the RK4/5 method by Cash and Karp

a_i	b_{ij}					
$\frac{1}{5}$	$\frac{1}{5}$					
$\frac{3}{10}$	$\frac{3}{40}$	$\frac{9}{40}$				
$\frac{3}{5}$	$\frac{3}{10}$	$-\frac{9}{10}$	$\frac{6}{5}$			
1	$-\frac{11}{54}$	$\frac{5}{2}$	$-\frac{70}{27}$	$\frac{35}{27}$		
$\frac{7}{8}$	$\frac{1631}{55296}$	$\frac{175}{512}$	$\frac{575}{13824}$	$\frac{44275}{110592}$	$\frac{253}{4096}$	
c_i	$\frac{37}{378}$	0	$\frac{250}{621}$	$\frac{125}{594}$	0	$\frac{512}{1771}$
c_i^*	$\frac{2825}{27648}$	0	$\frac{18575}{48384}$	$\frac{13525}{55296}$	$\frac{277}{14336}$	$\frac{1}{4}$

$$|\Delta\left(Y_{n+1}^{(\Delta t)}\right)| = |a|\Delta t^5 \approx |Y_{n+2}^{(\Delta t)} - Y_{n+2}^{(\Delta t/2)}|. \tag{14.96}$$

The step size Δt can now be adjusted to keep the local error within the desired limits.

Fehlberg [255–257] discovered an embedded RK method which uses six function evaluations to obtain a fifth order together with a fourth order method. The coefficients of this method have been improved later by Cash and Karp [258] and it has become a standard method. The general RK5 method has the form [259]

$$K_1 = f(Y(t_n), t_n) \tag{14.97}$$

$$K_2 = f(Y(t_n) + b_{21}\Delta t K_1, t_n + a_2\Delta t) \tag{14.98}$$

$$\vdots$$

$$K_6 = f(Y(t_n) + \Delta t(b_{61}K_1 + \ldots b_{65}K_5), t_n + a_6\Delta t) \tag{14.99}$$

$$Y(t_{n+1}) = Y(t_n) + \Delta t\,(c_1K_1 + \cdots c_6K_6) + O(\Delta t^6). \tag{14.100}$$

The embedded RK4 method uses the same function evaluations to obtain

$$Y^*(t_{n+1}) = Y(t_n) + \Delta t\,\left(c_1^*K_1 + \cdots c_6^*K_6\right) + O(\Delta t^5) \tag{14.101}$$

and the local error is estimated from

$$|Y(t_{n+1}) - Y^*(t_{n+1})| = \Delta t|\sum_{i=1}^{6}(c_i - c_i^*)K_i|. \tag{14.102}$$

The coefficients by Cash and Karp are shown in Table 14.1 as a Butcher tableau.

14.8 Extrapolation Methods

Application of the extrapolation method to calculate the integral $\int_{t_n}^{t_{n+1}} f(t)dt$ produces very accurate results but can also be time consuming. The famous Gragg-Bulirsch- Stoer method [2] starts from an explicit midpoint rule with a special starting procedure. The interval Δt is divided into a sequence of N sub-steps

$$h = \frac{\Delta t}{N}. \tag{14.103}$$

First a simple Euler step is performed

$$u_0 = Y(t_n) \tag{14.104}$$

$$u_1 = u_0 + h\, f(u_0, t_n) \tag{14.105}$$

and then the midpoint rule is applied repeatedly to obtain

$$u_{j+1} = u_{j-1} + 2h\, f(u_j, t_n + jh) \qquad j = 1, 2 \cdots N-1. \tag{14.106}$$

Gragg [260] introduced a smoothing procedure to remove oscillations of the leading error term by defining

$$v_j = \frac{1}{4}u_{j-1} + \frac{1}{2}u_j + \frac{1}{4}u_{j+1}. \tag{14.107}$$

He showed that both approximations (14.106) and (14.107) have an asymptotic expansion in powers of h^2 and are therefore well suited for an extrapolation method. The modified midpoint method can be summarized as follows:

$$u_0 = Y(t_n) \tag{14.108}$$

$$u_1 = u_0 + h\, f(u_0, t_n) \tag{14.109}$$

$$u_{j+1} = u_{j-1} + 2h\, f(u_j, t_n + jh) \quad j = 1, 2, \cdots N-1 \tag{14.110}$$

$$Y(t_n + \Delta t) \approx \frac{1}{2}\left(u_N + u_{N-1} + h\, f(u_N, t_n + \Delta t)\right) \tag{14.111}$$

The number of sub-steps N is increased according to a sequence like

$$N = 2, 4, 6, 8, 12, 16, 24, 32, 48, 64 \cdots \quad N_j = 2N_{j-2} \quad \text{Bulirsch - Stoer sequence} \tag{14.112}$$

or

$$N = 2, 4, 6, 8, 10, 12 \cdots \quad N_j = 2j \quad \text{Deuflhard sequence.}$$

After each successive N is tried, a polynomial extrapolation is attempted. This extrapolation returns both the extrapolated values and an error estimate. If the error is still too large then N has to be increased further. A more detailed discussion can be found in [261, 262].

14.9 Linear Multistep Methods

All methods discussed so far evaluated one or more values of the gradient $f(Y(t), t)$ only within the interval $t_n \cdots t_n + \Delta t$. If the state vector changes sufficiently smooth then multistep methods can be applied. Linear multistep methods use a combination of function values Y_n and gradients f_n from several steps

$$Y_{n+1} = \sum_{j=1}^{k} \left(\alpha_j Y_{n-j+1} + \beta_j f_{n-j+1}\Delta t\right) + \beta_0 f_{n+1}\Delta t. \tag{14.113}$$

The method is explicit if $\beta_0 = 0$ and implicit otherwise. Multistep methods have a small local error and need fewer function evaluations. On the other hand, they have to be combined with other methods (like Runge Kutta) to start and end properly and it can be rather complicated to change the step size during the calculation. Three families of linear multistep methods are commonly used: explicit Adams-Bashforth methods, implicit Adams-Moulton methods and backward differentiation formulas (also known as Gear formulas [263]).

The first order Taylor series expansion around t_n

$$\begin{aligned} Y_{n+1} &= \sum_{j=1}^{k} \left(\alpha_j (Y_n - (j-1)\Delta t f_n + \cdots) + \beta_j (f_n - (j-1)\Delta t \dot{f}_n + \cdots)\Delta t\right) \\ &\quad + \beta_0 (f_n + \Delta t \dot{f}_n + \cdots)\Delta t \\ &= \sum_{j=1}^{k} \alpha_j Y_n + \Delta t f_n \left[\beta_0 + \sum_{j=1}^{k} -\alpha_j (j-1) + \beta_j\right] + \cdots \end{aligned} \tag{14.114}$$

coincides with the exact expansion

$$Y_{n+1} = Y_n + \Delta t f_n + \cdots \tag{14.115}$$

only if

$$\sum_{j=1}^{k} \alpha_j = 1 \tag{14.116}$$

and

$$\beta_0 + \sum_{j=1}^{k} -\alpha_j(j-1) + \beta_j = 1 \tag{14.117}$$

which can be written as

$$\sum_{j=1}^{k} j\alpha_j = \sum_{j=0}^{k} \beta_j. \tag{14.118}$$

Conditions (14.116) and (14.118) are known as the conditions of consistency.

While (14.14) may be nonlinear in general, numerical methods are often analyzed applying them to a simple linear test equation[3]

$$f(t) = \frac{\mathrm{d}}{\mathrm{d}t} Y(t) = \lambda Y(t). \tag{14.119}$$

Equation (14.113) then simplifies to the recursion

$$Y_{n+1} = \frac{1}{1-\beta_0\lambda\Delta t} \sum_{j=1}^{k} \left(\alpha_j + \beta_j\lambda\Delta t\right) Y_{n-j+1} = \sum_{j=1}^{k} \gamma_j Y_{n-j+1} \tag{14.120}$$

with

$$\gamma_j = \frac{\alpha_j + \beta_j\lambda\Delta t}{1-\beta_0\lambda\Delta t}. \tag{14.121}$$

In matrix form this recursion can be written as

$$\begin{pmatrix} Y_{n+1-k} \\ Y_{n+2-k} \\ \vdots \\ Y_n \\ Y_{n+1} \end{pmatrix} = \begin{pmatrix} 0 & 1 & 0 & \cdots & 0 \\ \vdots & \ddots & 1 & \ddots & \vdots \\ \vdots & & \ddots & \ddots & 0 \\ 0 & \cdots & \cdots & 0 & 1 \\ \gamma_k & \cdots & \cdots & \gamma_2 & \gamma_1 \end{pmatrix} \begin{pmatrix} Y_{n-k} \\ Y_{n+1-k} \\ \vdots \\ Y_{n-1} \\ Y_n \end{pmatrix}. \tag{14.122}$$

To find the eigenvalues, we calculate the determinant

$$D_k = \det \begin{vmatrix} x & -1 & & & \\ & x & -1 & & \\ & & \ddots & \ddots & \\ & & & x & -1 \\ -\gamma_k & -\gamma_{k-1} & \cdots & -\gamma_2 & x-\gamma_1 \end{vmatrix}$$

[3] It may be assumed that for a not too large time interval the equation can be linearized.

$$= x \det \begin{vmatrix} x & -1 & & \\ & \ddots & \ddots & \\ & & x & -1 \\ -\gamma_{k-1} & \cdots & -\gamma_2 & x-\gamma_1 \end{vmatrix} + \det \begin{vmatrix} 0 & -1 & & \\ & x & \ddots & \\ & & \ddots & -1 \\ -\gamma_k & \cdots & -\gamma_2 & x-\gamma_1 \end{vmatrix} \quad (14.123)$$

and obtain the recursion

$$D_k = x D_{k-1} - \gamma_k \quad (14.124)$$

which, starting from

$$D_2 = x^2 - \gamma_1 x - \gamma_2 \quad (14.125)$$

leads to

$$D_k = x^k - \gamma_1 x^{k-1} - \gamma_2 x^{k-2} \cdots - \gamma_k = x^k - \sum_{j=1}^{k} \gamma_j x^{k-\gamma}. \quad (14.126)$$

We multiply by the common denominator and obtain the characteristic polynomial

$$\pi(x) = (1 - \beta_0 \lambda \Delta t)\, x^k - \sum_{j=1}^{k} \left(\alpha_j + \beta_j \lambda \Delta t\right) x^{k-j} = \rho(x) - \lambda \Delta t \sigma(x) \quad (14.127)$$

which can be divided into the first and second characteristic polynomial

$$\rho(x) = x^k - \sum_{j=1}^{k} \alpha_j x^{k-j} \quad (14.128)$$

$$\sigma(x) = \sum_{j=0}^{k} \beta_j x^{k-j}. \quad (14.129)$$

If the roots of $\rho(x)$ all have magnitude $|x| \le 1$ and there are no multiple roots with magnitude $|x| = 1$,[4] the method is called zero stable, which means that for the differential equation $\dot{Y} = 0$ numerical errors can not blow up in time [264].[5] Both, consistency and zero stability are required for a convergent method. To determine the region of stability, we look for roots with magnitude 1 which can be written

$$x = \mathrm{e}^{\mathrm{i}\varphi} \quad (14.130)$$

and solve the equation

$$\rho(\mathrm{e}^{\mathrm{i}\varphi}) - \lambda \Delta t \sigma(\mathrm{e}^{\mathrm{i}\varphi}) = 0. \quad (14.131)$$

[4] This is known as root condition.

[5] If there are multiple roots with magnitude 1, errors can grow algebraically.

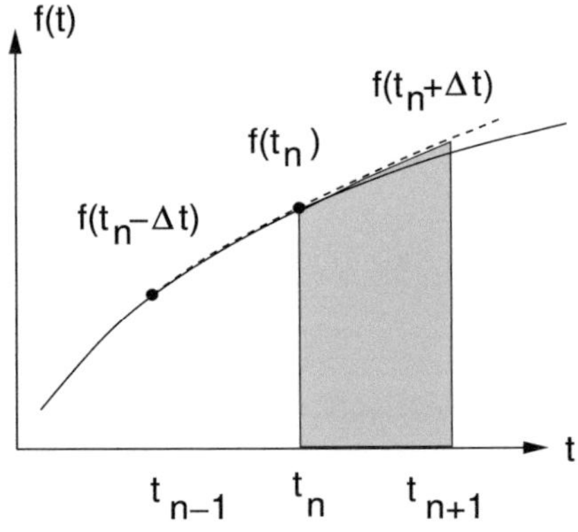

Fig. 14.9 Adams-Bashforth method

This defines a curve in the complex plane which contains all points, which are potentially on the boundary

$$z(\varphi) = \frac{\rho(\mathrm{e}^{\mathrm{i}\varphi})}{\sigma(\mathrm{e}^{\mathrm{i}\varphi})}. \tag{14.132}$$

For simpler methods this curve divides the plane into two regions and gives the boundary directly. For higher order methods, it may cross itself and define more than two regions. The interior of the stability region has to be identified by considering suitable points explicitly [265].

14.9.1 Adams-Bashforth Methods

The explicit Adams-Bashforth method of order r uses polynomial interpolation of the gradients from the last $r - 1$ steps (Fig. 14.9)[6]

$$p(t_n) = f(Y_n, t_n), \cdots p(t_{n-r+1}) = f(Y_{n-r+1}, t_{n-r+1}) \tag{14.133}$$

to calculate the approximation

$$Y_{n+1} - Y_n \approx \int_{t_n}^{t_{n+1}} p(t)dt \tag{14.134}$$

which is generally a linear combination of $f_n \cdots f_{n-r+1}$ of the form

$$Y_{n+1} - Y_n = \Delta t \sum_{j=1}^{r} \beta_j f_{n+1-j}. \tag{14.135}$$

[6] Coefficients for the implicit and explicit methods up to order 20 can be found in [266] and a method for efficient numerical calculation in [267].

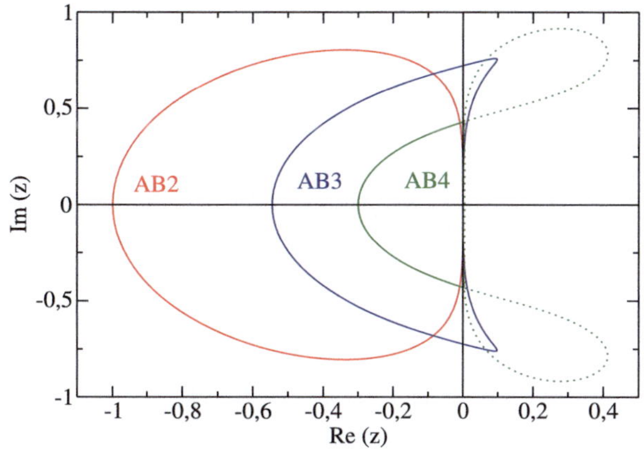

Fig. 14.10 Stability regions of explicit Adams-Bashforth methods. The curves defined by (14.132) are shown for the Adams-Bashforth method of 2nd order (14.137, red), 3rd order (14.138, blue) and 4th order (14.139, green). Stability is found only inside the areas enclosed by the full curves. For AB4, the curve intersects itself and encloses two more regions (dotted) of instability

The first characteristic polynomial

$$\rho = x^r - x^{r-1} = (x-1)x^{r-2} \tag{14.136}$$

always has a single root $x = 1$, hence Adams-Bashforth methods are zero stable. For example (Fig. 14.10), the Adams-Bashforth formulas of order r=2 to 4 are[7, 8]:

$$Y_{n+1} - Y_n = \frac{\Delta t}{2}(3f_n - f_{n-1}) + O(\Delta t^3) \tag{14.137}$$

$$Y_{n+1} - Y_n = \frac{\Delta t}{12}(23f_n - 16f_{n-1} + 5f_{m-2}) + O(\Delta t^4) \tag{14.138}$$

$$Y_{n+1} - Y_n = \frac{\Delta t}{24}(55f_n - 59f_{n-1} + 37f_{n-2} - 9f_{n-3}) + O(\Delta t^5). \tag{14.139}$$

14.9.2 Adams-Moulton Methods

The implicit Adams-Moulton method also uses the yet not known value f_{n+1} (Fig. 14.11) to obtain the polynomial

---

[7] AB1 is the explicit Euler rule.

[8] Obviously these methods are consistent since $\sum_{j=1}^{r} \alpha_j = \alpha_1 = 1$ and $\sum_{j=1}^{r} j\alpha_j = \sum_{j=0}^{r} \beta_j = 1$.

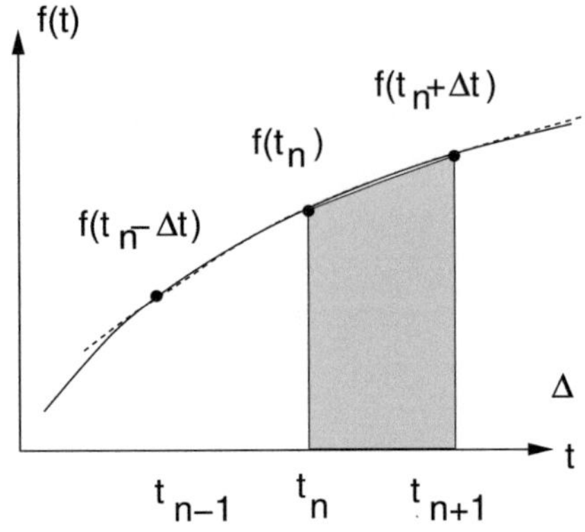

Fig. 14.11 Adams-Moulton method

$$p(t_{n+1}) = f_{n+1}, \cdots p(t_{n-r+2}) = f_{n-r+2}. \tag{14.140}$$

This leads to expressions of the form

$$Y_{n+1} - Y_n = \Delta t \left[\beta_0 f_{n+1} + \sum_{j=1}^{r-1} \beta_j f_{n+1-j} \right]. \tag{14.141}$$

The corresponding Adams-Moulton formulas of order r = 2–5 (Fig. 14.12) are[9, 10]:

$$Y_{n+1} - Y_n = \frac{\Delta t}{2}(f_{n+1} + f_n) + O(\Delta t^3) \tag{14.142}$$

$$Y_{n+1} - Y_n = \frac{\Delta t}{12}(5f_{n+1} + 8f_n - f_{n-1}) + O(\Delta t^4) \tag{14.143}$$

$$Y_{n+1} - Y_n = \frac{\Delta t}{24}(9f_{n+1} + 19f_n - 5f_{n-1} + f_{n-2}) + O(\Delta t^5) \tag{14.144}$$

$$Y_{n+1} - Y_n = \frac{\Delta t}{720}(251f_{n+1} + 646f_n - 264f + 106f_{n-2} - 19f_{n-3}) + O(\Delta t^6). \tag{14.145}$$

[9] AM1 coincides with the backward Euler method and AM2 with the implicit trapezoidal rule.

[10] Obviously these methods are consistent since $\sum_{j=1}^{r-1} \alpha_j = \alpha_1 = 1$ and $\sum_{j=1}^{r-1} j\alpha_j = \sum_{j=0}^{r-1} \beta_j = 1$.

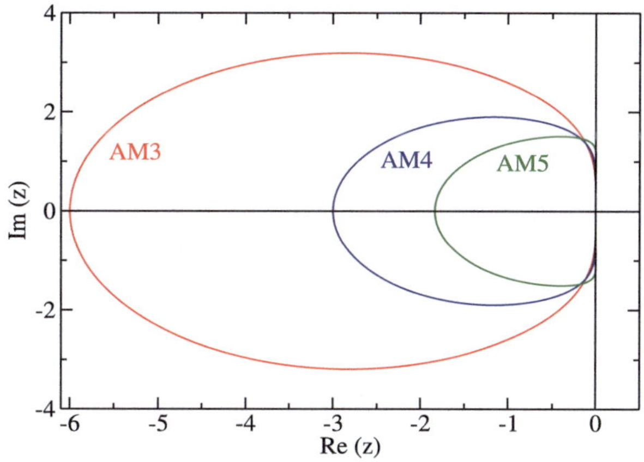

Fig. 14.12 Stability regions of implicit Adams-Moulton methods. The curves defined by (14.132) are shown for the Adams-Moulton method of 3rd order (14.143,red), 4th order (14.144,blue) and 5th order (14.145,green). Stability is found inside the areas enclosed by the curves. Stability regions are smaller for higher order methods

14.9.3 Adams-Bashforth-Moulton Methods

The Adams-Bashforth-Moulton methods, often simply called "implicit Adams methods" combine the explicit method as a predictor step to calculate an estimate Y_{n+1}^{p} with a corrector step using the implicit method of same order. The predictor step

$$Y_{n+1}^{p} = Y_n + \Delta t \sum_{j=1}^{r} \beta_j^p f_{n+1-j} \tag{14.146}$$

is used to approximate the new gradient

$$f_{n+1}^{p} = f(Y_{n+1}^{p}) \tag{14.147}$$

which is needed to apply the implicit corrector step

$$Y_{n+1} = Y_n + \Delta t \left[\beta_0^c f_{n+1}^{p} + \sum_{j=1}^{r-1} \beta_j^c f_{n+1-r} \right]. \tag{14.148}$$

The corrector step has to be repeated until convergence is achieved.

For non-stiff problems often higher order methods provide maximum accuracy. Figure 14.13 shows results from exercise Sect. 14.9.5 where the implicit Adams method of order 6–14 is used to simulate a circular Kepler orbit in the xy-plane. Since the exact solution is known, it is easy to start the calculation and to obtain the global error.

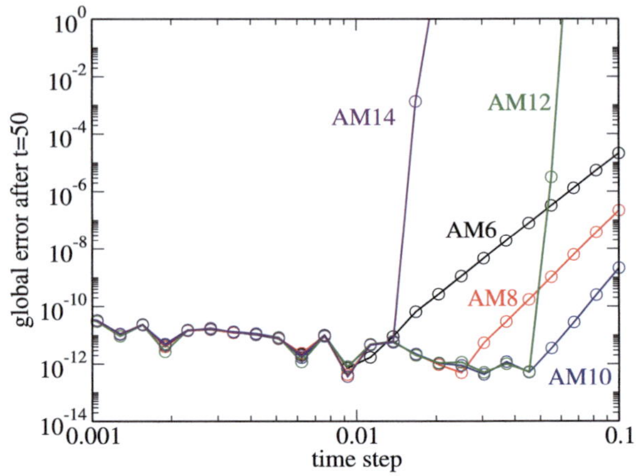

Fig. 14.13 Global error of higher order implicit Adams methods. For small time steps $\Delta t < 0.01$ numerical errors dominate which increase with the number of steps and therefore $\propto 1/\Delta t$. The minimum error is comparable but AM10 and AM12 allow for larger time steps. AM14 and even higher orders become unstable since their stability regions are small

14.9.4 Backward Differentiation (Gear) Methods

Gear methods [263] are popular for stiff problems.[11] They are implicit and usually combined with a modified Newton method. They make use of several previous function values $Y_n, Y_{n-1} \ldots$ and the gradient f_{n+1} at time $t + \Delta t$. Only methods of order $r \leq 6$ are stable and useful. The general formula (14.113) is

$$Y_{n+1} = \sum_{j=1}^{r} \alpha_j Y_{n-j+1} + \beta_0 f_{n+1} \Delta t. \tag{14.149}$$

The coefficients are determined such, that a polynomial of order r is integrated exactly. For $r = 1$ this becomes

$$Y_{n+1} = \alpha_1 Y_n + \beta_0 f_1 \Delta t \tag{14.150}$$

and all linear polynomials

$$p = p_0 + p_1(t - t_n), \quad \frac{dp}{dt} = p_1 \tag{14.151}$$

are integrated exactly if

$$p_0 + p_1 \Delta t = \alpha_1 p_0 + \beta_0 p_1 \tag{14.152}$$

which is the case for

$$\alpha_1 = 1, \quad \beta_0 = \Delta t. \tag{14.153}$$

Hence the first order Gear method is

---

[11] Differential equations with strongly varying eigenvalues which require very small time steps with standard methods.

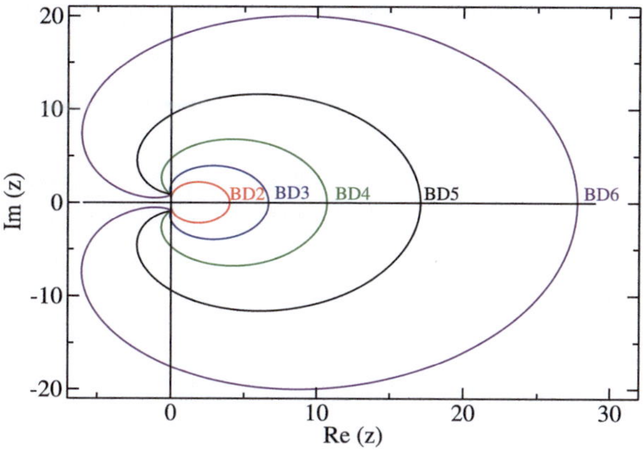

Fig. 14.14 Stability regions of implicit Backward differentiation formulas. The curves defined by (14.132) are shown for the BDF methods of order 3 to 6. Stability is found outside the areas enclosed by the curves. All methods are stable on the negative real axis

$$Y_{n+1} = Y_n + f_{n+1}\Delta t + O(\Delta t^2) \tag{14.154}$$

which coincides with the implicit Euler method. The higher order stable Gear methods (Fig. 14.14) are given by [268]

$$r = 2: \quad Y_{n+1} = \frac{4}{3}Y_n - \frac{1}{3}Y_{n-1} + \frac{2}{3}f_{n+1}\Delta t + O(\Delta t^3) \tag{14.155}$$

$$r = 3: \quad Y_{n+1} = \frac{18}{11}Y_n - \frac{9}{11}Y_{n-1} + \frac{2}{11}Y_{n-2} + \frac{6}{11}f_{n+1}\Delta t + O(\Delta t^4) \tag{14.156}$$

$$r = 4: \quad Y_{n+1} = \frac{48}{25}Y_n - \frac{36}{25}Y_{n-1} + \frac{16}{25}Y_{n-2} - \frac{3}{25}Y_{n-3} + \frac{12}{25}f_{n+1}\Delta t + O(\Delta t^5) \tag{14.157}$$

$$\begin{aligned} r = 5: \quad Y_{n+1} = {} & \frac{300}{137}Y_n - \frac{300}{137}Y_{n-1} + \frac{200}{137}Y_{n-2} - \frac{75}{137}Y_{n-3} \\ & + \frac{12}{137}Y_{n-4} + \frac{60}{137}f_{n+1}\Delta t + O(\Delta t^6) \end{aligned} \tag{14.158}$$

$$\begin{aligned} r = 6: \quad Y_{n+1} = {} & \frac{120}{49}Y_n - \frac{150}{49}Y_{n-1} + \frac{400}{147}Y_{n-2} - \frac{75}{49}Y_{n-3} \\ & + \frac{24}{49}Y_{n-4} - \frac{10}{147}Y_{n-5} + \frac{20}{49}f_{n+1}\Delta t + O(\Delta t^7). \end{aligned} \tag{14.159}$$

14.9.5 General Predictor-Corrector Methods

The general class of linear multistep predictor corrector methods [269] uses a predictor step

$$Y_{n+1}^{(0)} = \sum_{j=1}^{k} \left(\alpha_j^{(p)} Y_{n-j+1} + \beta_j^{(p)} f_{n-j+1} \Delta t \right) \tag{14.160}$$

which is corrected using the formula

$$Y_{n+1}^{(1)} = \sum_{j=1}^{k} \left(\alpha_j^{(c)} Y_{n-j+1} + \beta_j^{(c)} f_{n-j+1} \Delta t \right) + \beta_0 f(Y_{n+1}^{(0)}, t_{n+1}) \Delta t \tag{14.161}$$

and further iterations

$$Y_{n+1}^{(m+1)} = Y_{n+1}^{(m)} - \beta_0 \left[f(Y_{n+1}^{(m-1)}, t_{n+1}) - f(Y_{n+1}^{(m)}, t_{n+1}) \right] \Delta t \quad m = 1 \ldots M-1 \tag{14.162}$$

$$Y_{n+1} = Y_{n+1}^{(M)}, \quad \dot{Y}_{n+1} = f(Y_{n+1}^{(M-1)}, t_{n+1}). \tag{14.163}$$

The coefficients α, β have to be determined to optimize accuracy and stability.

14.10 Verlet Methods

For classical molecular dynamics simulations it is necessary to calculate very long trajectories. Here a family of symplectic methods often is used which conserve the phase space volume [270–275]. The equations of motion of a classical interacting N-body system are

$$m_i \ddot{\mathbf{x}}_i = F_i \tag{14.164}$$

where the force acting on atom i can be calculated once a specific force field is chosen. Let us write these equations as a system of first order differential equations

$$\begin{pmatrix} \dot{\mathbf{x}}_i \\ \dot{\mathbf{v}}_i \end{pmatrix} = \begin{pmatrix} \mathbf{v}_i \\ \mathbf{a}_i \end{pmatrix} \tag{14.165}$$

where $\mathbf{x}(t)$ and $\mathbf{v}(t)$ are functions of time and the forces $m\mathbf{a}(\mathbf{x}(t))$ are functions of the time dependent coordinates.

14.10.1 Liouville Equation

We rewrite (14.165) as

$$\begin{pmatrix} \dot{\mathbf{x}} \\ \dot{\mathbf{v}} \end{pmatrix} = \mathcal{L} \begin{pmatrix} \mathbf{x} \\ \mathbf{v} \end{pmatrix} \tag{14.166}$$

where the Liouville operator $\mathcal{L}$ acts on the vector containing all coordinates and velocities:

$$\mathcal{L} \begin{pmatrix} \mathbf{x} \\ \mathbf{v} \end{pmatrix} = \left(\mathbf{v}\frac{\partial}{\partial \mathbf{x}} + \mathbf{a}\frac{\partial}{\partial \mathbf{v}} \right) \begin{pmatrix} \mathbf{x} \\ \mathbf{v} \end{pmatrix}. \tag{14.167}$$

The Liouville equation (14.166) can be formally solved by

$$\begin{pmatrix} \mathbf{x}(t) \\ \mathbf{v}(t) \end{pmatrix} = \mathrm{e}^{\mathcal{L}t} \begin{pmatrix} \mathbf{x}(0) \\ \mathbf{v}(0) \end{pmatrix}. \tag{14.168}$$

For a better understanding let us evaluate the first members of the Taylor series of the exponential:

$$\mathcal{L} \begin{pmatrix} \mathbf{x} \\ \mathbf{v} \end{pmatrix} = \left(\mathbf{v}\frac{\partial}{\partial \mathbf{x}} + \mathbf{a}\frac{\partial}{\partial \mathbf{v}} \right) \begin{pmatrix} \mathbf{x} \\ \mathbf{v} \end{pmatrix} = \begin{pmatrix} \mathbf{v} \\ \mathbf{a} \end{pmatrix} \tag{14.169}$$

$$\mathcal{L}^2 \begin{pmatrix} \mathbf{x} \\ \mathbf{v} \end{pmatrix} = \left(\mathbf{v}\frac{\partial}{\partial \mathbf{x}} + \mathbf{a}\frac{\partial}{\partial \mathbf{v}} \right) \begin{pmatrix} \mathbf{v} \\ \mathbf{a}(\mathbf{x}) \end{pmatrix} = \begin{pmatrix} \mathbf{a} \\ \mathbf{v}\frac{\partial}{\partial \mathbf{x}}\mathbf{a} \end{pmatrix} \tag{14.170}$$

$$\mathcal{L}^3 \begin{pmatrix} \mathbf{x} \\ \mathbf{v} \end{pmatrix} = \left(\mathbf{v}\frac{\partial}{\partial \mathbf{x}} + \mathbf{a}\frac{\partial}{\partial \mathbf{v}} \right) \begin{pmatrix} \mathbf{a} \\ \mathbf{v}\frac{\partial}{\partial \mathbf{x}}\mathbf{a} \end{pmatrix} = \begin{pmatrix} \mathbf{v}\frac{\partial}{\partial \mathbf{x}}\mathbf{a} \\ \mathbf{a}\frac{\partial}{\partial \mathbf{x}}\mathbf{a} + \mathbf{v}\mathbf{v}\frac{\partial}{\partial \mathbf{x}}\frac{\partial}{\partial \mathbf{x}}\mathbf{a} \end{pmatrix}. \tag{14.171}$$

But since

$$\frac{\mathrm{d}}{\mathrm{d}t}\mathbf{a}(\mathbf{x}(t)) = \mathbf{v}\frac{\partial}{\partial \mathbf{x}}\mathbf{a} \tag{14.172}$$

$$\frac{\mathrm{d}^2}{\mathrm{d}t^2}\mathbf{a}(\mathbf{x}(t)) = \frac{\mathrm{d}}{\mathrm{d}t}(\mathbf{v}\frac{\partial}{\partial \mathbf{x}}\mathbf{a}) = \mathbf{a}\frac{\partial}{\partial \mathbf{x}}\mathbf{a} + \mathbf{v}\mathbf{v}\frac{\partial}{\partial \mathbf{x}}\frac{\partial}{\partial \mathbf{x}}\mathbf{a} \tag{14.173}$$

we recover

$$(1 + t\mathcal{L} + \frac{1}{2}t^2\mathcal{L}^2 + \frac{1}{6}t^3\mathcal{L}^3 + \cdots) \begin{pmatrix} \mathbf{x} \\ \mathbf{v} \end{pmatrix} = \begin{pmatrix} \mathbf{x} + \mathbf{v}t + \frac{1}{2}t^2\mathbf{a} + \frac{1}{6}t^3\dot{\mathbf{a}} + \cdots \\ \mathbf{v} + \mathbf{a}t + \frac{1}{2}t^2\dot{\mathbf{a}} + \frac{1}{6}t^3\ddot{\mathbf{a}} + \cdots \end{pmatrix}. \tag{14.174}$$

14.10.2 Split Operator Approximation

We introduce a small time step $\Delta t = t/N$ and write

$$e^{\mathcal{L}t} = \left(e^{\mathcal{L}\Delta t}\right)^N \tag{14.175}$$

For the small time step Δt the split-operator approximation can be used which approximately factorizes the exponential operator. For example, write the Liouville operator as the sum of two terms

$$\mathcal{L}_A = \mathbf{v}\frac{\partial}{\partial \mathbf{x}} \qquad \mathcal{L}_B = \mathbf{a}\frac{\partial}{\partial \mathbf{v}} \tag{14.176}$$

and make the approximation

$$e^{\mathcal{L}\Delta t} = e^{\mathcal{L}_A\Delta t}e^{\mathcal{L}_B\Delta t} + \cdots . \tag{14.177}$$

Each of the two factors simply shifts positions or velocities

$$e^{\mathcal{L}_A\Delta t}\begin{pmatrix}\mathbf{x}\\ \mathbf{v}\end{pmatrix} = \begin{pmatrix}\mathbf{x}+\mathbf{v}\Delta t\\ \mathbf{v}\end{pmatrix} \qquad e^{\mathcal{L}_B\Delta t}\begin{pmatrix}\mathbf{x}\\ \mathbf{v}\end{pmatrix} = \begin{pmatrix}\mathbf{x}\\ \mathbf{v}+\mathbf{a}\Delta t\end{pmatrix} \tag{14.178}$$

since these two steps correspond to either motion with constant velocities or constant coordinates and forces.

14.10.3 Position Verlet Method

Often the following approximation is used which is symmetrical in time

$$e^{\mathcal{L}\Delta t} = e^{\mathcal{L}_A\Delta t/2}e^{\mathcal{L}_B\Delta t}e^{\mathcal{L}_A\Delta t/2} + \cdots . \tag{14.179}$$

The corresponding algorithm is the so called position Verlet method (Fig. 14.15):

$$\mathbf{x}_{n+1/2} = \mathbf{x}_n + \mathbf{v}_n\frac{\Delta t}{2} \tag{14.180}$$

$$\mathbf{v}_{n+1} = \mathbf{v}_n + \mathbf{a}_{n+1/2}\Delta t = \mathbf{v}(t_n+\Delta t) + O(\Delta t^3) \tag{14.181}$$

$$\mathbf{x}_{n+1} = \mathbf{x}_{n+1/2} + \mathbf{v}_{n+1}\frac{\Delta t}{2} = \mathbf{x}_n + \frac{\mathbf{v}_n+\mathbf{v}_{n+1}}{2}\Delta t = \mathbf{x}(t_n+\Delta t) + O(\Delta t^3). \tag{14.182}$$

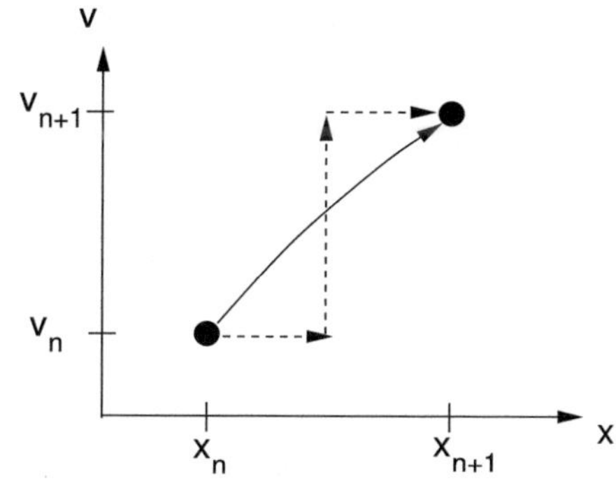

Fig. 14.15 Position Verlet method. The exact integration path is approximated by two half-steps with constant velocities and one step with constant coordinates

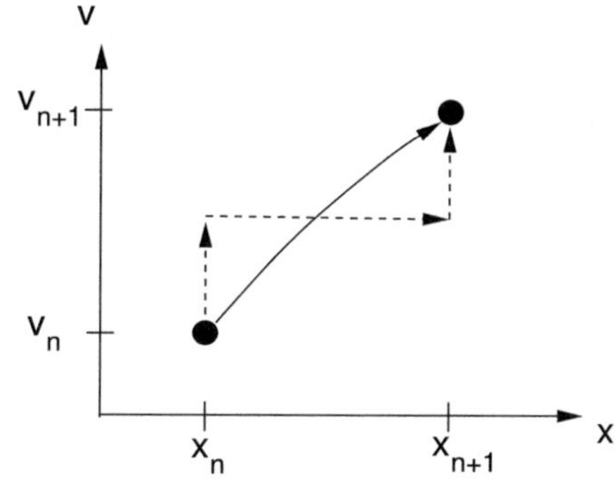

Fig. 14.16 Velocity Verlet method. The exact integration path is approximated by two half-steps with constant coordinates and one step with constant velocities

14.10.4 Velocity Verlet Method

If we exchange operators A and B we have

$$e^{\mathcal{L}\Delta t} = e^{\mathcal{L}_B \Delta t/2} e^{\mathcal{L}_A \Delta t} e^{\mathcal{L}_B \Delta t/2} + \cdots \tag{14.183}$$

which produces the velocity Verlet algorithm (Fig. 14.16):

$$\mathbf{v}_{n+1/2} = \mathbf{v}_n + \mathbf{a}_n \frac{\Delta t}{2} \tag{14.184}$$

$$\mathbf{x}_{n+1} = \mathbf{x}_n + \mathbf{v}_{n+1/2}\Delta t = \mathbf{x}_n + \mathbf{v}_n \Delta t + \mathbf{a}_n \frac{\Delta t^2}{2} = \mathbf{x}(t_n + \Delta t) + O(\Delta t^3) \tag{14.185}$$

$$\mathbf{v}_{n+1} = \mathbf{v}_{n+1/2} + \mathbf{a}_{n+1}\frac{\Delta t}{2} = \mathbf{v}_n + \frac{\mathbf{a}_n + \mathbf{a}_{n+1}}{2}\Delta t = \mathbf{v}(t_n + \Delta t) + O(\Delta t^3). \tag{14.186}$$

14.10.5 Stoermer-Verlet Method

The velocity Verlet method is equivalent to Stoermer's version [276] of the Verlet method which is a two step method given by

$$\mathbf{x}_{n+1} = 2\mathbf{x}_n - \mathbf{x}_{n-1} + \mathbf{a}_n \Delta t^2 \tag{14.187}$$

$$\mathbf{v}_n = \frac{\mathbf{x}_{n+1} - \mathbf{x}_{n-1}}{2\Delta t}. \tag{14.188}$$

To show the equivalence we add two consecutive position vectors

$$\mathbf{x}_{n+2} + \mathbf{x}_{n+1} = 2\mathbf{x}_{n+1} + 2\mathbf{x}_n - \mathbf{x}_n - \mathbf{x}_{n-1} + (\mathbf{a}_{n+1} + \mathbf{a}_n)\Delta t^2 \tag{14.189}$$

which simplifies to

$$\mathbf{x}_{n+2} - \mathbf{x}_n - (\mathbf{x}_{n+1} - \mathbf{x}_n) = (\mathbf{a}_{n+1} + \mathbf{a}_n)\Delta t^2. \tag{14.190}$$

This can be expressed as the difference of two consecutive velocities:

$$2(\mathbf{v}_{n+1} - \mathbf{v}_n) = (\mathbf{a}_{n+1} + \mathbf{a}_n)\Delta t \tag{14.191}$$

Now we substitute

$$\mathbf{x}_{n-1} = \mathbf{x}_{n+1} - 2\mathbf{v}_n \Delta t \tag{14.192}$$

to get

$$\mathbf{x}_{n+1} = 2\mathbf{x}_n - \mathbf{x}_{n+1} + 2\mathbf{v_n} \Delta t + \mathbf{a}_n \Delta t^2 \tag{14.193}$$

which simplifies to

$$\mathbf{x}_{n+1} = \mathbf{x}_n + \mathbf{v}_n \Delta t + \frac{\mathbf{a}_n}{2} \Delta t^2. \tag{14.194}$$

Thus the equations of the velocity Verlet algorithm have been recovered. However, since the Verlet method is a 2-step method, the choice of initial values is important. The Stoermer-Verlet method starts from two coordinate sets x_0, x_1. The first step is

$$\mathbf{x}_2 = 2\mathbf{x_1} - \mathbf{x_0} + a_1 \Delta t^2 \tag{14.195}$$

$$\mathbf{v}_1 = \frac{\mathbf{x}_2 - \mathbf{x}_0}{2\Delta t} = \frac{\mathbf{x}_1 - \mathbf{x}_0}{\Delta t} + \frac{\mathbf{a}_1}{2} \Delta t^2. \tag{14.196}$$

The velocity Verlet method, on the other hand, starts from one set of coordinates and velocities $\mathbf{x}_1, \mathbf{v}_1$. Here the first step is

$$\mathbf{x}_2 = \mathbf{x}_1 + \mathbf{v}_1 \Delta t + \mathbf{a}_1 \frac{\Delta t^2}{2} \tag{14.197}$$

$$\mathbf{v}_2 = \mathbf{v}_1 + \frac{\mathbf{a}_1 + \mathbf{a}_2}{2} \Delta t. \tag{14.198}$$

The two methods give the same resulting trajectory if we choose

$$\mathbf{x}_0 = \mathbf{x}_1 - \mathbf{v}_1 \Delta t + \frac{\mathbf{a}_1}{2} \Delta t^2. \tag{14.199}$$

If, on the other hand, $\mathbf{x}_0$ is known with higher precision, the local error order of Stoermer's algorithm changes as can be seen from addition of the two Taylor series

$$\mathbf{x}(t_n + \Delta t) = \mathbf{x}_n + \mathbf{v}_n \Delta t + \frac{\mathbf{a}_n}{2} \Delta t^2 + \frac{\dot{\mathbf{a}}_n}{6} \Delta t^3 + \cdots \tag{14.200}$$

$$\mathbf{x}(t_n - \Delta t) = \mathbf{x}_n - \mathbf{v}_n \Delta t + \frac{\mathbf{a}_n}{2} \Delta t^2 - \frac{\dot{\mathbf{a}}_n}{6} \Delta t^3 + \cdots \tag{14.201}$$

which gives

$$\mathbf{x}(t_n + \Delta t) = 2\mathbf{x}(t_n) - \mathbf{x}(t_n - \Delta t) + \mathbf{a}_n \Delta t^2 + O(\Delta t^4) \tag{14.202}$$

$$\frac{\mathbf{x}(t_n + \Delta t) - \mathbf{x}(t_n - \Delta t)}{2\Delta t} = \mathbf{v}_n + O(\Delta t^2). \tag{14.203}$$

14.10.6 Error Accumulation for the Stoermer-Verlet Method

Equation (14.202) gives only the local error of one single step. Assume the start values $\mathbf{x}_0$ and $\mathbf{x}_1$ are exact. The next value $\mathbf{x}_2$ has an error with the leading term $\Delta x_2 = \alpha \Delta t^4$. If the trajectory is sufficiently smooth and the time step not too large the coefficient α will vary only slowly and the error of the next few iterations is given by

$$\Delta x_3 = 2\Delta x_2 - \Delta x_1 = 2\alpha \Delta t^4$$

$$\Delta x_4 = 2\Delta x_3 - \Delta x_2 = 3\alpha \Delta t^4$$

$$\vdots$$

$$\Delta x_{n+1} = n\alpha \Delta t^4. \tag{14.204}$$

This shows that the effective error order of the Stoermer-Verlet method is only $O(\Delta t^3)$ similar to the velocity Verlet method.

14.10.7 Beeman's Method

Beeman and Schofield [277, 278] introduced a method which is very similar to the Stoermer-Verlet method but calculates the velocities with higher precision. This is important if, for instance, the kinetic energy has to be calculated. Starting from the Taylor series

$$\mathbf{x}_{n+1} = \mathbf{x}_n + \mathbf{v}_n \Delta t + \mathbf{a}_n \frac{\Delta t^2}{2} + \dot{\mathbf{a}}_n \frac{\Delta t^3}{6} + \ddot{\mathbf{a}}_n \frac{\Delta t^4}{24} + \ldots \tag{14.205}$$

the derivative of the acceleration is approximated by a backward difference

$$\begin{aligned}\mathbf{x}_{n+1} &= \mathbf{x}_n + \mathbf{v}_n \Delta t + \mathbf{a}_n \frac{\Delta t^2}{2} + \frac{\mathbf{a}_n - \mathbf{a}_{n-1}}{\Delta t} \frac{\Delta t^3}{6} + O(\Delta t^4) \\ &= \mathbf{x}_n + \mathbf{v}_n \Delta t + \frac{4\mathbf{a}_n - \mathbf{a}_{n-1}}{6} \Delta t^2 + O(\Delta t^4).\end{aligned} \tag{14.206}$$

This equation can be used as an explicit step to update the coordinates or as a predictor step in combination with the implicit corrector step

$$\begin{aligned}\mathbf{x}_{n+1} &= \mathbf{x}_n + \mathbf{v}_n \Delta t + \mathbf{a}_n \frac{\Delta t^2}{2} + \frac{\mathbf{a}_{n+1} - \mathbf{a}_n}{\Delta t} \frac{\Delta t^3}{6} + O(\Delta t^4) \\ &= \mathbf{x}_n + \mathbf{v}_n \Delta t + \frac{\mathbf{a}_{n+1} + 2\mathbf{a}_n}{6} \Delta t^2 + O(\Delta t^4)\end{aligned} \tag{14.207}$$

which can be applied repeatedly (usually two iterations are sufficient). Similarly, the Taylor series of the velocity is approximated by

$$\begin{aligned}\mathbf{v}_{n+1} &= \mathbf{v}_n + \mathbf{a}_n \Delta t + \dot{\mathbf{a}}_n \frac{\Delta t^2}{2} + \ddot{\mathbf{a}}_n \frac{\Delta t^3}{6} + \ldots \\ &= \mathbf{v}_n + \mathbf{a}_n \Delta t + \left(\frac{\mathbf{a}_{n+1} - \mathbf{a}_n}{\Delta t} + O(\Delta t) \right) \frac{\Delta t^2}{2} + \ldots \\ &= \mathbf{v}_n + \frac{\mathbf{a}_{n+1} + \mathbf{a}_n}{2} \Delta t + O(\Delta t^3).\end{aligned} \tag{14.208}$$

Inserting the velocity from (14.207) we obtain the corrector step for the velocity

$$\begin{aligned}\mathbf{v}_{n+1} &= \frac{\mathbf{x}_{n+1} - \mathbf{x}_n}{\Delta t} - \frac{\mathbf{a}_{n+1} + 2\mathbf{a}_n}{6} \Delta t + \frac{\mathbf{a}_{n+1} + \mathbf{a}_n}{2} \Delta t + O(\Delta t^3) \\ &= \frac{\mathbf{x}_{n+1} - \mathbf{x}_n}{\Delta t} + \frac{2\mathbf{a}_{n+1} + \mathbf{a}_n}{6} \Delta t + O(\Delta t^3).\end{aligned} \tag{14.209}$$

In combination with (14.206) this can be replaced by

$$\mathbf{v}_{n+1} = \mathbf{v}_n + \frac{4\mathbf{a}_n - \mathbf{a}_{n-1}}{6}\Delta t + \frac{2\mathbf{a}_{n+1} + \mathbf{a}_n}{6}\Delta t + O(\Delta t^3)$$

$$= \mathbf{v}_n + \frac{2\mathbf{a}_{n+1} + 5\mathbf{a}_n - \mathbf{a}_{n-1}}{6}\Delta t + O(\Delta t^3). \tag{14.210}$$

Together, (14.206) and (14.210) provide an explicit method which is usually understood as Beeman's method. Inserting the velocity (14.209) from the previous step

$$\mathbf{v}_n = \frac{\mathbf{x}_n - \mathbf{x}_{n-1}}{\Delta t} + \frac{2\mathbf{a}_n + \mathbf{a}_{n-1}}{6}\Delta t + O(\Delta t^3) \tag{14.211}$$

into (14.206) gives

$$\mathbf{x}_{n+1} = 2\mathbf{x}_n - \mathbf{x}_{n-1} + \mathbf{a}_n \Delta t^2 + O(\Delta t^4) \tag{14.212}$$

which coincides with the Stoermer-Verlet method (14.187). We conclude that Beeman's method should produce the same trajectory as the Stoermer-Verlet method if numerical errors can be neglected and comparable initial values are used. In fact, the Stoermer-Verlet method may suffer from numerical extinction and Beeman's method provides a numerically more favorable alternative.

14.10.8 The Leapfrog Method

Closely related to the Verlet methods is the so called leapfrog method [275]. It uses the simple decomposition

$$e^{\mathcal{L}\Delta t} \approx e^{\mathcal{L}_A \Delta t} e^{\mathcal{L}_B \Delta t} \tag{14.213}$$

but introduces two different time grids for coordinates and velocities which are shifted by $\Delta t/2$ (Fig. 14.17).

The leapfrog algorithm is given by

$$\mathbf{v}_{n+1/2} = \mathbf{v}_{n-1/2} + \mathbf{a}_n \Delta t \tag{14.214}$$

$$\mathbf{x}_{n+1} = \mathbf{x}_n + \mathbf{v}_{n+1/2}\Delta t. \tag{14.215}$$

Due to the shifted arguments the order of the method is increased as can be seen from the Taylor series:

$$\mathbf{x}(t_n) + \left(\mathbf{v}(t_n) + \frac{\Delta t}{2}\mathbf{a}(t_n) + \cdots\right)\Delta t = \mathbf{x}(t_n + \Delta t) + O(\Delta t^3) \tag{14.216}$$

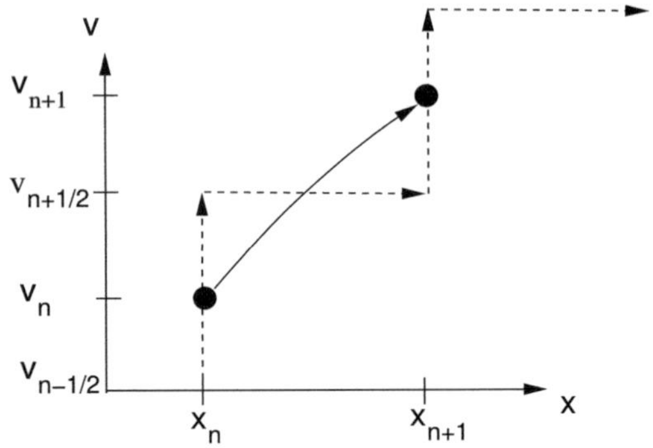

Fig. 14.17 Leapfrog method. The exact integration path is approximated by one step with constant coordinates and one step with constant velocities. Two different grids are used for coordinates and velocities which are shifted by $\Delta t/2$

$$\mathbf{v}(t_n + \frac{\Delta t}{2}) - \mathbf{v}(t_n - \frac{\Delta t}{2}) = \mathbf{a}(t_n)\Delta t + O(\Delta t^3). \tag{14.217}$$

One disadvantage of the leapfrog method is that some additional effort is necessary if the velocities are needed. The simple expression

$$\mathbf{v}(t_n) = \frac{1}{2}(\mathbf{v}(t_n - \frac{\Delta t}{2}) + \mathbf{v}(t_n + \frac{\Delta t}{2})) + O(\Delta t^2) \tag{14.218}$$

is of lower error order than (14.217).

14.11 Direct Integration Methods for Higher Order ODE

While higher order ODE can be always brought to first order form by introducing the derivatives as new variables (14.15), there are direct integration methods which can be more efficient in certain cases. In the following we concentrate on second order ODE of the form[12]

$$\frac{d^2}{dt^2}Y(t) = f(t, Y(t), \frac{d}{dt}Y(t)) \tag{14.219}$$

which often describe periodic phenomena in mechanics and astronomy. Introducing discrete times t_n we want to advance from[13] t_n to $t_{n+1} = t_n + \Delta t$ and to calculate new coordinates Y_{n+1} and velocities $\dot{Y}_{n+1}$[14] starting from Y_n and Y_{n+1}. While (14.219) may be nonlinear in general, numerical methods are often analyzed applying them to a simple linear oscillator model

---

[12] We concentrate on single equations but the methods can be generalized to systems of equations.

[13] Δt is not necessarily constant.

[14] For a simple one-dimensional oscillator, $x = Y$ is the coordinate, $v = \dot{Y}$ the velocity and $a = \ddot{Y} = F(t, x, v)/m$ the acceleration.

$$\frac{d^2}{dt^2}Y(t) = -Y(t). \tag{14.220}$$

In this case, one step forward in time is described by the linear equation

$$\begin{pmatrix} Y_{n+1} \\ \dot{Y}_{n+1} \end{pmatrix} = M(\Delta t)\begin{pmatrix} Y_n \\ \dot{Y}_n \end{pmatrix} \tag{14.221}$$

with a 2×2 matrix M for a one-step method and by

$$\begin{pmatrix} Y_{n+1} \\ Y_n \end{pmatrix} = M(\Delta t)\begin{pmatrix} Y_n \\ Y_{n-1} \end{pmatrix} \tag{14.222}$$

for a two step method. An algorithm is stable for (14.220) if the two eigenvalues of M have absolute values $|\lambda_{1,2}| \leq 1$ and are different if both have absolute value $|\lambda_1| = |\lambda_2| = 1$. If at least one eigenvalue has modulus $|\lambda| > 1$, the solution grows exponentially. If the eigenvalues are degenerate and have modulus 1, the solutions grow linearly in time. In the periodic regime, the two eigenvalues are complex conjugates and have the form

$$\lambda_{1,2} = |\lambda| e^{\pm i\varphi}. \tag{14.223}$$

14.11.1 Numerov Type Methods

Numerov [279, 280] type methods are often used to integrate second order initial value problems where the first order derivative is absent:

$$\ddot{Y}(t) = f(t, Y) \quad Y(t_0) = Y_0 \quad \dot{Y}(t_0) = \dot{Y}_0. \tag{14.224}$$

We expand the second order difference quotient as a Taylor series

$$Y_{n+1} - 2Y_n + Y_{n-1} = Y_n + \dot{Y}_n \Delta t + \frac{(\Delta t)^2}{2}\ddot{Y}_n + \frac{(\Delta t)^3}{6}\dddot{Y}_n + \frac{(\Delta t)^4}{24}\ddddot{Y}_n + \cdots$$

$$- 2Y_n + Y_n - \dot{Y}_n \Delta t + \frac{(\Delta t)^2}{2}\ddot{Y}_n - \frac{(\Delta t)^3}{6}\dddot{Y}_n + \frac{(\Delta t)^4}{24}\ddddot{Y}_n + \cdots$$

$$= (\Delta t)^2 Y_n'' + \frac{(\Delta t)^4}{12}\ddddot{Y}_n + \cdots . \tag{14.225}$$

Numerov takes the second derivative

$$\ddot{Y}_{n+1} - 2\ddot{Y}_n + \ddot{Y}_{n-1} = (\Delta t)^2 \ddddot{Y}_n + \frac{(\Delta t)^4}{12}\frac{d^6}{dt^6}Y_n + \cdots \tag{14.226}$$

to eliminate the fourth order with the help of

$$(\Delta t)^2 \ddddot{Y}_n = \ddot{Y}_{n+1} - 2\ddot{Y}_n + \ddot{Y}_{n-1} - \frac{(\Delta t)^4}{12} \frac{d^6}{dt^6} Y_n + \cdots$$

$$= f_{n+1} - 2f_n + f_{n-1} - \frac{(\Delta t)^4}{12} \frac{d^6}{dt^6} Y_n + \cdots \tag{14.227}$$

where all second derivatives have been replaced by the values of f. Finally we have

$$Y_{n+1} - 2Y_n + Y_{n-1} = (\Delta t)^2 f_n + \frac{(\Delta t)^2}{12} \left[f_{n+1} - 2f_n + f_{n-1} \right] - \frac{(\Delta t)^6}{144} \frac{d^6}{dt^6} Y_n + \cdots \tag{14.228}$$

from which a two-step method results[15]

$$Y_{n+1} = 2Y_n - Y_{n-1} + \frac{(\Delta t)^2}{12} \left[f_{n+1} + 10 f_n + f_{n-1} \right] + O(\Delta t^6). \tag{14.229}$$

The original Numerov method is not zero-stable since its characteristic polynomial $z^2 - 2z + 1 = (z - 1)^2$ has a double root with magnitude 1. It can be conditionally stable for finite time step Δt. As the simplest example we consider the linear ODE $\ddot{Y} = f(Y) = -Y$. With

$$p = \frac{1 - \frac{5}{12}(\Delta t)^2}{1 + \frac{(\Delta t)^2}{12}} \tag{14.230}$$

(14.229) becomes a recursion

$$Y_{n+1} = 2pY_n - Y_{n-1} \tag{14.231}$$

which can be formulated as

$$\begin{pmatrix} Y_{n+1} \\ Y_n \end{pmatrix} = \begin{pmatrix} 2p & -1 \\ 1 & 0 \end{pmatrix} \begin{pmatrix} Y_n \\ Y_{n-1} \end{pmatrix} \tag{14.232}$$

with the solution

$$\begin{pmatrix} Y_{n+1} \\ Y_n \end{pmatrix} = \begin{pmatrix} 2p & -1 \\ 1 & 0 \end{pmatrix}^n \begin{pmatrix} Y_1 \\ Y_0 \end{pmatrix}. \tag{14.233}$$

The eigenvalues of the matrix are the roots of

$$-\lambda(2p - \lambda) + 1 \tag{14.234}$$

which is nothing but the characteristic polynomial of (14.231). We find

[15] This is implicit for nonlinear f.

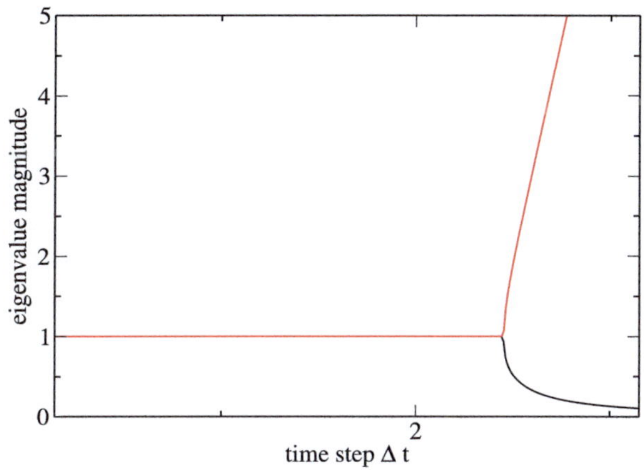

Fig. 14.18 Eigenvalue magnitude for Numerov's method. Numerov's method is stable for $\Delta t < \sqrt{6}$

$$\lambda_{\pm} = p \pm \sqrt{p^2 - 1}. \tag{14.235}$$

For $0 < \Delta t < \sqrt{6}$ it follows $p < 1$ and the two eigenvalues are complex conjugated on the unit circle,

$$\lambda_{\pm} = e^{\pm i\varphi} \tag{14.236}$$

hence the algorithm is p-stable [281].[16] For $\Delta t > \sqrt{6}$ both eigenvalues are real valued and one of them is larger than 1, hence the algorithm becomes unstable (Fig. 14.18). The region of stability can be increased by making the algorithm explicit [282]. The general solution is found as linear combination of the solutions corresponding to the eigenpairs as

$$Y_n = \frac{1}{2}\left[Y_0 + \frac{Y_1 - pY_0}{\sqrt{p^2-1}}\right]\left[p + \sqrt{p^2-1}\right]^n + \frac{1}{2}\left[Y_0 - \frac{Y_1 - pY_0}{\sqrt{p^2-1}}\right]\left[p - \sqrt{p^2-1}\right]^n. \tag{14.237}$$

For $p < 1$ this describes a harmonic oscillation with phase

$$n\varphi = n \arccos(p) = n\left[\Delta t + \frac{1}{480}\Delta t^5 + \cdots\right] \tag{14.238}$$

hence there is a relative phase lag of

$$\frac{n\varphi - n\sqrt{12h}}{n\sqrt{12h}} \approx \frac{3}{10}h^2 = \frac{1}{480}\Delta t^4. \tag{14.239}$$

Chawla [284] improved Numerov's method by adding a predictor step

[16] Which is required for the integration of periodic problems. [281, 283].

$$\tilde{Y}_n = Y_n - \alpha\Delta t^2\left[f_{n+1} - 2f_n + f_{n-1}\right] \tag{14.240}$$

$$\tilde{f}_n = f(\tilde{Y}_n, t_n) \tag{14.241}$$

$$Y_{n+1} = 2Y_n - Y_{n-1} + \frac{(\Delta t)^2}{12}\left[f_{n+1} + 10\tilde{f}_n + f_{n-1}\right]. \tag{14.242}$$

For the linear example of an harmonic oscillation $f = -Y$ this leads to

$$\tilde{Y}_n = Y_n + \alpha\Delta t^2\left[Y_{n+1} - 2Y_n + Y_{n-1}\right] \tag{14.243}$$

$$\tilde{f}_n = -\tilde{Y}_n = -Y_n - \alpha\Delta t^2\left[Y_{n+1} - 2Y_n + Y_{n-1}\right] \tag{14.244}$$

$$Y_{n+1} = 2Y_n - Y_{n-1} - \frac{(\Delta t)^2}{12}\left[Y_{n+1} + 10\tilde{Y}_n + Y_{n-1}\right] \tag{14.245}$$

$$Y_{n+1} = 2Y_n - Y_{n-1} - \frac{(\Delta t)^2}{12}\left[Y_{n+1} + 10Y_n + Y_{n-1} + 10\alpha\Delta t^2\left(Y_{n+1} - 2Y_n + Y_{n-1}\right)\right] \tag{14.246}$$

$$Y_{n+1}(1 + \frac{\Delta t^2}{12} + \frac{10\alpha}{12}\Delta t^4) = Y_n(2 - \frac{10}{12}\Delta t^2 + \frac{20\alpha}{12}\Delta t^4) + Y_{n-1}(-1 - \frac{\Delta t^2}{12} - \frac{10\alpha}{12}\Delta t^4) \tag{14.247}$$

$$Y_{n+1} = 2pY_n - Y_{n-1} \tag{14.248}$$

with

$$p = \frac{1 - \frac{5}{12}\Delta t^2 + \frac{5\alpha}{6}\Delta t^4}{1 + \frac{\Delta t^2}{12} + \frac{5\alpha}{6}\Delta t^4}. \tag{14.249}$$

The stability condition $|p| < 1$ requires

$$0 < 2 - \frac{1}{3}\Delta t^2 + \frac{5}{3}\alpha\Delta t^4 = \frac{5\alpha}{3}\left[\left(\Delta t^2 - \frac{1}{10\alpha}\right)^2 + \frac{6}{5\alpha}\left(1 - \frac{1}{120\alpha}\right)\right] \tag{14.250}$$

which, for $\alpha > 1/120$ holds for arbitrary Δt.

$$\arccos(p) = \Delta t + \Delta t^5(\frac{1}{480} - \frac{5\alpha}{12}) + \frac{1}{12096}\Delta t^7 + O(\Delta t^9). \tag{14.251}$$

The phase lag

$$\frac{\arccos(p) - \Delta t}{\Delta t} = \Delta t^4(\frac{1}{480} - \frac{5\alpha}{12}) + \frac{1}{12096}\Delta t^6 + O(\Delta t^8) \tag{14.252}$$

becomes minimal for

$$\alpha = \frac{1}{200}. \tag{14.253}$$

Higher order methods with minimal phase lag have been described in [283, 285–290]. Second order equations including also a first order derivative have been treated in [291]. An extension to two-dimensional problems was given in [292] and fourth order equations are treated in [293].

14.11.2 Direct Runge-Kutta Methods

Second order ODE without the presence of the first order derivative can be solved directly very efficiently with Runge-Kutta-Nyström methods.

A fourth order method was presented by Nyström [294].

He approximates

$$Y_{n+1} = Y_n + \dot{Y}_n \Delta t + \frac{1}{6} K_0 \Delta t^2 + \frac{1}{3} K_1 \Delta t^2 \tag{14.254}$$

$$\dot{Y}_{n+1} = \dot{Y}_n + \frac{\Delta t}{6} \left[K_0 + 4K_1 + K_2 \right] \tag{14.255}$$

with

$$K_0 = f(t_n, Y_n) \tag{14.256}$$

$$K_1 = f(t_n + \frac{\Delta t}{2}, Y_n + \frac{1}{2}\dot{Y}_n \Delta t + \frac{1}{8} K_0 \Delta t^2) \tag{14.257}$$

$$K_2 = f(t_n + \Delta t, Y_n + \dot{Y}_n \Delta t + \frac{1}{2} K_1 \Delta t^2). \tag{14.258}$$

For the linear oscillator equation $\ddot{Y} = -Y$ we obtain

$$K_0 = -Y_n \tag{14.259}$$

$$K_1 = -Y_n - \frac{1}{2}\dot{Y}_n \Delta t - \frac{1}{8} Y_n \Delta t^2 \tag{14.260}$$

$$K_2 = -Y_n - \dot{Y}_n \Delta t + \frac{1}{2} Y_n \Delta t^2 + \frac{1}{4} \dot{Y}_n \Delta t^3 - \frac{1}{16} Y_n \Delta t^4 \tag{14.261}$$

$$Y_{n+1} = Y_n + \dot{Y}_n \Delta t - \frac{1}{2} Y_n \Delta t^2 - \frac{1}{6} \dot{Y}_n \Delta t^3 + \frac{1}{24} Y_n \Delta t^4 \tag{14.262}$$

$$\dot{Y}_{n+1} = \dot{Y}_n - Y_n \Delta t - \frac{1}{2} \dot{Y}_n \Delta t^2 + \frac{1}{6} Y_n \Delta t^3 + \frac{1}{24} \dot{Y}_n \Delta t^4 - \frac{1}{96} Y_n \Delta t^5 \tag{14.263}$$

$$\begin{pmatrix} Y_{n+1} \\ \dot{Y}_{n+1} \end{pmatrix} = \begin{pmatrix} 1 - \frac{\Delta t^2}{2} + \frac{\Delta t^4}{24} & 1 - \frac{\Delta t^3}{6} \\ -1 + \frac{\Delta t^3}{6} - \mathbf{\frac{t^5}{96}} & 1 - \frac{\Delta t^2}{2} + \frac{\Delta t^4}{24} \end{pmatrix} \begin{pmatrix} Y_n \\ \dot{Y}_n \end{pmatrix}. \tag{14.264}$$

Comparing with the Taylor series of the cosine and sine functions we find that the error is of order $O(\Delta t^5)$ as expected for a fourth order method.

The eigenvalues are

$$\lambda = p \pm \sqrt{-r} \tag{14.265}$$

with

$$p = 1 - \frac{\Delta t^2}{2} + \frac{\Delta t^4}{24} \tag{14.266}$$

$$r = \Delta t^2 - \frac{\Delta t^4}{3} + \frac{11\Delta t^6}{288} - \frac{\Delta t^8}{576}. \tag{14.267}$$

For $0 < \Delta t < \sqrt{6}$ the root is imaginary and the two eigenvalues are complex conjugated with

$$|\lambda|^2 = p^2 - r = 1 - \frac{\Delta t^6}{288} < 1 \tag{14.268}$$

hence the algorithm is stable for (14.220). The phase angle is

$$\arg(\lambda) = \arctan\left(\pm\frac{\sqrt{r}}{p}\right) = \pm\left(\Delta t - \frac{\Delta t^5}{320} + \cdots\right) \tag{14.269}$$

hence the phase lag is of order $O(\Delta t^4)$.

Albrecht [295] derived the sixth order RKN method

$$Y_{n+1} = Y_n + \dot{Y}_n \Delta t + \frac{\Delta t^2}{90}\left[7K_0 + 24K_1 + 6K_2 + 8K_3\right] \tag{14.270}$$

$$\dot{Y}_{n+1} = \dot{Y}_n + \frac{\Delta t}{90}\left[7K_0 + 32K_1 + 12K_2 + 32K_3 + 7K_4\right] \tag{14.271}$$

with intermediate values

$$K_0 = f(t_n, Y_n) \tag{14.272}$$

$$K_1 = f(t_n + \frac{\Delta t}{4}, Y_n + \frac{\Delta t}{4}\dot{Y}_n + \frac{\Delta t^2}{32}K_0) \tag{14.273}$$

$$K_2 = f(t_n + \frac{\Delta t}{2}, Y_n + \frac{\Delta t}{2}\dot{Y}_n + \frac{\Delta t^2}{24}(4K_1 - K_0)) \tag{14.274}$$

$$K_3 = f(t_n + \frac{3\Delta t}{4}, Y_n + \frac{3\Delta t}{4}\dot{Y}_n + \frac{\Delta t^2}{32}(2K_2 + 4K_1 + 3K_0)) \tag{14.275}$$

$$K_4 = f(t_n + \Delta t, Y_n + \Delta t \dot{Y}_n + \frac{\Delta t^2}{14}(2K_3 - K_2 + 6K_1)). \tag{14.276}$$

For the linear oscillator equation $\ddot{Y} = -Y$ we obtain

$$K_0 = -Y_n \tag{14.277}$$

$$K_1 = \left(-1 + \frac{\Delta t^2}{32}\right) Y_n - \frac{\Delta t}{4} \dot{Y}_n \tag{14.278}$$

$$K_2 = \left(-1 + \frac{\Delta t^2}{8} - \frac{\Delta t^4}{192}\right) Y_n + \left(-\frac{\Delta t}{2} + \frac{\Delta t^3}{24}\right) \dot{Y}_n \tag{14.279}$$

$$K_3 = \left(-1 + \frac{9\Delta t^2}{32} - \frac{3\Delta t^4}{256} + \frac{\Delta t^6}{3072}\right) Y_n + \left(-\frac{3\Delta t}{4} + \frac{\Delta t^2}{16} - \frac{\Delta t^4}{384}\right) \dot{Y}_n \tag{14.280}$$

$$K_4 = \left(-1 + \frac{\Delta t^2}{2} - \frac{5\Delta t^4}{112} + \frac{\Delta t^6}{768} - \frac{\Delta t^8}{21504}\right) Y_n + \left(-\Delta t + \frac{5}{28}\Delta t^3 - \frac{\Delta t^5}{168} + \frac{\Delta t^7}{2688}\right) \dot{Y}_n \tag{14.281}$$

$$\begin{pmatrix} Y_{n+1} \\ \dot{Y}_{n+1} \end{pmatrix} = \begin{pmatrix} 1 - \frac{\Delta t^2}{2} + \frac{\Delta t^4}{24} - \frac{\Delta t^6}{720} + \frac{\boldsymbol{\Delta t^8}}{\mathbf{34560}} & 1 - \frac{\Delta t^3}{6} + \frac{\Delta t^5}{120} - \frac{\boldsymbol{\Delta t^7}}{\mathbf{4320}} \\ -1 + \frac{\Delta t^3}{6} - \frac{\Delta t^5}{120} + \frac{\boldsymbol{\Delta t^7}}{\mathbf{4608}} - \frac{\boldsymbol{\Delta t^9}}{\mathbf{276480}} & 1 - \frac{\Delta t^2}{2} + \frac{\Delta t^4}{24} - \frac{\Delta t^6}{720} + \frac{\boldsymbol{\Delta t^8}}{\mathbf{34560}} \end{pmatrix} \begin{pmatrix} Y_n \\ \dot{Y}_n \end{pmatrix}. \tag{14.282}$$

Comparing with the Taylor series of the cosine and sine functions we find that the error is of order $O(\Delta t^7)$ as expected for a sixth order method.

The eigenvalues are

$$\lambda = p \pm \sqrt{-r} \tag{14.283}$$

with

$$p = 1 - \frac{\Delta t^2}{2} + \frac{\Delta t^4}{24} - \frac{\Delta t^6}{720} + \frac{\Delta t^8}{34560} \tag{14.284}$$

$$r = \Delta t^2 - \frac{\Delta t^4}{3} + \frac{2\Delta t^6}{45} - \frac{223\Delta t^8}{69120} + \frac{613\Delta t^{10}}{4147200} - \frac{\Delta t^{12}}{230400} + \frac{\Delta t^{14}}{12441600} - \frac{\Delta t^{16}}{1194393600}. \tag{14.285}$$

For $\Delta t < 3.118$ the eigenvalues are complex conjugated with

$$|\lambda|^2 = p^2 + r = 1 - \frac{\Delta t^8}{23040} + \frac{13\Delta t^{10}}{4147200} \le 1 \tag{14.286}$$

$$\arg(\lambda) = \arctan\left(\pm\frac{\sqrt{r}}{p}\right) = \pm\left(\Delta t - \frac{5}{193536}\Delta t^7 + \cdots\right) \tag{14.287}$$

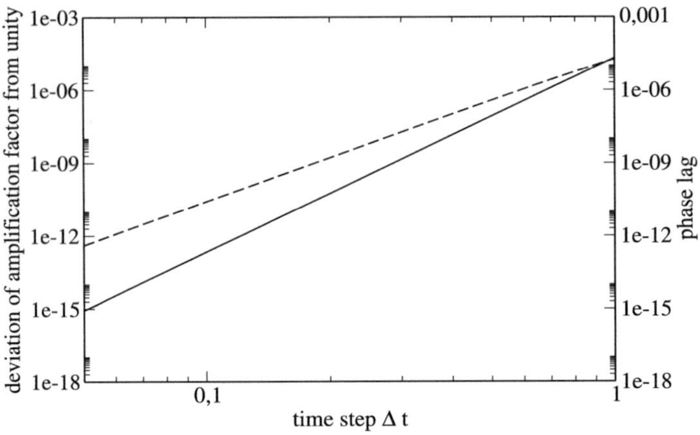

Fig. 14.19 Amplification factor and phase lag of Albrecht's RKN method. The RK method is stable for $0 < \Delta t < 3.118$, where the amplification factor $|\lambda| < 1$. The solid curve shows the deviation $1 - |\lambda|$ (14.286), which is dominated by a Δt^8dependence whereas the broken curve shows the phase lag (14.287) which is of order $O(\Delta t^6)$

hence the relative phase lag is of $O(\Delta t^6)$ (Fig. 14.19).

14.11.3 Direct Predictor-Corrector Methods

Kostek [296] uses Hermite polynomial interpolation for a predictor-corrector method to solve second order ODEs of the general form

$$\ddot{Y}(t) = f(t, Y(t), \dot{Y}(t)). \tag{14.288}$$

He approximates the second derivative in the interval $[t_n, t_{n+1}]$ with the help of the Hermite method 2.2.5 by a polynomial which is determined from the values of $\ddot{Y}(t)$ and its derivatives at the interval boundaries. Cubic interpolation for instance approximates the acceleration[17]

$$a(t) = a_n B_{0,3}(t) + a_{n+1} B_{3,3}(t) + \left(\frac{\dot{a}_n \Delta t}{3} + a_n\right) B_{1,3}(t) + \left(a_{n+1} - \frac{\dot{a}_{n+1}\Delta t}{3}\right) B_{2,3}(t) \tag{14.289}$$

which is obviously an implicit method. Integration then gives approximations to velocity and coordinate

$$v(t) = v_n + \int_{t_n}^{t} a(t)dt$$

[17] We prefer to represent the polynomial with generalized Bernstein polynomials and make use of the generalized relations [17] $\frac{d}{dt} B_{k,n}(t) = \frac{n}{\Delta t}(B_{k-1,n-1}(t) - B_{k,n-1}(t))$ $\int B_{k,n}(t)dt = \frac{\Delta t}{n+1} \sum_{j=k+1}^{n+1} B_{j,n+1}(t)$ and $\int_t^{t+\Delta t} B_{k,n} = \frac{\Delta t}{n+1}$.

$$= v_n + \frac{\Delta t}{4}\left\{ a_n \sum_{j=1}^{4} B_{j,4}(t) + a_{n+1} B_{4,4}(t) + \left(\frac{\dot{a}_n \Delta t}{3} + a_n\right) \sum_{j=2}^{4} B_{j,4}(t) \right.$$

$$\left. + \left(a_{n+1} - \frac{\dot{a}_{n+1}\Delta t}{3}\right) \sum_{j=3}^{4} B_{j,4}(t) \right\} \tag{14.290}$$

$$v_{n+1} = v(t_{n+1}) = v_n + a_n \frac{\Delta t}{2} + a_{n+1}\frac{\Delta t}{2} + \frac{\Delta t^2 \dot{a}_n}{12} - \frac{\Delta t^2 \dot{a}_{n+1}}{12} \tag{14.291}$$

$$x_{n+1} = x_n + \int_{t_n}^{t_{n+1}} v(t)dt = x_n + v_n \Delta t + \left[\frac{7}{20}a_n + \frac{3}{20}a_{n+1}\right]\Delta t^2 + \frac{\Delta t^3 \dot{a}_n}{20} - \frac{\Delta t^3 \dot{a}_{n+1}}{30}. \tag{14.292}$$

Starting from some initial guess for $a_{n+1}$and $\dot{a}_{n+1}$[18] this procedure is repeated several times to improve the approximation $a_{n+1} = f(t_{n+1}, x_{n+1}, v_{n+1})$ and finally extrapolation predicts new initial values for the next time interval

$$v_{n+2} = v_n + \int_{t_n}^{t_n+2\Delta t} a(t)dt$$

$$= v_n + 2a_n\Delta t + \left(\frac{2}{3}\dot{a}_n + \frac{4}{3}\dot{a}_{n+1}\right)\Delta t^2 \tag{14.293}$$

$$x_{n+2} = x_n + 2v_n\Delta t + \int_{t_n}^{t_n+2\Delta t} dt' \int_{t_n}^{t'} a(t)dt$$

$$= x_n + 2v_n\Delta t + (\frac{6}{5}a_n + \frac{4}{5}a_{n+1})\Delta t^2 + \frac{4}{15}(\dot{a}_n + \dot{a}_{n+1})\Delta t^3. \tag{14.294}$$

An algebra program is helpful to calculate the integrals for integrands outside the interval $[t_n, t_{n+1}]$.

To analyze the accuracy, we consider again a simple linear oscillator with

$$a = -x, \quad \dot{a} = -v, \quad \ddot{a} = x, \quad \dddot{a} = v, \cdots .$$

With the help of an algebra program, (14.291) and (14.292) can be solved for $x_{n+1}$and v_{n+1}as a function of x_n and v_n:

$$\begin{pmatrix} x_{n+1} \\ v_{n+1} \end{pmatrix} = \begin{pmatrix} \frac{\Delta t^4 - 104\Delta t^2 + 240}{\Delta t^4 + 16\Delta t^2 + 240} & \frac{\Delta t^5 - 72\Delta t^3 + 720\Delta t}{3(\Delta t^4 + 16\Delta t^2 + 240)} \\ \frac{24\Delta t^3 - 240\Delta t}{\Delta t^4 + 16\Delta t + 240} & \frac{3\Delta t^4 - 104\Delta t^2 + 240}{\Delta t^4 + 16\Delta t + 240} \end{pmatrix} \begin{pmatrix} x_n \\ v_n \end{pmatrix}$$

[18] Which could be obtained e.g. with the help of a Taylor series or RK method.

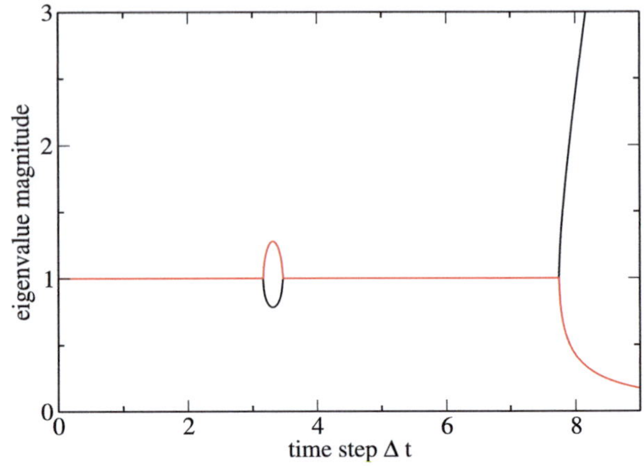

Fig. 14.20 Eigenvalue magnitude there are two stability regions $[0, \sqrt{10}]$and $[2\sqrt{3}, 2\sqrt{15}]$ with a conjugate pair of eigenvalues on the unit circle

$$= \begin{pmatrix} 1 - \frac{\Delta t^2}{2} + \frac{\Delta t^4}{24} - \frac{\boldsymbol{\Delta t^6}}{\mathbf{1440}} + \cdots & \Delta t - \frac{\Delta t^3}{6} + \frac{\Delta t^5}{120} + \frac{\Delta t^7}{7200} + \cdots \\ -\Delta t + \frac{\Delta t^3}{6} - \frac{\boldsymbol{\Delta t^5}}{\mathbf{144}} + \cdots & 1 - \frac{\Delta t^2}{2} + \frac{\Delta t^4}{124} - \frac{\Delta t^6}{1440} + \end{pmatrix} \begin{pmatrix} x_n \\ v_n \end{pmatrix}. \tag{14.295}$$

Comparison with the Taylor series expansion

$$\begin{pmatrix} x_{n+1} \\ v_{n+1} \end{pmatrix} = \begin{pmatrix} x_n + v_n \Delta t - \frac{x_n}{2}\Delta t^2 - \frac{v_n}{6}\Delta t^3 + \frac{x_n}{24}\Delta t^4 + \frac{v_n}{120}\Delta t^5 - \frac{\boldsymbol{x_n}}{\mathbf{720}}\boldsymbol{\Delta t^6} - \frac{v_n}{5040}\Delta t^7 \cdots \\ v_n - x_n \Delta t - \frac{v_n}{2}\Delta t^2 + \frac{x_n}{6}\Delta t^3 + \frac{v_n}{24}\Delta t^4 - \frac{\boldsymbol{x_n}}{\mathbf{120}}\boldsymbol{\Delta t^5} - \frac{v_n}{720}\Delta t^6 + \frac{x_n}{5040}\Delta t^7 \cdots \end{pmatrix} \tag{14.296}$$

shows that the local error is of order $O(\Delta t^5)$, hence the method is of order 4.

The two eigenvalues

$$\lambda_{\pm} = \frac{3\Delta t^4 - 104\Delta t^2 + 240 \pm 2\sqrt{2\Delta t^8 - 164\Delta t^6 + 2880\Delta t^4 - 14400\Delta t^2}}{\Delta t^4 + 16\Delta t^2 + 240} \tag{14.297}$$

are complex conjugates on the unit circle in the intervals $[0, \sqrt{10}]$and $[2\sqrt{3}, 2\sqrt{15}]$. For $d > 2\sqrt{15}$ and in the interval $[\sqrt{10}, 2\sqrt{3}]$ one eigenvalue has magnitude $|\lambda| > 1$ (Fig. 14.20). In the two stable regions harmonic oscillations result with phase

$$p = \arccos\left(\frac{3\Delta t^4 - 104\Delta t^2 + 240}{\Delta t^4 + 16\Delta t^2 + 240}\right) = \Delta t - \frac{1}{1440}\Delta t^5 + \cdots \tag{14.298}$$

and a phase lag of

$$\frac{p - \Delta t}{\Delta t} = -\frac{1}{1440}\Delta t^4 + \cdots. \tag{14.299}$$

So far we assumed that the number of iterations was sufficient to achieve convergence. If the number of iterations is limited, the precision of the method depends on this number [296]. The accuracy can be improved by using 5th order interpolation polynomials which are determined from the values of a_n, a_{n+1}, $\dot{a}_n$, $\dot{a}_{n+1}$, $\ddot{a}_n$ and $\ddot{a}_{n+1}$ to obtain a sixth order method [296].

Exercises

Exercise 14.1 Circular Orbits

In this computer experiment we consider a mass point moving in a central field. The equation of motion can be written as the following system of first order equations:

$$\begin{pmatrix} \dot{x} \\ \dot{y} \\ \dot{v}_x \\ \dot{v}_y \end{pmatrix} = \begin{pmatrix} 0 & 0 & 1 & 0 \\ 0 & 0 & 0 & 1 \\ -\frac{1}{(x^2+y^2)^{3/2}} & 0 & 0 & 0 \\ 0 & -\frac{1}{(x^2+y^2)^{3/2}} & 0 & 0 \end{pmatrix} \begin{pmatrix} x \\ y \\ v_x \\ v_y \end{pmatrix}. \tag{14.300}$$

For initial values

$$\begin{pmatrix} x \\ y \\ v_x \\ v_y \end{pmatrix} = \begin{pmatrix} 1 \\ 0 \\ 0 \\ 1 \end{pmatrix} \tag{14.301}$$

the exact solution is given by

$$x = \cos t \quad y = \sin t. \tag{14.302}$$

The following methods are used to calculate the position $x(t)$, $y(t)$ and the energy

$$E_{tot} = E_{kin} + E_{pot} = \frac{1}{2}(v_x^2 + v_y^2) - \frac{1}{\sqrt{x^2 + y^2}}. \tag{14.303}$$

- The explicit Euler method (Sect. 14.3)

$$\begin{aligned} x(t_{n+1}) &= x(t_n) + v_x(t_n)\Delta t \\ y(t_{n+1}) &= y(t_n) + v_y(t_n)\Delta t \\ v_x(t_{n+1}) &= v_x(t_n) - \frac{x(t_n)}{R(t_n)^3}\Delta t \\ v_y(t_{n+1}) &= v_y(t_n) - \frac{y(t_n)}{R(t_n)^3}\Delta t. \end{aligned} \tag{14.304}$$

- The 2nd order Runge Kutta method (Sect. 14.7.1)

which consists of the predictor step

$$x(t_n+\Delta t/2)=x(t_n)+\frac{\Delta t}{2}v_x(t_n) \tag{14.305}$$

$$y(t_n+\Delta t/2)=y(t_n)+\frac{\Delta t}{2}v_y(t_n) \tag{14.306}$$

$$v_x(t_n+\Delta t/2)=v_x(t_n)-\frac{\Delta t}{2}\frac{x(t_n)}{R(t_n)^3} \tag{14.307}$$

$$v_y(t_n+\Delta t/2)=v_y(t_n)-\frac{\Delta t}{2}\frac{y(t_n)}{R(t_n)^3} \tag{14.308}$$

and the corrector step

$$x(t_{n+1})=x(t_n)+\Delta t\ v_x(t_n+\Delta t/2) \tag{14.309}$$

$$y(t_{n+1})=y(t_n)+\Delta t\ v_y(t_n+\Delta t/2) \tag{14.310}$$

$$v_x(t_{n+1})=v_x(t_n)-\Delta t\frac{x(t_n+\Delta t/2)}{R^3(t_n+\Delta t/2)} \tag{14.311}$$

$$v_y(t_{n+1})=v_y(t_n)-\Delta t\frac{y(t_n+\Delta t/2)}{R^3(t_n+\Delta t/2)}. \tag{14.312}$$

- The fourth order Runge Kutta method (Sect. 14.7.3)
- The Verlet method (Sect. 14.10.5)

$$x(t_{n+1})=x(t_n)+(x(t_n)-x(t_{n-1}))-\Delta t\frac{x(t_n)}{R^3(t_n)} \tag{14.313}$$

$$y(t_{n+1})=y(t_n)+(y(t_n)-y(t_{n-1}))-\Delta t\frac{y(t_n)}{R^3(t_n)} \tag{14.314}$$

$$v_x(t_n)=\frac{x(t_{n+1})-x(t_{n-1})}{2\Delta t}=\frac{x(t_n)-x(t_{n-1})}{\Delta t}-\frac{\Delta t}{2}\frac{x(t_n)}{R^3(t_n)} \tag{14.315}$$

$$v_y(t_n)=\frac{y(t_{n+1})-y(t_{n-1})}{2\Delta t}=\frac{y(t_n)-y(t_{n-1})}{\Delta t}-\frac{\Delta t}{2}\frac{y(t_n)}{R^3(t_n)}. \tag{14.316}$$

To start the Verlet method we need additional coordinates at time $-\Delta t$ which can be chosen from the exact solution or from the approximation

$$x(t_{-1})=x(t_0)-\Delta t\ v_x(t_0)-\frac{\Delta t^2}{2}\frac{x(t_0)}{R^3(t_0)} \tag{14.317}$$

$$y(t_{-1}) = y(t_0) - \Delta t\, v_y(t_0) - \frac{\Delta t^2}{2} \frac{y(t_0)}{R^3(t_0)}. \tag{14.318}$$

- The leapfrog method (Sect. 14.10.8)

$$x(t_{n+1}) = x(t_n) + v_x(t_{n+1/2}) \Delta t \tag{14.319}$$

$$y(t_{n+1}) = y(t_n) + v_y(t_{n+1/2}) \Delta t \tag{14.320}$$

$$v_x(t_{n+1/2}) = v_x(t_{n-1/2}) - \frac{x(t_n)}{R(t_n)^3} \Delta t \tag{14.321}$$

$$v_y(t_{n+1/2}) = v_y(t_{n-1/2}) - \frac{y(t_n)}{R(t_n)^3} \Delta t \tag{14.322}$$

where the velocity at time t_n is calculated from

$$v_x(t_n) = v_x(t_{n+1/2}) - \frac{\Delta t}{2} \frac{x(t_{n+1})}{R^3(t_{n+1})} \tag{14.323}$$

$$v_y(t_n) = v_y(t_{n+1/2}) - \frac{\Delta t}{2} \frac{y(t_{n+1})}{R^3(t_{n+1})}. \tag{14.324}$$

To start the leapfrog method we need the velocity at time $t_{-1/2}$ which can be taken from the exact solution or from

$$v_x(t_{-1/2}) = v_x(t_0) - \frac{\Delta t}{2} \frac{x(t_0)}{R^3(t_0)} \tag{14.325}$$

$$v_y(t_{-1/2}) = v_y(t_0) - \frac{\Delta t}{2} \frac{y(t_0)}{R^3(t_0)}. \tag{14.326}$$

Compare the conservation of energy for the different methods as a function of the time step Δt. Study the influence of the initial values for leapfrog and Verlet methods.

Exercise 14.2 N-body System

In this computer experiment we simulate the motion of three mass points under the influence of gravity. Initial coordinates and velocities as well as the masses can be varied. The equations of motion are solved with the 4th order Runge Kutta method with quality control for different step sizes. The local integration error is estimated using the step doubling method. Try to simulate a planet with a moon moving round a sun!

Exercise 14.3 Adams-Bashforth Methods

In this computer experiment we simulate a circular orbit with explicit and implicit Adams methods of order $3 \cdots 14$. The absolute error at time T

$$\Delta(T) = |x(T) - \cos(T)| + |y(t) - \sin(T)| + |v_x(T) + \sin(T)| + |v_y(T) - \cos(T)| \quad (14.327)$$

is shown as a function of the time step Δt in a log-log plot. From the slope

$$s = \frac{d(\log_{10}(\Delta))}{d(\log_{10}(\Delta t))} \quad (14.328)$$

the leading error order s can be determined. For very small step sizes rounding errors become dominating which leads to an increase $\Delta \sim (\Delta t)^{-1}$.

Determine maximum precision and optimal step size for different orders of the method. Compare with the explicit Euler, Verlet and RK4 methods.

Appendix A: About the Computer Experiments

The computer exercises are realized as Java programs which can be run on almost any platform if a JDK≥11 is available. I use the NetBeans environment and OpenJDK which are both free and open-source. The screenshot in Fig. A.1 shows an example from Exercise 8.2.

With proper permission rights double clicking is sufficient. Alternatively you can run a program *prog.jar* from the command line with *java -jar prog.jar*.

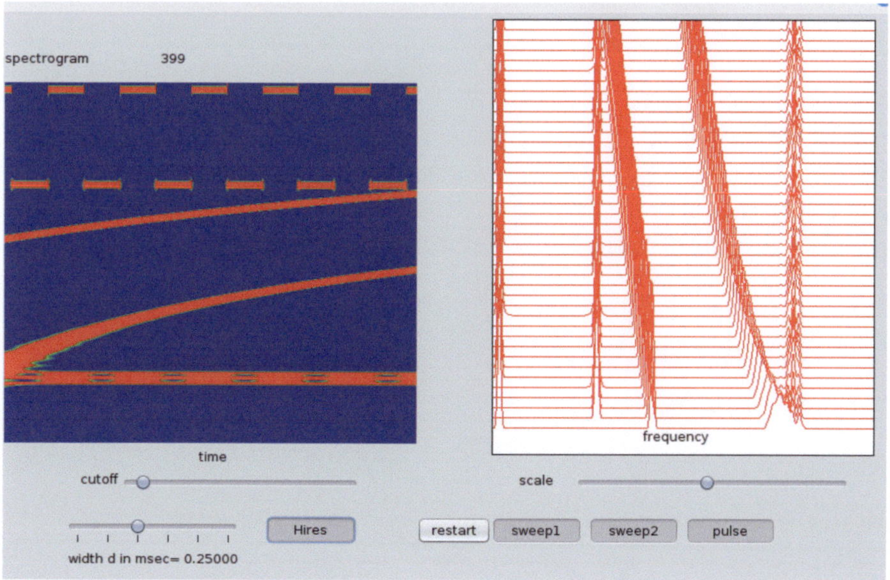

Fig. A.1 Screenshot from Exercise 8.2

P. O. J. Scherer, *Computational Physics I*, Graduate Texts in Physics,
https://doi.org/10.1007/978-3-032-07856-8

Appendix B: Methods and Algorithms

Purpose	Method	Comments
Interpolation	Lagrange polynomial	Explicit form, easy to evaluate
	Barycentric Lagrange polynomial	For evaluation at many points
	Newton's divided differences	New points added easily
	Neville method	For evaluation at one point
	Bernstein polynomials	For Hermite splines and Beziér curves
	Casteljau's algorithm	Calculation of Bernstein polynomials
	Spline interpolation	Smoother, less oscillatory
	Rational interpolation	Smoother, less oscillatory, often less coefficients necessary
	Pade approximation	Often better than Taylor series
	Barycentric rational interpolation	Easy to evaluate
	Rational interpolation without poles	Alternative to splines, analytical
	Multivariate interpolation	Multidimensional
	Trigonometric interpolation	Periodic functions
Differentiation	One-sided difference quotient	Low error order
	Central difference quotient	Higher error order
	Extrapolation	High accuracy
	Higher derivatives	Finite difference methods
	Partial derivatives	Finite difference methods
	Algorithmic differentiation	Machine learning and artificial intelligence
Integration	Newton-Cotes formulas	Equally spaced points
	Trapezoidal rule	Simple, closed interval
	Midpoint rule	Simple, open interval
	Simpson's rule	More accurate
	Composite Newton-Cotes rules	For larger intervals
	Extrapolation (Romberg)	High accuracy
	Clenshaw-Curtis expressions	Suitable for adaptive and multidimensional quadrature
	Gaussian integration	High accuracy if polynomial approximation possible
	Monte Carlo integration	High dimensional integrals

P. O. J. Scherer, *Computational Physics I*, Graduate Texts in Physics,
https://doi.org/10.1007/978-3-032-07856-8

Purpose	Method	Comments
Linear equations	Gaussian elimination (LU reduction)	Standard method for linear equations and matrix inversion
	QR decomposition	Numerically more stable
	Iterative solution	Large sparse systems
	Richardson iteration	Simplest iterative method
	Jacobi relaxation	Iterative matrix-splitting method, converges for diagonally dominant matrices, parallel computation possible
	Gauss-Seidel relaxation	Iterative matrix-splitting method, converges for symmetric positive definite or diagonal dominant matrices, no extra storage
	Chessboard (black-red)	Two independent subgrids, especially for Poisson equation
	Damping and successive over-relaxation	Speeds up convergence for proper relaxation parameter
	Multigrid method	Fast convergence but more complicated
	Conjugate gradients method (CG)	Krylov space method for symmetric positive definite matrices, preconditioning often necessary
	General minimum residual method (GMRES)	Krylov space method for nonsymmetric systems
	Special LU decomposition	Tridiagonal linear equations
	Sherman-Morrison formula	Cyclic tridiagonal systems
Root finding	Bisection	Reliable but slow continuous functions
	Regula falsi (false position)	Speed and robustness between bisection and interpolation
	Newton-Raphson	Continuous derivative necessary, converges fast if starting point is close to a root
	Interpolation (secant)	No derivative necessary, but slower than Newton
	Inverse interpolation	Mainly used by combined methods
	Dekker's combined method	Combination of bisection and secant method
	Brent's combined method	Combination of bisection, secant, and quadratic inverse interpolation methods, very popular
	Chandrupatla's combined method	Uses quadratic interpolation whenever possible, faster than Brent's method, especially for higher order roots
Multidimensional root finding	Newton-Raphson	Needs full Hessian
	Quasi-Newton (Broyden)	Hessian not needed, no matrix inversion
Function minimization	Ternary search	No gradient needed, very simple, for unimodal functions
	Golden section search (Brent)	Faster than ternary search but more complicated
Multidimensional minimization	Gradient descent, steepest descent	simple but slow
	Stochastic gradient descent	Very large dimension, ANN
	Gradient descent with momentum	Escapes local minima, Adam, AMAdam
	Stochastic reconfiguration	Energy minimasation for NN quantum states
	Conjugate gradients	Faster than steepest descent
	Newton-Raphson	Fast, if starting point close to minimum, needs full Hessian
	Quasi-Newton (BFGS, DFP)	Hessian not needed, very popular

Purpose	Method	Comments
Fourier transformation	Goertzel's algorithm	Efficient if only some Fourier components are needed
	Fast Fourier transform	Much faster than direct discrete Fourier transform
Time-frequency analysis	Short time Fourier Transform (STFT)	Constant resolution for all frequencies, often used for audio signals
	Gabor transform	STFT with Gaussian window represents signal by elementary signals localized in time and frequency
	Discrete STFT	Reduced redundancy, still invertible
	Continuous wavelet transform	Constant relative frequency resolution, better time resolution for high frequencies,very time consuming convolution integral
	Discrete wavelet transform	Uses orthogonal or biorthogonal wavelets, fast scalar product
	Multiresolution analysis	Represents a signal by a basic approximation and a series of details with increasing resolution
	Fast wavelet transform	Recursive filtering, very fast
Random numbers	Linear congruent mapping (LC)	Simple pseudo-random number generator
	Xorshift	Fast, maximum possible period
	Multiply with carry (MWC)	Similar to LC but uses a varying carry
	Complementary multiply with carry (CMWC)	Improves MWC, passes many tests
	RN with given distribution	Inverse of cumulative distribution function needed
	Random points on unit sphere	Random directions
	Gaussian RN (Box-Muller)	Gaussian random numbers
Thermal average	Simple sampling	Inefficient
	Importance sampling	Samples preferentially important configurations
	Metropolis-hastings MCMC	Generates configurations according to a canonical distribution
	Gibbs sampling	Scans coordinates sequentially, machine learning, Ising models
Eigenvalue problems	Direct solution	only for very small dimension
	Tridiagonal matrices	Explicit solutions for some special tridiagonal matrices
	Jacobi	Simple but not very efficient
	Power iteration	Finds dominant eigenvector
	QL and QR	Efficient power iteration method for not too large matrices, especially in combination with tridiagonalization by Householder transformations
	Lanczos	Iterative method for very large matrices or if only a few eigenvalues are needed
	Singular value decomposition (SVD)	Generalization for arbitrary matrices
Data fitting	Least square fit	Fit a model function to a set of data
	Linear least square fit with normal equations	Simple but less accurate
	Linear fit with orthogonalisation	Better numerical stability
	Linear fit with SVD	Expensive but more reliable, also for rank deficient matrices
	Low rank matrix approximation	Data compression, total linear least squares

Purpose	Method	Comments
Data analysis with ANN	Linear classification	Perceptron, Support vector machine
	Nonlinear classification	
	Forward and Backward propagation	Automatic differentiation for ANN
	Regression and interpolation with feed forward NN	
	Time series analysis with recurrent NN	
	Principle component analysis with NN	Dimensionality reduction for large data sets
Discretization	Method of lines	Continuous time, discretized space
	Eigenvector expansion	
	Finite differences	Simplest discretization, uniform grids
	Finite volumes	Partial differential equations with a divergence term (conservation laws), flux conservative, allows unstructured meshes and discontinuous material parameters
	Finite elements	Very flexible and general discretization method but also more complicated
	Spectral methods	Expansion with global basis functions, mostly polynomials and Fourier sums, less expensive than finite elements but not as accurate for discontinuous material parameters and complicated geometries
	Dual grid	For finite volumes
	Weighted residuals	General method to determine the expansion coefficients
	Point collocation	Simplest criterion, often used for nonlinear problems and spectral methods
	Sub-domains	More general than finite volumes
	Least square	Popular for computational fluid dynamics and electrodynamics
	Galerkin	Most widely used criterion, leads often to symmetric matrices
	Fourier pseudo-spectral method	Very useful whenever a Laplacian is involved, reduces dispersion
	Boundary elements	If the Green's function is available
	ANN as solvers	Analytic and differentiable trial function
Time evolution	Explicit forward Euler	low error order and unstable, mainly used as predictor step
	Implicit backward Euler	Low error order but stable, used for stiff problems and as corrector step
	Improved Euler (Heun, predictor-corrector)	Higher error order
	Nordsieck predictor-corrector	Implicit method, has been used for molecular dynamics
	Gear predictor-corrector	Optimized for molecular dynamics
	Explicit Runge Kutta (2nd, 3rd, 4th)	General and robust methods, easy step size and quality control
	Extrapolation (Gragg-Bulirsch-Stoer)	Very accurate and very slow

Purpose	Method	Comments
	Explicit Adams-Bashforth	High error order but not self-starting, for smooth functions, can be used as predictor
	Implicit Adams-Moulton	Better stability than explicit method, can be used as corrector
	Backward differentiation (Gear)	Implicit, especially for stiff problems
	Linear multistep predictor-corrector	General class, includes Adams-Bashforth-Moulton and Gear methods
	Verlet integration	Symplectic, time reversible, for molecular dynamics
	Position Verlet	Less popular
	Velocity Verlet	Often used
	Stoermer-Verlet	If velocities are not needed
	Beeman's method	Velocities more accurate than for Stoermer-Verlet
	Leapfrog	Simple but two different grids
	Direct integration of second order ODE	Numerov, Kostek, Runge-Kutta-Nyström no first order derivative
	Crank-Nicolson	Implicit, stable, diffusion and Schroedinger equation
	Lax-Wendroff	Hyperbolic differential equations
	Two-step	Differential equation with second order time derivative
	Reduction to a first order equation	Derivatives treated as additional variables
	Two-variable	Transforms wave equation into a system of two first order equations
	Split operator	Approximates an operator by a product
Unitary time evolution	Rational approximation	implicit,unitary
	Second order differencing	Explicit, not exactly unitary
	Split operator Fourier	Low dispersion, needs fast Fourier transformation
	Real space product formula	Fast but less accurate, useful for wavepackets in coupled states
Rotation	Reorthogonalization	Restore orthogonality of rotation matrix
	Quaternions	Optimum parametrization of the rotation matrix
	Euler angles	Numerical singularities
	Explicit method	Low accuracy, reorthogonalization needed
	Implicit method	Higher accuracy, orthogonal transformation
Molecular dynamics	Force field gradients	Needed for molecular dynamics
	Normal mode analysis	Small amplitude motion around an equilibrium
	Behrendsen thermostat	Simple method to control temperature
	Langevin dynamics	Brownian motion
Many body quantum systems	Variational Quantum Monte-Carlo method (VQMC)	Calculates energy for non separable trial wavefunctions
	Time dependent variational principle	Time dependent trial wavefunction, ANN

Purpose	Method	Comments
Physics informed machine learning	Physics informed neural network, PINN	Data driven solution and discovery of differential equations
	Hamiltonian NN	Solves Hamilton's equations while conserving energy
	Boltzmann machine	Stochastic network, learns a probability distribution and generates samples, nonlinear optimization, trial functions for VQMC
	Neural network quantum states	Trial functions for many-body quantum systems
Quantum simulation	Adiabatic quantum simulation AQS	Finds the ground state of a spin Hamiltonian
	Quadratic unconstrained binary optimization QUBO	Important for current analog quantum computing
	Boolean satisfiability problems SAT	NP-complete
	Quantum annealing QA	
	Tensor networks and matrix product states	Trial wavefunctions with reduced dimensionality for simulated QA
	Path integral method for spin systems	Classical simulation of quantum annealing

References

1. IEEE 754-2008, *Standard for Floating-Point Arithmetic* (IEEE Standards Association, 2008)
2. J. Stoer, R. Bulirsch, *Introduction to Numerical Analysis*, 3rd revised edn. (Springer, New York, 2010). ISBN 978-1441930064
3. H. Jeffreys, B.S. Jeffreys, Lagrange's interpolation formula 9.011, in *Methods of Mathematical Physics*, 3rd edn. (Cambridge University Press, Cambridge, England, 1988), p. 260
4. J.-P. Berrut, L.N. Trefethen, Barycentric lagrange interpolation. SIAM Rev. **46**(3), 501–517 (2004)
5. H. Jeffreys, B.S. Jeffreys, Divided differences 9.012, in *Methods of Mathematical Physics* 3rd edn. (Cambridge University Press, Cambridge, England, 1988), pp. 260–264
6. W. Werner, Math. Comp. **43**, 205 (1984)
7. E.H. Neville, Indian Math. Soc. **20**, 87 (1933)
8. S. Bernstein, Comm. Kharkov Math. Soc. **13**, 1–2 (1912)
9. Y. Şimşek, (2010) arXiv preprint arXiv:1012.5538
10. Y. Şimşek, Hacettepe J. Math. Stat. **43**(1), 1–14 (2014)
11. D. Barrera, S. Eddargani, M.J. Ibáñez, S. Remogna, J. Comput. Appl. Math. **424**, 11501 (2023)
12. N. Ezhov, F. Neitzel, S. Petrovic, Mathematics **9**, 2198 (2021)
13. L. Allen, (Baylor University ProQuest Dissertations Publishing, 2021), p. 28776447
14. M. Amirfakhrian, Int. J. Comput. Math. **88**(6), 1294–1304
15. M. Ainsworth, G. Andriamaro, O. Davydov, SIAM J. Sci. Comput. **33**(6) (2011)
16. M. Gaĕlle Andriamaro, Doctoral thesis T13753 (University of Strathclyde, 2013). https://doi.org/10.48730/ftgh-7068
17. E.H. Doha, A.H. Bhrawy, M.A. Saker, Appl. Math. Lett. **24**, 559–565 (2011)
18. W. Boehm, A. Muller, Comput. Aided Geomet. Des. **16**, 587–605 (1999)
19. C.T.H. Baker, Numer. Math. **15**, 315 (1970)
20. I.J. Schoenberg, Q. Appl. Math. **4**, pp. 45–99 and 112–141 (1946)
21. G. Nürnberger, *Approximation by Spline Functions* (Springer Verlag, 1989). ISBN 3-540-51618-2
22. M.S. Floater, K. Hormann, Numer. Math. **107**, 315 (2007)
23. J.P. Berrut, R. Baltensperger, H.D. Mittelmann, Int. Series Num. Math. **151**, 27 (2005)
24. C. Schneider, W. Werner, Math. Comp. **47**, 285 (1986)

P. O. J. Scherer, *Computational Physics I*, Graduate Texts in Physics,
https://doi.org/10.1007/978-3-032-07856-8

25. G.A. Baker Jr., P. Graves-Morris, *Pad Approximants* (Cambridge University Press, New York, 1996)
26. J.P. Berrut, Comput. Math. Appl. **14**, 1 (1988)
27. L.F. Richardson, Philos. Trans. R. Soc. Lond. Ser. Á **210**, 307–357 (1911)
28. A.G. Baydin, B.A. Pearlmutter, A.A. Radul, J.M. Siskind, J. Mach. Learn. Res. **18**, 1–43 (2018)
29. https://www.tensorflow.org/guide/autodiff
30. https://pytorch.org/tutorials/beginner/basics/autogradqs_tutorial.html
31. https://www.mathworks.com/help/deeplearning/ug/deep-learning-with-automatic-differentiation-in-matlab.html
32. F. Šrajera, Z. Kukelova, A. Fitzgibbon (2018) arXiv:1807.10129v1 [cs.MS]
33. F.L. Bauer, SIAM. J. Numer. Anal. **11**(1), 87 (1974)
34. A.L. Davis, R.M. Keller, Keller, IEEE Comput. **15**(2), 26–41 (1982)
35. M. Bartholomew-Biggs, S. Brown, B. Christianson, L. Dixon, J. Comput. Appl. Math. **124**, 171–190 (2000)
36. A.B. Kahn, Commun. ACM **5**(11), 558–562 (1962)
37. G. Di Battista, R. Tamassia, Theoret. Comput. Sci. **61**, 175–19 (1988)
38. R.E. Wengert, Commun. ACM **7**(8), 463 (1964)
39. C.P. Fries, Quant. Finan. **19**(6), 1043–1059 (2019)
40. L.N. Trefethen, SIAM Rev. **50**, 67–87 (2008)
41. M. Novelinkova, in *WDS'11 Proceedings of Contributed Papers*, Part I (2011), 67. ISBN 978-80-7378-184-2
42. C.W. Clenshaw, A.R. Curtis, Numer. Math. **2**, 197 (1960)
43. J. Waldvogel, BIT Numer. Math. **46**, 195 (2006)
44. A. Sommariva, Comput. Math. Appl. **65**(4), 682–693 (2013)
45. G.H. Golub, J.H. Welsch, Math. Comput. **23**, 221 (1969)
46. R.Z. Iqbal, Master Thesis (University of Birmingham, School of Mathematics, 2008)
47. H. Wilf, *Mathematics for the Physical Sciences* (Wiley, New York, 1962)
48. M.R. Hestenes, E. Stiefel, J. Res. Natl. Bur. Stand. **49**, 435 (1952)
49. Y. Saad, M. Schultz, SIAM J. Sci. Stat. Comput. **7**, 856–869 (1986)
50. R. Fletcher, Conjugate gradient methods for indefinite systems, in *Numerical Analysis*, ed. by W.G. Alistair. Lecture Notes in Mathematics, vol. 506 (Springer, Berlin/Heidelberg, 1976), pp. 73–89
51. H.A. Van der Vorst, SIAM J. Sci. Stat. Comput. **13**(2), 631–644 (1992)
52. Y. Saad, *Iterative Methods for Sparse Linear Systems*, 2nd edn. (SIAM, p. 194)
53. C. Paige, S.I.A.M.J. Saunders, Numer. Anal. **12**, 617–629 (1975)
54. R. Freund, N. Nachtigal, Numer. Math. **60**, 315–339 (1991)
55. C. Paige, B. Parlett, H. van der Vorst, Numer. Lin. Alg. Appl. **29**, 115–134 (1995)
56. C. Brezinski, H. Sadok, Appl. Numer. Math. **11**, 443–473 (1993)
57. M.H. Gutknecht, Acta Numer. **6**, 271–397 (1997)
58. S.-L. Zhang, SIAM J. Sci. Stat. Comput. **18**, 537–551 (1997)
59. W.H. Press, S.A. Teukolsky, W.T. Vetterling, B.P. Flannery, LU decomposition and its applications, in *Numerical Recipes, The Art of Scientific Computing*, 3rd edn. (Cambridge University Press, 2007), pp. 48–55
60. W.H. Press, S.A. Teukolsky, W.T. Vetterling, B.P. Flannery, Cholesky decomposition, in *Numerical Recipes, The Art of Scientific Computing*, 3rd edn. (Cambridge University Press, 2007), pp. 100–101
61. L.N. Trefethen, D. Bau III, *Numerical Linear Algebra* (Society for Industrial and Applied Mathematics, 1997)
62. G.H. Golub, C.F. Van Loan, *Matrix Computations*, 3rd edn. (Johns Hopkins, 1976). ISBN 978-0-8018-5414-9
63. J.W. Daniel, W.B. Gragg, L. Kaufmann, G.W. Stewart, Math. Comput. **30**, 772 (1976)
64. W.H. Press, S.A. Teukolsky, W.T. Vetterling, B.P. Flannery, Cyclic tridiagonal systems, in *Numerical Recipes, The Art of Scientific Computing*, 3rd edn. (Cambridge University Press, 2007), p. 79

65. W.W. Hager, SIAM Rev. **31**, 221 (1989)
66. J. Sherman, W.J. Morrison, Ann. Math. Stat. **20**, 621 (1949)
67. I.N. Bronshtein, K.A. Semendyayev, *Handbook of Mathematics*, 3rd edn. (Springer-Verlag, New York, 1997), p. 892
68. W.H. Press, S.A. Teukolsky, W.T. Vetterling, B.P. Flannery, Relaxation methods for boundary value problems, in *Numerical Recipes, The Art of Scientific Computing*, 3rd edn. (Cambridge University Press, 2007), pp. 1059–1065
69. H. Jeffreys, B.S. Jeffreys, *Methods of Mathematical Physics,* 3rd edn. (Cambridge University Press, Cambridge, England, 1988), pp. 305–306
70. J.H. Wilkinson, J. ACM **8**, 281 (1961)
71. D. Hilbert, L. Nordheim, J. von Neumann, Mathematische Annalen **98**, 1 (1927)
72. A. Bermann, S. Gueron, Math. Gaz. **86**, 274 (2002)
73. B. Beckermann, Numer. Math. **85**, 553 (2000)
74. I.R. Savage , E. Lukacs, in *Contributions to the Solution of Systems of Linear Equations and the Determination of Eigenvalues*, ed. by O. Taussky. National Bureau of Standards, Applied Mathematics Series, vol. 39 (1954)
75. R.L. Burden, J.D. Faires, *Numerical Analysis* (Brooks Cole Pub Co, Boston, 2010). (chapter 2)
76. J.A. Ford, *Improved Algorithms of Illinois-type for the Numerical Solution of Nonlinear Equations*. Technical Report CSM-257 (University of Essex Press, 1995)
77. J.Y. Tjalling, Historical development of the Newton-Raphson method. SIAM Rev. **37**, 531 (1995)
78. J.M. Papakonstantinou, A historical development of the (n+1)-point secant method. M.A. thesis, Rice University Electronic Theses and Dissertations, 2007
79. D.E. Muller, Mathematical tables and other aids to computation, vol. 10, 208 (1956)
80. R.P. Brent, Comput. J. **14**, 422 (1971)
81. T.R. Chandrupatla, Adv. Eng. Softw. **28**, 145 (1997)
82. T.J. Dekker, Finding a zero by means of successive linear interpolation, in *Constructive Aspects of the Fundamental Theorem of Algebra*, eds. by B. Dejon, P. Henrici. (Wiley-Interscience, London, 1969)
83. J.C.P. Bus, T.J. Dekker, ACM Trans. Math. Softw. **1**, 330 (1975)
84. R.P. Brent, *Algorithms for Minimization without Derivatives, Chap. 4* (Prentice-Hall, Englewood Cliffs, NJ, 1973)
85. A. Quarteroni, R. Sacco, F. Saleri, *Numerical Mathematics*, 2nd edn. (Springer, Berlin Heidelberg, 2007)
86. C.G. Broyden, Math. Comput. **19**, 577 (1965)
87. W.H. Teukolsky, S. Vetterling, W. Flannery, Section 10.2. Golden section search in one dimension, in *Numerical Recipes: The Art of Scientific Computing*, 3rd edn. (Cambridge University Press, New York, 2007). ISBN 978-0-521-88068-8
88. J. Kiefer, Proc. Am. Math. Soc. **4**, 502–506 (1953)
89. L. Armijo, Pac. J. Math. **16**(1) (1966)
90. P. Wolfe, SIAM Rev. **11**(2), 226–235 (1969)
91. P. Wolfe, SIAM Rev. **13**(2), 185–188 (1971)
92. J. Nocedal, S.J. Wright, *Numerical Optimization* (Springer Science & Business Media, 2006)
93. J.C. Spall, *Introduction to Stochastic Search and Optimization: Estimation, Simulation, and Control* (Wiley, Hoboken, NJ, 2003)
94. H. Robbins, S. Monro, Ann. Math. Stat. **22**(3), 400 (1951)
95. F. Rosenblatt, Psychol. Rev. **65**, 386–408 (1958) (Reprinted in Neurocomputing (MIT Press, 1988))
96. L. Bottou, Large-scale machine learning with stochastic gradient descent, in *Proceedings of COMPSTAT'2010* (Springer, 2010), pp. 177–186
97. B.T. Poliak, Z. Vych, Math. **4**(5), 791–803 (1964)
98. B.T. Poliak, *Introduction to Optimization* (Optimization Software Inc Publications Division, New York, 1987)

99. Y. Nesterov, Soviet Math. Doklady **27**(2), 372–376 (1983)
100. N.J. Walkington, SIAM Rev. **65**(2), 539–562 (2023)
101. J. Duchi, E. Hazan, Y. Singer, J. Mach. Learn. Res. **12**, 2121–2159 (2011)
102. T. Tieleman, G. Hinton, Lecture 6.5—rmsprop, divide the gradient by a running average of its recent magnitude, in *Coursera: Neural Networks for Machine Learning*, vol. 4 (2012)
103. M.D. Zeiler (2012) arXiv:1212.5701v1
104. D.P. Kingma, J.L. Ba (2017) arXiv:1412.6980v9
105. H. Kabiri, Y. Ghanou, H. Khalifi, G. Casalino, Knowl. Inf. Syst. **66**(6), 3427–3458 (2024)
106. R. Fletcher, C. Reeves, Comput. J. **7**, 149, 1–3 (1964)
107. C.G. Broyden, J. Inst. Math. Appl. **6**, 76 (1970)
108. R. Fletcher, Comput. J. **13**, 317 (1970)
109. D. Goldfarb, Math. Comput. **24**, 23 (1970)
110. D.F. Shanno, Math. Comput. **24**, 647 (1970)
111. Z. Wang, B.R. Hunt, Appl. Math. Comput. **16**, 19 (1985)
112. N. Ahmed, T. Natarajan, K.R. Rao, IEEE Trans. Comput. **23**, 90 (1974)
113. K.R. Rao, P. Yip, *Discrete Cosine Transform: Algorithms, Advantages, Applications* (Academic Press, Boston, 1990)
114. F.J. Harris, Proc. IEEE **66**, 51 (1978)
115. G. Goertzel, Am. Math. Mon. **65**, 34 (1958)
116. E.I. Jury, *Theory and Application of the Z-Transform Method* (Krieger Pub Co, 1973). ISBN 0-88275-122-0
117. P. Duhamel, M. Vetterli, Signal Process. **19**, 259 (1990)
118. H.J. Nussbaumer, *Fast Fourier Transform and Convolution Algorithms* (Springer-Verlag, Berlin, 1990)
119. J.W. Cooley, J.W. Tukey, Math. Comput. **19**, 297 (1965)
120. D. Gabor, J. Inst. Electr. Eng. **93**, 429 (1946)
121. M.J. Bastiaans, Appl. Opt. **33**, 5241 (1994)
122. M.J. Bastiaans, Proc. IEEE **68**, 538 (1980)
123. R.S. Orr, IEEE Trans. Sign. Proc. **41**, 122 (1993)
124. S. Qian, IEEE Trans. Sign. Proc. **41**, 2429 (1993)
125. J. Ashmead, Quanta **1**, 58 (2012)
126. A. Grossmann, J. Morlet, SIAM J. Math. Anal. **15**, 723 (1984)
127. S.G. Mallat, Trans. Am. Math. Soc. **315**, 69 (1989)
128. S.G. Mallat, IEEE Trans. Pattern Anal. Mach. Intell. **11**, 674 (1989)
129. A. Haar, Math. Ann. **69**, 331 (1910)
130. I. Daubechies, A. Grossmann, Y. Meyer, J. Math. Phys. **27**, 1271 (1986)
131. Y. Meyer, *Wavelets and Operators*, vol. I (Cambridge University Press, 1995)
132. V. Vermehren Valenzuela, H.M. de Oliveira, OALib J. (2015). arxiv.org/abs/1502.00161v1
133. D. Esteban, C. Galand, *Proceedings of the International Conference Acoustics Speech, Signal Processing* (1977)
134. I. Daubechies, Comm. Pure Appl. Math. **16**, 909 (1988)
135. N. Metropolis, S. Ulam, J. Am. Stat. Assoc. **44**, 335 (1949)
136. G.S. Fishman, *Monte Carlo: Concepts, Algorithms, and Applications* (Springer, New York, 1996). ISBN 038794527X
137. C.P. Robert, G. Casella, G. *Monte Carlo Statistical Methods (2nd ed.).* New York: Springer (2004) ISBN 0387212396
138. R.E. Caflisch, Monte Carlo and quasi-Monte Carlo methods. Acta Numerica, vol. 7 (Cambridge University Press, 1998), pp. 1–49
139. R.D. Richtmyer, *Principles of Modern Mathematical Physics I* (Springer, New York, 1978)
140. H. Pollard, Quart. J. Pure Appl. Math. **49**, 1 (1920)
141. J. Rice, *Mathematical Statistics and Data Analysis*, 2nd edn. (Duxbury Press, 1995). ISBN 0-534-20934-3
142. True random number service at RANDOM.ORG

143. J. Kelsey, B. Schneier, N. Ferguson, Yarrow-160 notes on the design and analysis of the yarrow cryptographic pseudorandom number generator, in *Sixth Annual Workshop on Selected Areas in Cryptography* (Springer, 1999)
144. N. Ferguson, B. Schneier, *Practical Cryptography* (John Wiley & Sons, 2003)
145. The Marsaglia random number CDROM including the diehard battery of tests of randomness. http://stat.fsu.edu/pub/diehard/
146. G. Marsaglia, J. Mod. Appl. Stat. Methods **1**, 2 (2003)
147. The TestU01 website http://simul.iro.umontreal.ca/testu01/tu01.html
148. P. L'ecuyer, R. Simard, ACM Trans. Math. Softw. **33**, 22 (2007)
149. M. Matsumoto, T. Nishimura, ACM Trans. Model. Comput. Simul. **8**, 3 (1998)
150. K. Entacher, ACM Trans. Model. Comput. Simul. **8**, 61 (1998)
151. G.E.P. Box, M.E. Muller, Ann. Math. Stat. **29**, 610 (1958)
152. N. Metropolis, A. Rosenbluth, M. Rosenbluth, A. Teller, E. Teller, J. Chem. Phys. **21**, 1087 (1953)
153. W. Hastings, Monte Carlo sampling methods using Markov chains and their application. Biometrika **57**, 97–109 (1970)
154. L. Tierney, Ann. Stat. **22**(4), 1701 (1994)
155. S. Chib, E. Greenberg, Am. Stat. **49**(4), 327 (1995)
156. G.O. Roberts, J.S. Rosenthal, Probab. Surv. **1**, 20–71 (2004)
157. A.D. Kennedy, J. Kuti, S. Meyer, B. Pendelton, J. Comp. Phys. **64**, 133 (1986)
158. G. Bhanot, Rep. Prog. Phys. **51**, 429 (1988)
159. G.O. Roberts, A.F.M. Smith, Stoch. Process. Appl. **49**, 207–216 (1994)
160. G.O. Roberts, A.F.M. Smith, Stoch. Process. Appl. **49**(2), 207–216 (1994)
161. C.P. Robert, arXiv:1504.01896v3 [stat.CO] 27 Jan 2016
162. K. Brügge, A. Fischer, C. Igel, arXiv:2006.14999v1 [stat.ML] 26 Jun 2020
163. G. Casella, E.I. George, Am. Stat. **46**(3), 167 (1992). (August)
164. S. Geman, D. Geman, IEEE Trans. Pattern Anal. Mach. Intell. **6**, 721–741 (1984)
165. R.J. Glauber, J. Math. Phys. **4**, 294–307 (1963)
166. A.E. Gelfand, A.F.M. Smith, J. Am. Stat. Assoc. **85**(410), 398 (1990)
167. P. Diaconis, K. Khare, L. Saloff-Coste, Stat. Sci. **23**(2), 151–178 (2008)
168. E. Bonomi, J.-L. Lutton, SIAM Rev. **26**, 551 (1984)
169. S. Kirkpatrick, C. Gelatt Jr, M. Vecchi, Science **220**, 4598, 671 (1983)
170. B.N. Parlett, *The Symmetric Eigenvalue Problem* (Society for Industrial and Applied Mathematics, Philadelphia, 1998)
171. H.W. Chang, S.E. Liu, R. Burridge, Linear Algebra Appl. **430**, 999 (2009)
172. S. Kouachi, Electron. J. Linear Algebra **15**, 115 (2006)
173. C.M. da Fonseca, V. Kowalenko, Acta Math. Hungar. **160**(2), 376–389 (2020)
174. J.G.F. Francis, Comput. J. **4**, 265 (1961); **4**, 332 (1962)
175. C. Lanczos, J. Res. Natl. Bur. Stand. **45**, 255 (1950)
176. H.D. Simon, Linear Algebra Appl. **61**, 101 (1984)
177. B.N. Parlett, D.S. Scott, Math. Comput. **33**, 217 (1979)
178. D. Calvetti, L. Reichel, D.C. Sorensen, Electron. Trans. Numer. Anal. **2**, 1 (1994)
179. D.C. Sorenson, NASA contractor report 198342 (1996)
180. J. Wolberg , *Data Analysis Using the Method of Least Squares: Extracting the Most Information from Experiments.* (Springer, 2005)
181. J.A. Richards, *Remote Sensing Digital Image Analysis* (Springer, Berlin, Heidelberg, New York, 1993)
182. A.E. Garcia, Phys. Rev. Lett. **86**, 2696 (1992)
183. T.D. Romo et al., Proteins **22**, 311 (1995)
184. D.P. Derrarr et al., in: *A Practical Approach to Microarray Data Analysis*, (Kluwer, 2003), p. 91
185. DGESVD routine from the freely available software package LAPACK (http://www.netlib.org/lapack)
186. G. Golub, W. Kahan, J.S.I.A.M. Numer, Anal. B **2**, 205 (1965)

187. G.W. Stewart, SIAM Rev. **35**, 551 (1993)
188. J. Mandel, Am. Stat. **36**, 15 (1982)
189. A. Ben-Israel, T.N.E. Greville, *Generalized Inverses* (Springer-Verlag, 2003). ISBN 0-387-00293-6
190. R. Rochas, *Neural Networks: A Systematic Introduction* (Springer-Verlag, Berlin, 1966)
191. M. Minsky, S. Papert, *Perceptrons: An Introduction to Computational Geometry* (MIT Press, Cambridge, MA, 1969)
192. J.C. Spall, *Introduction to Stochastic Search and Optimization* (Wiley, New York, 2003)
193. T. Zhang, Solving large scale linear prediction problems using stochastic gradient descent algorithms, in *Proceedings of the Twenty-First International Conference on Machine Learning* (2004)
194. Q. Wang, Y. Ma, K. Zhao, Y. Tian, Ann. Data Sci. **9**, 187–212 (2022)
195. L. Ciampiconi, A. Elwood, M. Leonardi, A. Mohamed, A. Rozza, arXiv:2301.05579v1 [cs.LG] 13 Jan 2023
196. N. Cristianini, J. Shawe-Taylor, *An Introduction to Support Vector Machines* (Cambridge University Press, 2000)
197. C. Cortes, V. Vapnik, Mach. Learn. **20**, 273–297 (1995)
198. S. Shalev-Shwartz, Y. Singer, N. Srebro, A. Cotter, Math. Program. Ser. B **127**(1), 3–30 (2011)
199. G.X. Yuan, C.H. Ho, C.J. Lin, Proc. IEEE **100**(9) (2012)
200. T. Hofmann, B. Schölkopf, A.J. Smola, Ann. Stat. **36**(3), 1171–1220 (2008)
201. B.E. Boser, I.M. Guyon, V.N. Vapnik, A training algorithm for optimal margin classifiers, in *Proceedings of the 5th Annual Workshop Computing Learning Theory* (1992), pp. 144–152
202. R.E. Wengert, Commun. ACM **7**(8), 463 (1964)
203. F.L. Bauer, SIAM. J. Numer. Anal. **11**(1), 87 (1974)
204. J. Schmidhuber (2022) arXiv:2212.11279
205. K. Hornik, M. Stinchcombe, H. White, Neural Netw. **2**, 359 (1989)
206. P. Auer, H. Burgsteiner, W. Maass, Neural Netw. **21**(5), 786 (2008)
207. A.D. Jagtap, G. Em Karniadakis (2022) arXiv:2209.02681v5
208. Q.V. Le, N. Jaitly, G.E. Hinton (2015) arXiv preprint arXiv:1504.00941
209. G.B. Zhou, J. Wu, C.L. Zhang, Z.H. Zhou (2016) arXiv:1603.09420v1
210. R. Jozefowicz, W. Zaremba, I. Sutskever, An empirical exploration of recurrent network architectures, in *Proceedings of the International Conference on Machine Learning*, pp. 2342–2350 (2015)
211. S. Hochreiter, J. Schmidhuber, J. Neural Comput. **9**(8), 1735–1780 (1997)
212. F.A. Gers, J. Schmidhuber, F. Cummins, Learning to forget: continual prediction with LSTM, in *Proceedings of the International Conference Artificial Neural Networks*, vol. 2 (1999), pp. 850–855
213. J.C. Heck Jr., F.M. Salem, *IEEE 60th International Midwest Symposium on Circuits and Systems (MWSCAS)* (Boston, MA, USA, 2017), p. 1593
214. X. Kong, C. Hu, Z. Duan, Zhansheng, Neural networks for principal component analysis (2017). https://doi.org/10.1007/978-981-10-2915-8_3
215. D.O. Hebb, *The Organization of Behavior* (Wiley & Sons, New York, 1949)
216. T.D. Sanger, Neural Netw. **2**(6) (1989)
217. E. Oja, J. Math. Biol. **15**, 267 (1982)
218. E. Oja, J. Karhunen, J. Math. Anal. Appl. **106**(1), 69 (1985)
219. E. Oja, Neural Netw. **5**, 927 (1992)
220. P. Baldi, K. Hornik, Neural Netw. **2**, 53 (1989)
221. N. Migenda , R. Muller, W. Schenck , PLoS ONE **16**(3), e0248896. https://doi.org/10.1371/journal.pone.0248896
222. C. Fyfe, Neural Process. Lett. **6**, 33 (1997)
223. J. Peiro, S. Sherwin, in *Handbook of Materials Modeling*, vol.1, ed. by S. Yip (Springer, 2005), pp. 1–32
224. H.P. Langtangen, *Computational Partial Differential Equations: Numerical Methods and Diffpack Programming* (Springer, 2003)

225. W. Ames, *Numerical Methods for Partial Differential Equations*, (Academin Press Inc., 1992)
226. F. John, Duke Math. J. **4**, 300 (1938)
227. D.J. Acheson, *Elementary Fluid Dynamics* (Oxford University Press, 1990)
228. B.G. Galerkin, On electrical circuits for the approximate solution of the Laplace equation. Vestnik Inzh. **19**, 897–908 (1915)
229. R. Eymard, T. Galloue, R. Herbin , Finite Volume Methods, in *Handbook of Numerical Analysis*, vol. 7, ed. by P.G. Ciarlet, J.L. Lions (Elsevier, 2000), pp. 713–1020
230. E.L. Ortiz, SIAM J. Numer. Anal. **6**, 480 (1969)
231. B.N. Jiang, *The Least-Squares Finite Element Method* (Springer, 1998)
232. P. Bochev, M. Gunzburger, *Least Squares Finite Element Methods*, (Springer, 2009)
233. C.A.J. Fletcher, *Computational Galerkin Methods*, (Springer, 1984)
234. J.P. Boyd, *Chebyshev and Fourier Spectral Methods* (Dover publications N.Y., 2001)
235. J. Fish, T. Belytschko, *A First Course in Finite Elements* (John Wiley & Sons, 2007)
236. S.S. Rao, *The Finite Element Method in Engineering*, (Elsevier 2011)
237. C.A.J. Fletcher, *Computational Techniques For Fluid Dynamics*, vol. I, 2nd edn (Springer, 1991)
238. G. Beer, I. Smith, C. Duenser, *The Boundary Element Method with Programming: For Engineers and Scientists* (Springer, 2008)
239. L.C. Wrobel, M.H. Aliabadi, *The Boundary Element Method* (Wiley, New Jersey, 2002)
240. G. Wunsch, *Feldtheorie* (VEB Technik, 1973)
241. I.E. Lagaris, A. Likas, D.I. Fotiadis (1979) arXiv:physics/9705023v1
242. I.E. Lagaris, A. Likas, D.I. Fotiadis, IEEE Trans. Neural Netw. **9**(5), 987 (1998)
243. J.B. Pedro, J. Maroas, R. Paredes (2019) arXiv:1912.04737v1 [physics.comp-ph]
244. N. Yadav, A. Yadav, M. Kumar, *An Introduction to Neural Network Methods for Differential Equations* (Springer, 2015)
245. T.V.S.S. Chaitanya, R.P. Kumar, D. Chaudhary, S. Singh, P. Shrivastav, D. Yadav, Numerical solution of first and second order differential equations with deep neural networks, in *2023 IEEE World Conference on Applied Intelligence and Computing (AIC)* (Sonbhadra, India, 2023), pp. 180–186. https://doi.org/10.1109/AIC57670.2023.10263878
246. M.M. Chiaramonte, M. Kiener, Solving differential equations using neural networks (2017). http://cs229.stanford.edu/proj2013
247. A. Honchar, Neural networks for solving differential equations [WWW Document]. Becom. Hum. Artif. Intell. Mag. (2017). https://becominghuman.ai/neural-networks-for-solving-differential-equations-fa230ac5e04c. Last accessed 22 Apr 2019
248. M. Mattheakis, D. Sondak, A.S. Dogra, P. Protopapas, Phys. Rev. E **105**, 2022 (2022)
249. J.A. Sirignano, K. Spiliopoulos, J. Comput. Phys. **375**, 1339 (2018)
250. A. Nordsieck, Math. Comput. **16**, 22 (1962)
251. C.W. Gear, *Numerical Initial Value Problems in Ordinary Differential Equations* (Prentice-Hall, Englewood Cliffs, NJ, 1971)
252. G.J. Martyna, M.E. Tuckerman, J. Chem. Phys. **102**, 8071 (1995)
253. J.C. Butcher, *The Numerical Analysis of Ordinary Differential Equations: Runge-Kutta and General Linear Methods* (Wiley, New York, 1987)
254. C. W. Gear, *Numerical Initial Value Problems in Ordinary Differential Equations* (Prentice Hall, Englewood Cliffs, NJ, 1971), p 5.4
255. E. Fehlberg, Computing **4**, 93 (1969)
256. E. Fehlberg, Computing **6**, 61 (1970)
257. E. Fehlberg, Computing **10**, 305–315 (1972)
258. J.R. Cash, A.H. Karp, ACM Trans. Math. Softw. **16**, 201 (1990). https://doi.org/10.1145/79505.79507
259. W.H. Press, S.A. Teukolsky, Comput. Phys. **6**, 188 (1992). https://doi.org/10.1063/1.4823060
260. W.B. Gragg, SIAM J. Num. Anal. **2**, 384 (1965)
261. L.F. Shampine, IMA J. Num. Anal. **3**, 383 (1983)
262. L.F. Shampine, L.S. Baca, Numer. Math. **41**, 165 (1983)
263. C.W. Gear, IEEE Trans. Circuit Theory **18**, 89–95 (1971)

264. G. Dahlquist, Math. Scand. **4**, 33 (1956)
265. R.J. Leveque, Finite Difference Methods for Ordinary and Partial Differential Equations: Steady-State and Time-Dependent Problems (Cambridge University Press, 2007). ISBN-13 978-0898716290
266. J.C. Kirkpatrick, NASA report TM X-58182 (1976)
267. B. Seidu, Int. J. Comp. Appl. Math. **6**(3), 215 (2011)
268. A. Iserles, *A First Course in the Numerical Analysis of Differential Equations* (Cambridge University Press, 1996). ISBN 978-0-521-55655-2
269. C.W. Gear, Math. Comput. **21**, 146 (1967)
270. I.P. Omelyan, I.M. Mryglod, R. Folk, Comput. Phys. Commun. **151**, 272 (2003)
271. S.-H. Tsai et al., Braz. J. Phys. **34**, 384 (2004)
272. M. Tuckerman, B.J. Berne, J. Chem. Phys. **97**, 1990 (1992)
273. M.P. Allen, D.J. Tildesley, *Computer Simulation of Liquids* (Oxford University Press, 1989). ISBN 0-19-855645-4
274. L. Verlet, Phys. Rev. **159**, 98 (1967)
275. E. Hairer, C. Lubich, G. Wanner, Acta Numer. **12**, 399 (2003)
276. W.H. Press, S.A. Teukolsky, W.T. Vetterling, B.P.Flannery, Section 17.4. Second-order conservative equations, in *Numerical Recipes: The Art of Scientific Computing*, 3rd edn. (Cambridge University Press, 2007)
277. P. Schofield, Comp. Phys. Commun. **5**, 17 (1973)
278. D. Beeman, J. Comp. Phys. **20**, 130 (1976)
279. B.V. Noumerov, Publ. Observ. Cent. Astrophys. Russ. **2**, 188 (1923)
280. B.V. Noumerov, Mon. Not. R. Astron. Sot. **84**, 180 (1924); 592 (1924)
281. J.D. Lambert, I.A. Watson, J. Inst. Math. Appl. **18**, 189 (1976)
282. M.M. Chawla, BIT **24**, 117–118 (1984)
283. J.M. Franco, M. Palacios, J. Comp. Appl. Math. **30**, 1 (1990)
284. M.M. Chawla, BIT **23**, 541–542 (1983)
285. M.M. Chawla, P.S. Rao, J. Comput. Appl. Math. **11**, 277–281 (1984)
286. E. Hairer, Numer. Math. **32**, 373–379 (1979)
287. U. Ananthakrishnaiah, J. Comput. Appl. Math. **14**, 455–459 (1986)
288. T.E. Simos, A.D. Raptis, Computing **45**, 175–181 (1990)
289. Z.A. Anastassi, D.S. Vlachos, T.E. Simos, J. Math. Chem. **46**, 621–651 (2009)
290. A. Konguetsof, J. Math. Chem. **47**, 871–890 (2010)
291. A. Ndanusa, K.R. Adeboye, Sci. World J. **15**(4), 72–75 (2020)
292. Z. Kalogiratou, T. Monovasilis, T.E. Simos, J. Math. Chem. **37**(3), 271–279 (2005)
293. A. Ndanusa, K.R. Adeboye, A.U. Mustapha, R. Abdullahi, FUDMA J. Sci. (FJS) **4**(4), 355–362 (2020)
294. E.J. Nyström, Acta Soc. Sci. Fenn. **50**(13) (1925)
295. J. Albrecht, Z. Angew, Math. Mech. **35**, 100–110 (1955)
296. R. Kostek, Appl. Math. Comput. **219**, 10082–10095 (2013)

Index

P. O. J. Scherer, *Computational Physics I*, Graduate Texts in Physics,
https://doi.org/10.1007/978-3-032-07856-8

U

V

W

X

Z

MIX
Papier aus verantwortungsvollen Quellen
Paper from responsible sources
FSC® C105338

If you have any concerns about our products, you can contact us on
ProductSafety@springernature.com

In case Publisher is established outside the EU, the EU authorized representative is:
Springer Nature Customer Service Center GmbH
Europaplatz 3, 69115 Heidelberg, Germany

Printed by Libri Plureos GmbH
in Hamburg, Germany